Diseases of Horticultural Crops:
Nematode Problems and their Management

P. PARVATHA REDDY

Former Head, Division of Entomology & Nematology and Former Director
Indian Institute of Horticultural Research
Bangalore – 560 089

SCIENTIFIC PUBLISHERS (INDIA)

P.O. BOX 91

JODHPUR

Published by:
Pawan Kumar
Scientific Publishers (India)
5-A, New Pali Road, P.O. Box 91
Jodhpur - 342 001 (Raj.)
Tel.: +92-291-2433323, 2624154
E-mail: info@scientificpub.com
www.scientificpub.com

ISBN: 978-81-7233-543-4

Cover design: Tanay Sharma

Lasertype Set: Rajesh Ojha
Printed in India

*Affectionately dedicated
to the memory of my parents*
**Smt. SHIVAMMA
Shri. SANNA BASAPPA**

PREFACE

Horticulture in India is fast emerging as a major commercial venture, because of higher remuneration per unit area and the realization that consumption of fruits and vegetables is essential for health and nutrition. In the last one decade, export potential of horticultural crops has significantly increased attracting even multinationals into floriculture, processing and value added products. Horticulture development has been accorded high priority in the XI th Five Year Plan. The impact of enhanced investment in horticulture has been highly encouraging in terms of vastly improved quality production and export potential - an increase from 96.1 million tonnes in 1990-91 to 169.8 million tonnes during 2004-05. India is now the second largest producer of fruits and vegetables only next to China.

Nematodes continue to threaten horticultural crops throughout the world, particularly in tropical and sub-tropical regions. For century's man's essential crop plants have been plagued by these microscopic organisms that feed on roots, buds, stems, crowns, leaves and developing seed. Estimated overall average annual yield loss of the world's major horticultural crops due to damage by plant parasitic nematodes is 13.54%. For the seven life sustaining horticultural crops (banana, cassava, coconut, field bean, potato, sugar beet and sweet potato) that stand between man and starvation, an estimated annual yield loss of 12.77% is reported. The 16 economically important horticultural crops (cocoa, citrus, coffee, cowpea, eggplant, grapevine, guava, melons, okra, ornamentals, papaya, pepper, pineapple, tea, tomato and yam) that represent a group important for food or export value were reported to have an estimated annual yield loss of 14.31%. Monetary losses due to nematodes on 10 horticultural crops, six of which are life sustaining were estimated at US $ 19.37 billion annually based on 1984 production figures and prices.

The destructive plant-parasitic nematodes are one of the major limiting factors in the production of horticultural crops throughout the country. Roots damaged by the nematodes are not efficient in the utilization of available moisture and nutrients in the soil resulting in reduced functional metabolism. Furthermore, roots weakened and damaged by nematodes are easy prey to many types of fungi and bacteria which invade the roots and accelerate root decay. These deleterious effects on plant growth result in reduced yields and poor quality of horticultural crops. Nematode management is therefore, important for high yields and quality that are required by the high cost of modern crop production. Nematode management can be achieved by host resistance and by suppression of nematode population through regulatory, physical, cultural, chemical, biological and integrated methods.

The information on nematode management, especially crop-wise, is very much scattered and there is no book which deals with nematode diseases and their management entirely on horticultural crops. Hence, the present book is an attempt which comprehensively deals with the subject. This book deals with nematode diseases and their management in horticultural crops such as fruit, vegetable, ornamental, medicinal, aromatic, plantation, spice and tuber crops. Each nematode disease is described in adequate detail under the following heads: economic importance, crop losses, distribution, symptoms, hosts range, life cycle, races/biotypes, host-parasite relationship, histopathology, ecology, spread, survival, interaction with other pathogens and management. The management methods presented includes regulatory, physical, cultural, chemical, biological, host resistance and integrated approach. An entire chapter is devoted for sources of availability of critical inputs used for nematode management. Very useful information on nematode diseases and their management is provided in the Appedices. The book is adequately illustrated with about 50 figures.

This book is a practical guide to practicing farmers of horticultural crops. Further, it is a useful reference to policy makers, research and extension workers and students. The material can also be used for teaching undergraduate and post-graduate courses. Suggestions to improve the contents of the book are most welcome (E-mail: reddy_parvatha@yahoo.com). The publisher, Scientific publishers (India), Jodhpur deserves commendation for their professional contribution.

Bangalore
March 10, 2008

P. Parvatha Reddy

CONTENTS

Chapter 1

INTRODUCTION

1.1. Importance of Horticultural Crops

Agriculture is the mainstay of the Indian economy and Horticulture is a crucial component thereof. Horticulture development has been accorded high priority in the XIth Five Year Plan. The impact of enhanced investment in horticulture has been highly encouraging in terms of vastly improved quality production and export potential - an increase from 96.1 million tonnes in 1990-91 to 157.8 million tonnes in 2003-04 and to 169.8 million tonnes during 2004-05 (Table 1.1). India is now the second largest producer of fruits and vegetables only next to China and has first position in several horticultural crops, namely mango, banana, papaya, areca nut and okra.

Table 1.1. Area, production and productivity of horticultural crops in India (2004-05) (Chamber, 2006)

Horticultural crops	Area (million ha)	Production (million tonnes)	Productivity (tonnes/ha)
Fruits	4.964	49.295	9.9
Vegetables	6.756	101.434	15.0
Flowers-loose	0.116	0.655	---
Flowers-cut	---	1952.000*	---
Mushrooms	---	0.040	---
Spices	5.155	5.113	1.0
Plantation crops	3.102	13.161	4.2
Nuts	0.106	0.121	1.1
Total	20.200	169.800	8.4

*Numbers

Horticulture in India is fast emerging as a major commercial venture, because of higher remuneration per unit area and the realization that consumption of fruits and vegetables is essential for health and nutrition. In the last one decade, export potential of horticultural crops has significantly increased attracting even multinationals into floriculture, processing and value added products.

Horticulture has emerged as an important sector for diversification of agriculture. Presently, horticulture has established its credibility in improving income through increased productivity, generating employment and in enhancing exports besides providing protective food, household nutritional security, precious raw material for various processing and ancillary industries and finally in enhancing the aesthetic aspects of life. The focused attention on investment in horticulture during the last decade has been rewarding in terms of increased production and productivity of horticultural crops with manifold export potential. Horticulture has played a significant role in women empowerment, providing employment opportunities through mushroom cultivation, floriculture, processing, nursery raising, vegetable seed production etc. The present annual growth rate of horticulture is more than 6.5% which contributes 24.5% from the mere 8% of area to the GDP and 54.55% to the export earnings in the agriculture sector (Table 1.2).

Table 1.2. Export of horticultural produce from India (2004-05) (Chamber, 2006)

Horticultural product	Export quantity (tonnes)	Value (Rs. Crores)	% Share of value
Floriculture & seeds			
Floriculture	26262.35	210.99	6.9
Fruit & vegetable seeds	6307.33	62.94	2.1
Total	**32569.68**	**273.93**	**9.0**
Fruits & vegetables			
Fresh onions	833209.81	621.09	20.5
Other fresh vegetables	181956.66	224.38	7.4
Fresh grapes	35936.17	110.67	3.6
Other fresh fruits	131541.49	164.00	5.4
Total	**1240700.23**	**1299.94**	**42.8**
Processed fruits & vegetables			
Dried & preserved vegetables	351034.32	765.75	25.2
Other processed fruits and vegetables	80760.50	275.53	9.1
Total	**589976.71**	**1462.72**	**48.2**
Grand Total	1863246.62	3036.59	—

The major destinations for export of different horticultural products are presented in Table 1.3.

Table 1.3. Major destinations for export of different horticultural products (Chamber, 2006)

Horticultural crops	Major destinations for export
Fruits	UAE, Saudi Arabia, Sri Lanka, Bangla Desh, The Netherlands, USA, Malaysia
Vegetables	UAE, UK, Saudi Arabia, Sri Lanka, USA, Singapore
Floriculture	USA, Japan, The Netherlands, UK, Italy
Vegetable seeds	USA, The Netherlands, France, Italy, Japan, Pakistan
Fresh onion	UAE, Singapore, Sri Lanka, Bangla Desh, Malaysia
Fresh vegetables	UAE, Saudi Arabia, Sri Lanka, USA
Fresh grapes	UK, UAE, The Netherlands, Bangla Desh, Germany
Dried vegetables	UAE, Sri Lanka, USA, UK, Singapore

Potential international markets for fresh and processed products are presented in Table 1.4.

Table 1.4. Potential international markets for fresh and processed fruits and vegetables

Fresh & processed fruits & vegetables	Potential international markets
Fresh grapes	UK, Netherlands, Sri Lanka, UAE
Pomegranate	UAE, Saudi Arabia, Sri Lanka, Kuwait
Fresh guava	Saudi Arabia, Netherlands, USA, UAE, Kuwait
Gooseberries	UAE, Qatar
Fresh onion	Sri Lanka, Malaysia, UAE, Singapore
Gherkin	France, USA, Spain, Belgium
Fresh tomato	Bangle Desh, Nepal, Sri Lanka
Mango pulp	UK, Canada, Bangledesh, Bahrain, Netherlands, Japan, Saudi Arabia
Mango slices	UK, Saudi Arabia, Japan, USA
Dried grapes	UK, Malaysia, Netherlands, UAE
Mango pickles	Saudi Arabia, UAE, UK, Canada
Mango chutney	Oman, USA
Guava jam & jelly	UK, Germany, Hong Kong, USA, Netherlands, Saudi Arabia
Dried tamarind	Egypt, UAE, Pakistan, Saudi Arabia, UK
Dried figs	Nepal, Bangladesh
Dehydrated onion	Germany, Romania, Sri Lanka, USA, Malaysia, Netherlands, UK

Diversification in horticulture is a best option as there are several advantages of growing horticultural crops over field crops.

- Produce higher biomass than field crops per unit area resulting in efficient utilization of natural resources.

- Highly remunerative for replacing subsistence farming and thus alleviate poverty level in rain fed, dry land, hilly, arid and coastal agro eco-systems.

- Have potential for development of wastelands through planned strategies.

- Need comparatively less water than food crops.

- Provide higher employment opportunity.

- Important for nutritional security.

- Environment friendly.

- High value crops with high potential for value addition.

- High potential for foreign exchange earnings.

- Make higher contribution to GDP i.e. 24.5% from 8.5% area under these crops.

India is the second largest producer of fruits and vegetables next only to China contributing 10.10% and 14.40% of the total world production, respectively. Total production of fruits during 2004-05 has been estimated at 49.295 million tonnes from 4.964 million ha. Vegetables occupy an area of 6.756 million ha with a production of 101.434 million tonnes (Chamber, 2006). India is the largest producer of mango, banana, papaya, areca nut and okra. India produces 41.7% of the world mangoes, 25.7% of the bananas and 13.6% of the world onions. In grapes, India has recorded the highest productivity per unit area in the world. In vegetables, India occupies the first position in the production of okra, second in onion, brinjal, cabbage, cauliflower, peas, tomato and third in potato. The per capita consumption of fruits has increased from 40 to 85g/day and vegetables from 96 to 175g/day within the last one decade.

Floriculture is estimated to cover an area of 116,000 ha with production of 655,000 MT of loose flowers. The area under cut flowers has increased significantly in recent years and the production has reached 1952 million numbers.

India has emerged as the largest producer of areca nut. Coconuts sustain an area of 1.933 million ha with a production of 121.78 million tonnes. Areca nut is cultivated in 0.365 million ha with a production of 0.439 million tonnes. Cashew nuts are grown in an area of 0.706 million ha with a production of 0.46 million tonnes. The annual production of spices is 5.113 million tonnes from 5.155 million ha.

Horticulture today is not merely a means of diversification but forms an integral part of food and nutritional security, as also an essential ingredient of economic security. Adoption of horticulture both by small and marginal farmers brought prosperity in many regions of the country of which Maharashtra, Karnataka and Andhra Pradesh are the prime examples. A Golden Revolution through significant production increases in horticultural crops is already on the horizon.

Since India is now a signatory member of WTO, it will require its produce and products to be competitive both in the domestic and export markets. This would call for the use of improved HYV/F1 hybrids, recommended agro-techniques and plant protection measures, hi-tech intervention and scientific post-harvest management.

After the success of 'Green Revolution' in agriculture, India is now moving towards the **'Golden Revolution'** comprising fruits, vegetables and flowers. Commercialization of horticulture along with post harvest management, processing and marketing assume great significance and add value not only in the context of farmer's income, but also its contribution to the agriculture GDP. Commercialization of horticulture is largely dependent upon use of appropriate technology and market oriented production.

1.2. The Role of Plant Parasitic Nematodes in Horticultural Crops

Plant parasitic nematodes present some of the most difficult pest problems encountered in our horticultural economy. According to Thorne (1961) "Each year these minute organisms exact an ever-increasing toll from almost every cultivated acre in the world; a bag of potatoes in England, a bale of cotton in Georgia, a bushel of corn in Iowa, a box of apples in New York, a sack of wheat in Kansas, or a crate of oranges in California".

The destructive plant-parasitic nematodes are one of the major limiting factors in the production of horticultural crops throughout the world. For centuries, man has been plagued by these microscopic organisms feeding on the roots of crop plants essential to his survival and well being. Roots damaged by the nematodes are not efficient in the utilization of available moisture and nutrients in the soil resulting in reduced functional metabolism. Visible symptoms of nematode attack often include reduced growth of individual plants, varying degrees of chlorosis, wilting of the foliage and sometimes death of plants. Furthermore, roots weakened and damaged by nematodes are easy prey to many types of fungi and bacteria which invade the roots and accelerate root decay. These deleterious effects on plant growth result in reduced yields and poor quality of crops. Nematode management is therefore, important for high yields and quality that are required by the high cost of modern crop production.

1.2.1. Economic Importance

1.2.1.1. International Scenario: On a worldwide basis, the 10 most important genera of plant parasitic nematodes were reported to be as follows (Sasser and Freckman, 1987):

(i) *Meloidogyne*	(vi) *Tylenchulus*
(ii) *Pratylenchus*	(vii) *Xiphinema*
(iii) *Heterodera*	(viii) *Radopholus*
(iv) *Ditylenchus*	(ix) *Rotylenchulus*
(v) *Globodera*	(x) *Helicotylenchus*

Estimated overall average annual yield loss of the world's major horticultural crops due to damage by plant parasitic nematodes is 13.54% (Table 1.5). For the 7 horticultural crops (left-hand column) that stand between man and starvation (life sustaining horticultural crops), an estimated annual yield loss of 12.77% is reported.

The 16 economically important horticultural crops (right-hand column) that represent a miscellaneous group important for food or export value were reported to have an estimated annual yield loss of 14.31%.

Table 1.5. Estimated annual yield losses due to damage by plant parasitic nematodes – world basis (Sasser and Freckman, 1987).

Life sustaining horticultural crops	Loss (%)	Economically important horticultural crops	Loss (%)
Banana	19.7	Cocoa	10.5
Cassava	8.4	Citrus	14.2
Coconut	17.1	Coffee	15.0
Field bean	10.9	Cowpea	15.1
Potato	12.2	Eggplant	16.9
Sugar beet	10.9	Grapes	12.5
Sweet potato	10.2	Guava	10.8
		Melons	13.8
		Okra	20.4
		Ornamentals	11.1
		Papaya	15.1
		Pepper	12.2

		Pineapple	14.9
		Tea	8.2
		Tomato	20.6
		Yam	17.7
Average	12.77 %	Average	14.31 %

Overall average – 13.54%

Monetary losses due to nematodes on 10 horticultural crops, 6 of which are life sustaining, were estimated at US $ 19.37 billion annually based on 1984 production figures and prices (Table 1.6).

These figures are staggering and the real figure, when all horticultural crops are considered, probably exceeds US $ 50 billion annually. The losses are 5.8% greater in developing countries than in developed countries (Sasser and Freckman, 1987).

Table 1.6. Estimated annual losses due to nematodes for selected world horticultural Crops (Sasser and Freckman, 1987)

Crop	No. of estimates/ crop	FAO production estimates ('000 MT)	Estimated price / MT (US $)	Estimated yield loss due to nemas (%)	Estimated monetary loss due to nematodes (US $)
Banana	78	2097	431	19.7	178049979
Cassava	25	129020	90	8.4	975391200
Citrus	102	56100	505	14.2	4022931000
Cocoa	13	1660	2584	10.5	450391200
Coffee	36	5210	3175	15.0	2481262500
Field bean	70	19508	544	10.9	1156746300
Potato	141	312209	152	12.2	5789403696
Sugar beet	51	293478	37	10.9	1183596774
Sweet potato	67	117337	219	10.2	2621073906
Tea	16	2218	2807	8.2	510562300
				Total	19369408865

1.2.1.2. Indian Scenario: The ten major nematode pests in India have been identified and the priority list is as follows (Sharma *et al.*, 2002):

(i) *Meloidogyne incognita*

(ii) *M. javanica*

(vi) *Tylenchulus semipenetrans*

(vii) *Radopholus similis*

    (iii) *Rotylenchulus reniformis*   (viii) *Pratylenchus zeae*

    (iv) *Heterodera avenae*   (ix) *M. graminicola*

    (v) *H. cajani*   (x) *Hirschmanniella* spp.

The best guess rankings of nematode problems of important crops in India made on the basis of survey by highly experienced nematologists of the country are presented in Table 1.7.

Table 1.7. Best guess rankings of nematode problems of important horticultural crops in India (Sharma *et al.*, 2002)*.

State	Vegetable crops	Temperate fruit crops	Tropical fruit crops	Plantation crops
Andhra Pradesh	3	---	5	---
Assam	---	---	3	---
Bihar	3	---	7	---
Delhi	3	---	7	---
Gujarat	1	---	5	---
Haryana	1	---	5	---
Himachal Pradesh	1	3	---	---
Jammu & Kashmir	5	7	---	---
Karnataka	1	---	3	---
Kerala	5	---	3	3
Madhya Pradesh	3	---	---	---
Maharashtra	3	---	5	---
Manipur	---	---	7	---
Punjab	1	---	3	---
Rajasthan	3	---	---	---
Tamil Nadu	3	---	5	---
Uttar Pradesh	1	---	---	---
West Bengal	3	---	---	---

* Ranking on 1-9 scale: 1-Most important, 3-Very important, 5- Important, 7-Moderately important, 9-Least important.

The avoidable yield losses due to plant parasitic nematodes in horticultural crops have been presented in Table 1.8.

Table 1.8. Avoidable yield losses in horticultural crops due to plant parasitic nematodes.

Crop	Nematode(s)	Yield loss (%)	Reference(s)
Banana	*Radopholus similis*	38.00	Rajagopalan & Naganathan, 1977b
		32.00	Parvatha Reddy *et al.*, 1996d
		41.00	Nair, 1979
	Meloidogyne incognita	30.90	Jonathan & Rajendran, 2000
Sweet orange	*Tylenchulus semipenetrens*	69.00	Baghel & Bhatti, 1983a
Lemon	*T. semipenetrans*	29.00	Mukhopadhyaya & Suryanarayana, 1969
Sweet lime	*T. semipenetrans*	19.00	Mukhopadhyaya & Dalal, 1971
Grapevine	*M. incognita*	55.00	Rajagopalan & Naganathan, 1977a
	M. javanica	53.00	Baghel & Bhatti, 1983b
Papaya	*Rotylenchulus reniformis*	28.00	Rajendran & Naganathan, 1981
Pomegranate	*Meloidogyne* sp.	24.64-27.45	Singh *et al.*, 2003
Peach	*Macroposthonia xenoplax*	33.00	Anon, 1990a
Plum	*M. xenoplax*	10.00	Anon, 1990a
Potato	*M. incognita*	42.50	Prasad, 1989
	Globodera rostochiensis	99.50	Prasad, 1989
Tomato	*M. incognita*	30.57-46.92	Bhatti & Jain, 1977; Reddy, 1985; Darekar & Mahse, 1988
	M. javanica	77.50	Anon, 1993b
	R. reniformis	42.25 – 49.02	Subramanyam *et al.*, 1990
Brinjal	*M. incognita*	27.30-48.55	Bhatti & Jain, 1977; Parvatha Reddy & Singh, 1981; Darekar & Mahse, 1988
Chilli	*Meloidogyne* sp.	24.54 – 28.00	Singh *et al.*, 2003
Okra	*M. incognita*	90.90	Bhatti & Jain, 1977
		28.08	Parvatha Reddy & Singh, 1981
	M. javanica	20.20-41.20	Jain *et al.*, 1986
French bean	*M. incognita*	19.38 - 43.48	Das, 1994; Patel *et al.*, 2004

			Parvatha Reddy & Singh, 1981
	M. javanica	30-40	Sharma *et al.*, 2002
Cowpea	*M. incognita*	28.60	Parvatha Reddy & Singh, 1981
	R. reniformis	13.20-32.00	Palanisamy & Sivakumar, 1981
			Hasan & Jain, 1998
Peas	*M. incognita*	20.0-50.61	Parvatha Reddy, 1985
			Upadhyay & Dwivedi, 1987
			Sharma, 1989
	R. reniformis	15.8	Dalal & Vats, 1998
Carrot	*M. incognita*	56.64	Devi, 1993
Bitter gourd	*M. incognita*	36.72	Darekar & Mahse, 1988
Pointed gourd	*M. incognita*	30-40	Verma, 2001
Water melon	*M. incognita*	18-33	Hasan & Jain, 1998
Mushroom	*Ahelenchoides sacchari*	40.6-100.0	Singh *et al.*, 2003
	A. composticola	35-60	Laqman Khan, 2001
	A. avenae	25.8-53.5	Bajaj & Jain, 2001
Crossandra	*M. incognita*	21.64	Khan & Parvatha Reddy, 1994
Tuberose	*M. incognita*	13.78	Khan & Parvatha Reddy, 1994
Patchouli	*M. incognita*	47.00	Prasad & Reddy, 1984
Davana	*M. incognita*	50	Haseeb & Pandey, 1989a
Cymbopogon	*M. incognita*	20	Pandey, 1994a
Carnation	*M. incognita*	27	Nagesh & Parvatha Reddy, 2000
Gerbera	*M. incognita*	31	Nagesh & Parvatha Reddy, 2000
Coleus forskohlii	*M. incognita*	70.2	Senthamarai *et al.*, 2006c
Chethikodu-veli (*Plumbago rosea*)	*M. incognita*	29.00-43.96	Santhosh Kumar & Sheela, 2004
Kacholam (*Kaempferia galanga*)	*M. incognita*	18-64	Sheela & Rajani, 1998
Colocasia	*M. incognita*	24	Anon, 1990b
Menthol mint	*M. incognita*	30	Pandey, 2003
Cardamom	*M. incognita*	32-47	Ali, 1986

Betel vine	*M. incognita*	21.1 - 38	Saikia, 1992
			Jonathan *et al.,* 1990
Ginger	*M. incognita*	29.60-33.35	Ramana *et al.,* 1998
		74.10	Koshy, 2002
		46.40	Charles & Kuriyan, 1979
	R. similis	39-73	Sundararaju *et al.,* 1979
Turmeric	*M. incognita*	18.6-25.0	Poornima & Vadivelu, 1998
	R. similis	46-76	Sosamma *et al.,* 1979
Coriander	*M. incognita*	51	Midha & Trivedi, 1991
Cumin	*M. incognita*	34	Midha & Trivedi, 1991
Fennel	*M. incognita*	39	Midha & Trivedi, 1991
Black pepper	*R. similis* & *M. incognita*	38.5-64.6	Mohandas & Ramana, 1991
	R. similis	59	Mohandas & Ramana, 1991
	M. incognita	46	Mohandas & Ramana, 1991
Coconut	*R. similis*	30	Koshy and Geetha, 1992

1.3. Nematode Management Options

Realizing that nematodes cannot be eradicated, and that we must live with them, the overall goal is to keep the population density as low as possible. The economic threshold is the density of nematodes where the losses incurred exceed the cost of nematode management.

Horticultural ecosystem provides a congenial ecological niche for one or the other nematodes, the population of which if unchecked, may reach alarming proportions resulting in serious losses. Nematode management is therefore important for high yields and quality that are required by the high cost of modern crop production. The direct and indirect benefits of nematode management include – increased quantity and quality of the produce, improved health of plants thereby reducing their susceptibility to plant pathogens and increasing the ability to withstand adverse growing conditions, and a better utilization of nutrients and moisture.

In the field, several nematode management methods like regulatory, physical, cultural, chemical, biological and host resistance can be effectively employed to keep nematode population to a minimum level. It is not advisable to depend on a single method to control nematodes. Efficient management requires the carefully integrated combinations of several practices.

1.3.1. Regulatory Methods

If a harmful species of nematode is not present in an area, legal quarantine is the most effective method for prevention. Numerous attempts have been made to prevent the introduction of nematodes into countries or provinces by means of plant quarantine. Quarantines are established by legislative or other government actions and usually give quarantine authorities power to develop and enforce regulations to accomplish the purpose. Such regulations prohibit bringing infected plants into protected areas where similar crops are vulnerable to infection.

1.3.2. Physical Methods

Heat treatment is one of the oldest methods of nematode control. Plant parasitic nematodes can be killed by heat, desiccation, irradiation, high osmotic pressure, etc., but it is more difficult to employ physical methods in killing nematodes in soil and planting materials. It is virtually impossible to kill nematodes when they are in growing plants. Field scale treatment of soil is also difficult. Nematodes can be killed at 44 to 48°C. The nematode enzymes are inactivated at short time exposure to about 50°C *in vitro*. Bare root dip in hot water has to be specifically determined for different nematode species. Solarization is a modern technique to be developed for each situation.

1.3.3. Cultural Methods

Basically, cultural management involves depriving the nematode of a suitable host and thereby reducing the nematode population by starvation. There are many cultural practices used in nematode management. Probably the most common of these practices is crop rotation. Other cultural methods used for nematode management include fallowing, flooding, mulching, trap cropping, time of planting, use of organic amendments to the soil, intercropping with antagonistic plants and influence of fertilizers.

Botanicals and their products play an important role in the reduction of nematode population in soil. A large number of reports have been published showing that growing of antagonistic plants (marigold, mustard, sesame, asparagus, etc.) or addition of organic materials (green manures, oil cakes, crop residues, cellulosic soil amendments, etc.) to the nematode infested soils results in definite reduction of several plant parasitic nematodes.

1.3.4. Chemical Methods

The use of chemicals on a field scale for control of plant parasitic nematodes was not possible until early 1940's when effective and economical soil fumigants like D-D and EDB were discovered which made it possible to provide growers with spectacular differences in growth and yield through the effective control of nematodes and other soil pests. Several effective nematicides belonging to organophosphate and carbamate groups were developed and improvements in methods of application provided more economic control. In many parts of the world today, soil treatment for nematode control is an established farm practice.

The primary advantage of chemical control over other methods is the quick reduction of nematode population within days after the chemical is applied. The grower is able to plant a crop soon after treatment or, in some cases, at the time of treatment. Most crops are especially vulnerable to nematode attack during the seedling stage when the young root system is becoming established. Crops planted in treated soil develop extensive root systems, and the crop, especially annuals, mature before residual population of nematodes can increase to a damaging level.

1.3.5. Host Resistance

The use of resistant and tolerant cultivars is economically attractive management tactics because most infested fields have a dominant or *target* nematode. Furthermore, nematodes spread slowly, persist in the soils for long periods of time, and management by physical and chemical means is relatively expensive. Available information indicates that plant resistance to nematodes is genetically controlled by one gene in some cases, and by more than one gene in other plant-nematode combinations. Practical results have been obtained in breeding for resistance to a number of major nematodes.

1.3.6. Biological Methods

Biological suppression of plant parasitic nematodes with antagonistic fungi, mycorrhizae, bacteria and Actinomycetes is gaining increasing importance as a result of realization that many environmental and health hazards are linked with the use of chemicals. Standardization of methods for effective utilization of biocontrol agents is very important for evolving ecologically sound integrated nematode management strategies.

1.3.7. Integrated Approach

Integrated nematode management (INM) includes research, development, technology transfer and implementation needed to integrate two or more control measures to manage one or more nematode species. The aim of INM is to reduce and stabilize the nematode pest population below damaging levels, resulting in favourable long term socio-economic and environmental consequences. The components of INM systems include (i) biological monitoring; (ii) environmental monitoring; (iii) agricultural production systems; (iv) nematode crop ecosystem models; (v) system design; and (vi) implementation. The proceedings available for INM system design and implementation are based on the principles of exclusion, population reduction and tolerance.

1.3.8. Biointensive Integrated Nematode Management (BINM)

Biointensive integrated nematode management is a systems approach to pest management based on an understanding of nematode ecology. It begins with steps to accurately diagnose the nature and source of nematode problems, and then relies on a range of preventive tactics and biological controls to keep nematode populations within acceptable limits. Reduced-risk nematicides are used if other tactics have not been adequately effective, as a last resort, and with care to minimize risks.

An important difference between conventional and biointensive INM is that the emphasis of the latter is on proactive measures to redesign the agricultural ecosystem to the disadvantage of a nematode and to the advantage of its parasite and predator complex.

BINM options may be considered as proactive or reactive.

1.3.8.1. Proactive Options

Proactive options, such as crop rotations and creation of habitat for beneficial organisms, permanently lower the carrying capacity of the farm for the nematode. The carrying capacity is determined by the factors like food, shelter, natural enemy complex and weather, which affect the reproduction and survival of a nematode species. Cultural control practices are generally considered to be proactive strategies. Proactive practices include crop rotation, resistant crop cultivars including transgenic plants, disease-free seed and plants, crop sanitation, spacing of plants, altering planting dates, mulches, etc.

The proactive strategies (cultural controls) include:

- Healthy, biologically active soils (increasing below-ground diversity).
- Habitat for beneficial organisms (increasing above-ground diversity).
- Appropriate plant cultivars.

1.3.8.2. Reactive Options

The reactive options mean that the grower responds to a situation, such as an economically damaging population of nematodes, with some type of short-term suppressive action. Reactive methods generally include inundative releases of biological control agents, mechanical and physical controls, botanical pesticides and chemical controls.

Chapter 2

FRUIT CROPS

A large variety of fruits are grown in India. Of these, mango, banana, citrus, pineapple, papaya, guava, sapota, jack fruit, litchi and grape (tropical and subtropical fruits); apple, pear, peach, plum, apricot, almond and walnut (temperate fruits) and amla, ber, pomegranate, annona, fig and phalsa (arid zone fruits) are important. The leading fruit growing states are Maharashtra, Andhra Pradesh, Tamil Nadu, Karnataka, Gujarat, Uttar Pradesh, Bihar and West Bengal.

India is the second largest producer of fruits next only to China contributing 10.1% of the total world production. Total production of fruits has been estimated at 49.295 million tonnes from 4.964 million ha (Table 2.1). India is the largest producer of mango, banana and papaya. India produces 41.7% of the world mangoes and 25.7% of the bananas. In grapes, India has recorded the highest productivity per unit area in the world. The per capita consumption of fruits has increased from 40 to 85g/day within the last one decade.

Table 2.1. Area, production and productivity of fruit crops in India (2004-05) (Chamber, 2006)

Fruit crop	Area ('000 ha)	Production ('000 tonnes)	Productivity (tonnes/ha)
Apple	231	1739	7.5
Banana	530	16225	30.6
Citrus	712	5997	8.4
Grapes	60	1546	25.7
Litchi	60	369	6.1
Papaya	73	2568	35.2
Pineapple	81	1229	15.1
Pomegranate	113	793	7.0
Total	4964	49295	9.9

Fruits are amongst the most cherished bounties of nature. They are the major sources of Vitamins 'A' and 'C' and minerals like Calcium and Iron. Fresh fruits have low calorific values and are low in fats. The present area under fruit crops is estimated to be about 4.964 million hectares with an estimated production of 49.295 million tonnes. Of the several factors responsible for lower productivity of fruits, nematode parasites constitute an important limiting factor.

The destructive plant parasitic nematodes are one of the major limiting factors in fruit production throughout India. Fruit crops being perennial in nature, harbour and encourage the build up of nematode population. For centuries, man has been plagued by these microscopic organisms feeding on the roots of crop plants essential for his survival and well being. Roots damaged by the nematodes lose efficiency in the utilization of available soil moisture and nutrients resulting in reduced functional metabolism. Furthermore, the weakened and damaged roots become easy prey to many fungi and bacteria which accelerate root decay. Visible symptoms of nematode attack often include reduced growth, varying degrees of chlorosis, wilting and sometimes death of plants. These deleterious effects result in reduced yields and poor fruit quality in several fruits such as citrus, banana, pineapple and papaya. Nematode management is therefore important for high yields and quality that are required by the high cost of modern crop production.

The first report of nematode damage to fruit crops in India is that of *Meloidogyne* sp. to citrus in Andhra Pradesh (Thirumala Rao, 1956). Siddiqi (1961) reported the citrus nematode, *Tylenchulus semipenetrans* for the first time from India. Another important plant parasite, the burrowing nematode *Radopholus similis* was reported on banana from Kerala state by Nair *et al.* (1966). These are some of the early records of plant parasitic nematodes which indicated their importance in the production of fruit crops.

2.1. TROPICAL AND SUBTROPICAL FRUIT CROPS

2.1.1. Banana

Globally, banana (*Musa* spp.) is one of the maximum distributed fruit crop grown in more than 130 countries in 10 million hectares with an annual production of 95 million tonnes. It is the fourth most important food crop after rice, wheat and maize. It is a dessert fruit for millions and is used in different regions of the world as staple food. As a diet, banana is filling, easy to digest, nearly fat free, rich source of carbohydrate with calorific value of 67 calories/100g and is free from sodium, making it salt-free diet and also has zero cholesterol. Hence, regular use of it also considered to be good for heart. Banana also contains several minerals such as calcium, magnesium, potassium and phosphorus. Banana fruit is particularly rich in vitamin-C and contains significant amount of other vitamins such as vitamin-A.

Banana is an important fruit crop having great socio-economic significance in India, as it supports livelihood of millions of people. India is the largest producer of banana in the world with an annual production of 16.225 million tonnes from an area of 0.53 million ha. Although, banana occupies 10.49 per cent of the area under fruits it contributes 21.87 per cent of the total fruit production in the country. The major banana producing states include Maharashtra, Tamil Nadu, Gujarat, Karnataka, Andhra Pradesh, Bihar, Madhya Pradesh and Assam.

The national productivity of banana is very impressive (30.6 tonnes per ha). Its productivity in commercial zones especially in Maharashtra, Gujarat, Tamil Nadu, Karnataka, Andhra Pradesh and Bihar is highly impressive and comparable to the productivity in other countries. Although there is a regional preference for cultivars, "Dwarf Cavendish" and "Robusta"(AAA) are the major cultivars of commerce having substantial area. Other important banana cultivars grown in the country are Poovan, Rasthali, Virupakshi, Nendran (AAB), Chakrakeli (AAA) and Monthan (ABB).

Research and development efforts made in the last two decades have resulted in commendable increase in productivity (144 per cent in the last decade) and overall production of banana in the country. The improved production technologies include higher plant densities (4500 plants per ha), improved crop geometries coupled with adoption of drip irrigation, more judicious fertilization management through balanced use of organic and inorganic fertilizers, appropriately timed split applications of inorganics, use of proper sized rhizomes and well timed plantings, commercial production of *in vitro* plants and development of effective integrated pest management strategies. By adoption of these improved technologies, productivity of several individual farmers in different parts of the country has resulted in substantial increase (60 to 65 tonnes per ha).

Nematodes constitute one of the major limiting factors to banana production. They cause extensive root necrosis resulting in serious economic consequences viz., fertilizers are not effectively utilized, the period from planting to harvesting is lengthened, maximum bunch weights are not attained, the quality of fruit is impoverished and fields have to be replanted every 2 to 3 years because of drastic reduction in plant numbers. Furthermore, roots weakened and damaged by nematodes are easy prey to fungi which invade the roots and accelerate root decay. At present 71 species belonging to 33 genera of plant parasitic nematodes have been reported in association with banana roots from India. The burrowing nematode, *Radopholus similis* is the most important pest on banana. *Helicotylenchus multicinctus, Pratylenchus coffeae* and *Heterodera oryzicola* are the other nematodes that are of some economic importance.

2.1.1.1. The Burrowing Nematode, Radopholus similis

This nematode in India was first reported on banana from the Palghat district of Kerala by Nair *et al.* (1966). *R. similis* causes **'rhizome rot'** or **'toppling'** or **'black head'** disease of banana and is becoming a serious problem. It causes retarded growth and extensive root and rhizome necrosis.

(i) Economic Importance and Losses: The burrowing nematode is responsible for 30.76 to 41 pr cent yield loss in banana (Rajagopalan and Naganathan, 1977b; Nair, 1979; Parvatha Reddy *et al.*, 1996d; Vadivelu *et al.*, 1987). Root population of *R. similis* is indirectly correlated with the yield (Charles *et al.*, 1985).

(ii) Occurrence and Distribution: It has been reported from Tamil Nadu (Rajagopalan and Chinnarajan, 1976), Karnataka (Venkitesan, 1976), Maharashtra (Parvatha Reddy and Singh, 1980), Andhra Pradesh (Parvatha Reddy and Khan, 1987), Gujarat (Sethi *et al.*, 1981), Madhya Pradesh (Tewari and Dave, 1985), Bihar, Orissa (Mohanty *et al.*, 1972), Kerala, Uttar Pradesh, Goa, Assam, Manipur, Tripura, Nagaland and Andaman and Lakshadweep Islands.

(iii) Symptoms: Wounding of banana roots by the burrowing nematode usually induce reddish-brown cortical lesions which are diagnostic of the disease (Fig. 2.1). These lesions are clearly seen when an affected root is split longitudinally and examined immediately. Root and rhizome necrosis is manifested by varying degrees of retarded growth, leaf yellowing and falling of mature plants.

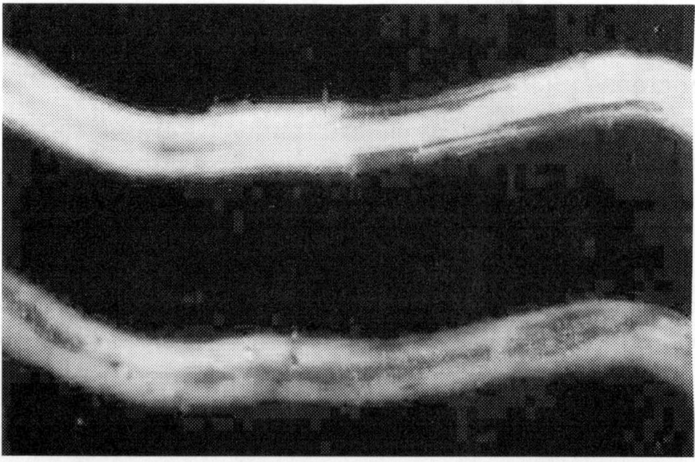

Fig. 2.1. Banana roots infected with *Radopholus similis*. Upper – Longitudinally cut root; Lower – Complete root (Courtesy: Union Carbide Agril. Products Co. Inc., 1986).

With the increase in nematode population, feeding roots are invaded and destroyed as fast as they are formed. The resulting setback in the uptake of plant nutrients leads to debility of the plant and production of smaller fruits. The lesioning of the primary roots together with the girdling and death of these anchor roots makes the plant prone to 'tip over' by wind action (Fig. 2.2).

Fig. 2.2. Premature fall of banana plants due to infection of *Radopholus similis*. Front – Toppled plants from non-treated plots; Back – Treated plants. (Courtesy: Union Carbide Agril. Products Co. Inc., 1986).

(iv) Life Cycle: The burrowing nematode has a migratory endoparasitic habit. Although the stages remain vermiform throughout, sexual dimorphism is apparent with adult males being somewhat degenerate and probably non-parasitic. Eggs are normally laid in infested tissue over 7-8 days at the rate of about 4 eggs per day. The life cycle from egg to egg extends over 20 to 25 days at 24-32°C with eggs taking 8-10 days to hatch and the larvae 10-13 days to mature. All larval stages and females except males are infective. They invade at any portion of the root, causing more root damage and are capable of spending their entire life in the roots.

(v) Host- parasite Relationship: On entering the root, the nematodes occupy an intercellular position in the cortical parenchyma where they feed on the cytoplasm of nearby cells destroying them and thus create cavities. These cavities coalesce and are continually enlarged by the nematodes feeding and tunneling laterally and towards the endodermis, producing the characteristic reddish brown lesions throughout the cortex. Three to four weeks after infection, when extensive cavities have formed, one or more deep cracks with raised margins appear on the root surface. The stele is not invaded. Eventually the root system may be reduced to a few short stubs.

(vi) Host Range: The banana biotype of *R. similis* is known to infect *Musa* spp., *Ipomoea batatas, Pueraria phaseoloides javanica*, pepper, areca nut and coconut.

(vii) Survival and Spread: The nematode does not seem to have any special adaptations for survival in the absence of a susceptible host. Free-living stages and eggs do not survive for more than 12 weeks. After a plantation is removed, the nematodes in the corms may probably survive for as long as these remain succulent.

The burrowing nematode is spread from one locality to another through planting material (corms). Trimming sets, which is often done before they are planted, is usually not sufficiently rigorous to remove infections that extend deeply into a set. *R. similis* may be disseminated when water that drains from infested areas gets recycled into irrigation system. Soil that adheres to implements, tyres of motor vehicles and shoes of plantation workers may also spread *R. similis*.

(viii) Sampling Techniques: Survey for *R. similis* may be carried out during September to December which is the peak period of nematode population. The maximum population is available at a distance of 25 to 50 cm from the base of the plant and at a depth of 20 to 40 cm. 200 ml of soil and 25 g of root samples should be collected. Tender portions of roots showing lesions yield maximum number of *R. similis* (Koshy, 1986). The roots should be cut to a length of 0.5 to 1 cm and split longitudinally into 4 pieces and incubated at 25 to 27°C to extract highest number of nematodes.

(ix) Interaction with Other Pathogens: The incidence of Panama wilt caused by *Fusarium oxysporum* f. sp. *cubense* was doubled in the presence of *R. similis* during an experimental period of 3 months. When Gros Michel bananas infected with *R. similis* were inoculated with *F. oxysporum* f. sp. *cubense*, the onset of wilt was considerably shortened. The nematodes caused the breakdown of resistance to Panama wilt in Lacatan bananas.

(x) Biotypes: Venkitesan (1976) reported that the population of *R. similis* isolated from pepper roots did not infect *Citrus sinensis, C. reticulata, C. aurantifolia, Coffea arabica*, coconut and areca nut; but infected banana only. The population isolated from banana infected pepper, coconut and areca nut; while those from areca nut infected pepper and banana but not coconut. And thus he concluded that these four isolates were different "host types".

Koshy and Jasy (1991) reported 10 races of *R. similis* in South India from among 28 populations on the basis of their multiplication on 17 differential hosts. Citrus, clove, pineapple and sugarcane cv. CO 997 were immune to all the 28 populations (Table 2.2).

Table 2.2. Races of *Radopholus similis* in India (Koshy and Jasy, 1991)

Differential hosts	Races									
	1	2	3	4	5	6	7	8	9	10
Coconut	+	+	+	+	+	+	+	+	+	+
Areca nut	+	+	+	+	+	+	+	+	-	-
Banana (Nijaln)	+	+	+	+	+	-	+	+	+	+
Banana (Robusta)	+	+	+	+	+	-	+	+	+	+
Black pepper	+	+	+	+	-	+	-	-	+	-
Sweet potato	+	+	+	+	+	+	+	+	+	+
Sugarcane (CO 449)	+	+	+	+	+	+	-	-	+	+
Sugarcane (CO 997)	-	-	-	-	-	+	-	-	-	-
Avocado	+	+	+	-	+	-	+	+	+	+
Cardamom	+	+	+	+	+	+	+	+	+	+
Coffee	+	+	+	+	+	+	+	+	+	+
Tea	+	+	-	+	+	+	-	-	-	+
Nutmeg	+	-	-	-	+	+	-	-	-	-
Clover	-	-	-	-	-	-	-	-	-	-
Pineapple	-	-	-	-	-	-	-	-	-	-
Citrus	-	-	-	-	-	-	-	-	-	-

(xi) Management Methods

(a) Physical methods: To avoid introducing inoculum into a new plantation, sets may be disinfected by paring and heat therapy. Corms of suckers should be trimmed carefully to remove the nematode affected necrotic tissue. Deep paring of infected banana suckers used for planting helps to reduce nematode population. The pared sets can be disinfected by dipping them in a hot water bath at 55°C for 25 minutes or 53°C for 30 minutes without causing significant losses. Sun drying the rhizomes before planting has been shown to be effective in the control of nematodes, which is being commonly practiced in Kerala.

(b) Cultural Methods: Nematode-free suckers should be selected for planting. Introduction of the nematode can be avoided by using nematode-free planting material grown by the meristem culture technique.

The land should be ploughed thoroughly during summer and left fallow or planted with cover crops/green manure crops (sunnhemp) not affected by nematodes for at least 6 months. Sugarcane has been shown to be a good alternate crop.

The number of nematodes in soil may be decreased either by fallowing, crop rotation or intercropping. The banana biotype of *R. similis* has a narrow host range and limited survival period in soil in the absence of host. Therefore, destruction of the infected area either mechanically with a bulldozer or by injection of an herbicide into the pseudo stem and replanting the area with nematode-free sets seems a practical approach to rehabilitate the infested plantations.

Cover Cropping: After destroying an infested plantation, planting of cover crops like *Panicum maximum* var. *trichoglume* and *Phaseolus atropurpureus* did not support *R. similis* population. Planting of sugarcane or pangola grass (*Digitaria decumbens*) following the destruction of bananas eliminated *R. similis* after 10 weeks (Loos, 1961).

Fallowing: Fallowing for a period of 3 months after banana effectively suppressed the burrowing nematode population. Flood fallowing for about 5 months destroyed not only the *Fusarium* but also the burrowing nematodes (Rajendran *et al.*, 1979).

Crop Rotation: Rotation of banana with paddy, sugarcane, green gram, cotton, pangola grass, sorghum, tobacco or cassava suppressed the nematode population and increased the yield (Rajendran *et al.*, 1979). Crop rotation with rice in wet lands and with vegetables, cotton and cereals in garden lands helps to reduce the nematode population. Rice or green gram grown after banana suppressed the population of *R. similis, P. coffeae* and *H. multicinctus* (Rajendran *et al.*, 1979).

Intercropping: The nematode population both in soil and roots were considerably reduced when banana was intercropped with nematode antagonistic plants such as marigold. *Crotalaria juncea* was most effective in reduction of soil nematodes and *Coriandrum sativum* was more effective in reduction of root nematodes. *C. juncea* gave maximum bunch weight (16.7 kg) recording 51% more than control (11.03 kg) followed by *Tagetes erecta* (16.2 kg), *Sesamum indica* (15.87 kg), *Acorus calamus* (15.67 kg) and carbofuran (15.23 kg) (Charles and Venkitesan, 1993).

Intercropping with sunnhemp, coriander, marigold, radish or lucerne reduced nematode population and increased yield (Vadivelu *et al.*, 1987). Intercropping sunnhemp in banana field and ploughing the green manure 45 days after sowing was found effective in reducing *R. similis, P. coffeae* and *H. multicinctus* population by 38.4%, followed by marigold and cowpea which recorded 29.0 and 22.3% reduction, respectively (Shanthi, 2003). Banana intercropped with *Crotalaria* reduced *R. similis* population in root and increased plant growth and yield (Charles *et al.*, 1985). Lakshmana Murthy (1983) reported that when banana was intercropped with papaya and marigold, *R. similis* was not encountered. Subramaniam and Selvaraj (1988) reported that intercropping with *Tagetes, Crotalaria* and radish significantly reduced the *R. similis* population.

Mulching: Mulching with black polythene at 20 per cent moisture depletion recorded highest yield with low population of *R. similis* (Vadivelu *et al.*, 1987). Covering soil with black polyethylene, sugarcane trash and banana trash are reported to reduce *R. similis* and *Pratylenchus* sp. in banana roots (Bhattacharya and Rao, 1984).

Botanicals: Application of neem cake at 800 g per plant (two split applications) or neem coated urea at 110 g per plant reduced the population of the burrowing nematode and also increased the yield of banana (Vadivelu *et al.*, 1987). Neem formulations like Econeem and Nimbicidine showed maximum efficacy in reducing the nematode population with increased plant growth and yield.

Application of press mud obtained from sugarcane waste treated plants has shown significant reduction in nematode populations (Sundararaju *et al.*, 2004).

(c) Chemical methods

Sucker Treatment: Paring and pralinage with carbofuran at 1.2 g a.i. per sucker recorded low incidence of *R. similis* and highest yield (Rajendran and Naganathan, 1977a). Dipping of banana suckers in 0.1 per cent aldicarb solution increased plant height, bunch weight, fingers per bunch, induced early flowering and effectively reduced the burrowing nematode population (Venkitesan and Charles, 1983).

Bare Root-dip Treatment: Various chemicals tried as bare root dip treatment for the management of *R. similis* on banana is summarized in Table 2.3.

Table 2.3. Effect of various chemicals as bare root dip treatment for the management of *R. similis* on banana

Chemical	Dosage	Duration & type of treatment	Effectiveness
Carbofuran	750 ppm (75% WP)	10 min. immersion	Eliminated infection
	1.2 g a.i./sucker	Immersion of suckers in clay slurry, uniform sprinkling of nematicide	Reduced population significantly and increased bunch weight
DBCP	Combined with Bordeaux mixture	Root dip	Reduced infection from 89 to 1% after 8 months

	0.05% emulsion	20 min. dip after paring	Eliminated infection , healthy root growth & nematode management
	Solution based on suspension (70%)	Root dip	Nematode management
	1% suspension	5 min. dip	Eliminated infection
	550 ml + 40 l clay soil + 50 l water	Dip in suspension for few seconds (sets kept moist for 24 hr.)	No attack or spread or multiplication till 14 months
	0.5% emulsion	5 min. dip	Eliminated infection
Ethoprophos	5000 ppm	1 min. dip	Eliminated infection
	600 ppm	20 min. dip	Eliminated infection
Fensulfothion	750 ppm	10 min. immersion	Eliminated infection
	1.7g a.i./sucker	Immersion of suckers in clay slurry, uniform sprinkling of nematicide	Reduced population significantly and increased bunch weight
Oxamyl	1500 ppm	10 min. immersion	Eliminated infection
	1000 ppm	20 min. immersion	Eliminated infection
Parathion	0.03% solution	20 min. dip after paring	Nematode management
Phenamiphos	50 ppm (40% emulsion), 8-11 ml/plant	15-30 min. dip	Eliminated infection
	100 ppm (40% emulsion)	5 min. dip	Eliminated infection
	250 ppm (40% emulsion)	30 min. dip	Eliminated infection
	2500 ppm (40% emulsion)	30 min. dip followed by 30 min. water rinse	Reduced infection from 71% to 14%
Phorate	50 ppm (40% emulsion), 8 to 11 ml/plant	15-30 min. dip	Eliminated infection

Main Field Treatment: Soil application of fenamiphos at 3 kg a.i. per ha was effective in increasing plant height, plant girth, number of leaves and yield and in reducing the nematode population (Parvatha Reddy *et al.*, 1991).

Various chemicals tried for the management of *R. similis* on banana under field conditions is presented in Table 2.4.

Table 2.4. Effect of various chemicals for the management of *R. similis* on banana under field conditions

Chemical	Dosage	Duration & type of treatment	Effectiveness
Aldicarb	6 g a.i./plant	4 times a year	Reduced the nematode population significantly
	3 g a.i./plant	3-4 times a year	Reduced the nematode population significantly
	4 g a.i./plant	Applied at planting time & 4 months later	Reduced the nematode population, produced early crop & improved yields
Aldicarb sulfone	6.75 g a.i./plant	3-4 times a year	Reduced the nematode population & increased yields
	9 g a.i./plant	4 times a year	Reduced the nematode population & increased yields
	4 g a.i./plant	Thrice a year in a 35 cm band around crown/plant	Doubled the yield
Carbofuran	4 g a.i./plant	Applied at planting time & 4 months later	Reduced the population, increased growth & yields
	1 g a.i./plant	Applied at planting time & 3 months later	Reduced the population & increased yields
DBCP	40 litres/ha	At planting (May-June), 25 l/ha in Oct. & 15 l/ha in March	Controlled nematodes & gave heavy bunches
DD	300 litres/ha	At planting	Increased yields by 30-50%
EDB	150 kg/ha	At planting	Increased yields by 30-50%
Ethoprophos	4-5 g a.i./ha	3-4 times a year	Reduced the nematode population & doubled yields
Oxamyl	3 g a.i./plant	3-4 times a year	Reduced the population, increased growth & yields
Phenamiphos	2-3 g a.i./plant	Applied at 4 monthly intervals (April, August & December)	Reduced the population, increased growth & yields

	3 g a.i./plant	Thrice a year in a 35 cm band around crown/plant	Doubled the yield
Phorate	3-4.5 g a.i./plant	At planting after paring	Eliminated infection
Vapam	900 litres/ha	Soil fumigation before planting	Reduced the nematode population

(d) Biological Methods

Antagonistic Bacteria: Soil application of *Pseudomonas fluorescens* at 20 g/plant (2.5 x 10^8 cfu/g) was effective in increasing the plant growth parameters (pseudo stem height and girth, number of leaves, root length and weight) and in reducing the nematode population in soil (62.34, 67.12 and 58.87% of *R. similis, P. coffeae* and *H. multicinctus*, respectively) and roots (60.35, 62.00 and 56.54% of *R. similis, P. coffeae* and *H. multicinctus*, respectively). Rhizosphere treatment with *P. fluorescens* at 20 g/250 ml water along with soil treatment with 100 g powder/rhizome gave 56.5% reduction in nematode population. During the second month the nematode reduction was 63.8% (Nirmal Johnson and Devarajan, 2004).

Antagonistic Fungi: Application of fungal antagonists like *Paecilomyces lilacinus, Trichoderma viride, Pochonia chlamydosporia* and bacterial antagonists like *Pasteuria penetrans* are effective in reducing the nematode population in soil and roots of banana.

Arbuscular Mycorrhizal Fungi: In an interaction study of arbuscular mycorrhizal fungus (AMF), *Glomus fasciculatum* and *R. similis*, the multiplication of nematodes in the root was suppressed and soil population was significantly lower. The banana plants inoculated with mycorrhizae also recorded higher N, P, K, Ca, Mg, sugars, phenols and amino acids in their root system suggesting AMF as a potential biocontrol agent in the management of *R. similis* (Umesh *et al.*, 1988). Inoculation of *G. mosseae* during raising of banana nursery was found effective in reducing the nematode populations of *R. similis* and *M. incognita* besides enhancing plant growth and bunch weight under field conditions. There was 35.7% increase in bunch weight, 30% increase in number of hands and 39.6% increase in number of fingers in the mycorrhizae colonized plants (Sosamma *et al.*, 1998).

Biofertilizers: Soil application of biofertilizer, *Azospirillum* at 20 g/plant had significant effect in enhancing the plant growth and reduced the population of *R. similis, P. coffeae* and *H. multicinctus* in soil to an extent of 50.1, 45.5 and 48.9%, respectively over control (Shanthi *et al.*, 2004).

(e) Host Resistance

Banana cvs./hybrids viz., Gold Finger, Kadali, Octoman, Then Kunnan, Kunnan, Dulmal Bhog, Malbhug, Tongat, Pedalimoongil, Pisang Lilin, Pisang Jari Buaya, Poovan, Anil Vazhai, Velatothan, Dudhsagar, Chinachampa, Chinia Kurlan, Benid, Ayiramkapoovan, Pisang Seribu, Vennettukunnan, Veneetu Mannan, Palayankodan, Beecha, Brilechinia, Local Dwarf, Manick Champa, Singhlal, Pey Kunnan, Elakkibale, Paka, Annaikomban, Bodles Altafort and Hybrid-74 (Matti x Pisang Lilin) were found tolerant/ resistant to *R. similis* (Parvatha Reddy *et al.*, 1989; Ravichandra and Krishnappa, 1985; Charles *et al.*, 1983; Vadivelu *et al.*, 1987; Sundararaju *et al.*, 2004). Pisang Jari Buaya from Malaysia, is known to be resistant to *R. similis* (Pinochet and Rowe, 1979).

(f) Integrated Methods

Bioagents and Botanicals: Integration of neem cake at 400 g per plant with *P. penetrans* at 100 g soil (300 spores/g)/*T. harzianum* / *T. viride* at 250 g/plant while planting was found effective in reducing nematode population both in soil and roots of banana by more than 50 per cent and increased plant growth parameters. The treatment should be repeated 4 months after planting. The root-knot and the burrowing nematode on banana were effectively managed by integration of neem cake at 500 g and FYM enriched with *T. harzianum* at 2 kg/plant. The above treatment increased the fruit yield to 45 kg/plant and bunches came to harvest 65 to 75 days earlier.

Soil application of 2 kg FYM with *P. fluorescens* (with 1×10^9 spores/g) and *P. chlamydosporia* (with 1×10^6 spores/g) per plant at the time of planting and at an interval of 4 months significantly reduced the burrowing nematode by 64% compared to control.

Bioagents and AMF: Integration of neem cake (at 200 g/plant) with *Glomus mosseae* (100 g/plant with 25-30 chlamydospores/g) was most effective in reducing the burrowing nematode population both in soil and roots, while karanj cake with *G. mosseae* gave maximum increase in plant growth parameters and fruit yield of banana. Mycorrhizal root colonization and number of chlamydospores of *G. mosseae* were maximum in neem cake amended soil (Parvatha Reddy *et al.*, 2002).

The following INM measures are suggested for the effective management of the burrowing nematode:

- Mulching with black polythene at 20 per cent moisture can reduce the populations of nematodes.
- Intercropping with sunnhemp, coriander, marigold or lucerne can reduce nematode population in the soil.

- Fallowing for a period of 3 months after banana can suppress the burrowing nematode population.

- Flood fallowing for about 5 months destroy not only the Panama wilt fungus - *Fusarium* but also the nematodes.

- Use of resistant/tolerant cultivars.

- Use of nematode-free certified planting material.

- Increased use of green manure and other organic amendments.

- Crop rotation with paddy, sugarcane and coriander.

- Intercropping with sunnhemp or marigold.

- Application of neem cake at 800 grams per plant (two split applications - while planting and 1 month after planting) can reduce the population of the burrowing nematode and also increase the yield of banana.

- Paring and pralinage with carbofuran at 5 grams per sucker reduce the incidence of nematode. For doing this, chop the rhizome portion thoroughly. Prepare mud slurry and dip this chopped portion of rhizome of the sucker in the mud slurry and sprinkle 5 grams of carbofuran on the rhizome.

- Dipping of banana suckers in 0.1 per cent aldicarb solution increase plant height, bunch weight, fingers per bunch, induce early flowering and effectively reduce the burrowing nematode population.

- Application of bio-control agents like *Paecilomyces lilacinus* and *Trichoderma harzianum* to the plants along with neem cake and farm yard manure (FYM) is beneficial. One kg of each of these bio-agents can be used to enrich 1 tonne of farm yard manure. Farm yard manure added with these bio-agents should be left under shade for a period of 15 days and in between it is advisable to mix FYM thoroughly at an interval of 3 days to enrich FYM with the bio-agents. This enriched FYM can be added at 1 kg per pit at the time of planting and thrice after planting with an interval of 2 months at the same dosage.

2.1.1.2. The Spiral Nematode, Helicotylenchus multicinctus

The spiral nematode is considered to be the important nematode pest in banana causing severe economic loss.

(i) Economic Importance and Losses: The spiral nematode causes serious decline of banana yield to the tune of 34 to 56% with delayed flowering. It is responsible for 33.83 per cent loss in yield, 55.88 per cent loss in number of fruits per bunch and delayed fruiting by 134 days (Vadivelu *et al.*, 1987).

(ii) Symptoms: The spiral nematode incites discrete, relatively shallow, necrotic lesions on banana roots. *H. multicinctus* causes extensive root necrosis, die-back and dysfunction leading eventually to debility of the entire plant.

(iii) Life Cycle: Groups of 8-26 eggs were observed in discoloured cortical tissues. Eggs hatch in 48-51 hr in tap water at 30°C and the first molt is believed to take place outside the egg shortly after hatching. The second stage larvae are recognized by their ventral digitate tail process. The female gonad primordium is 2 and 6 celled in the early third stage larvae and at the third molt, respectively. During the fourth molt, the male and the female gonads complete their development and the vulva and vagina are seen in female which perhaps complete entire life cycle within the roots.

(iv) Host Range: This nematode is known to infect banana, cocoa, sweet potato, citrus, sugarcane, cassava, coffee, maize, mango, oil palm, rice, rubber, tea, yam, avocado, Bermuda grass and grapevine.

(v) Host-parasite Relationship: The adults and larvae, penetrate the epidermis of the root within 36 hr of inoculation. The females and the males feed directly on parenchyma cells and after 4 days, the nematodes get usually wholly within the cortex to a depth of 4-6 cells. The head of each nematode is usually oriented parallel to the long axis of the root, but the posterior portion is curved and occupies several contiguous cells. The nematode withdraws cytoplasm from surrounding cells, the cell walls are distorted or ruptured, and the nucleus if intact, is enlarged. Evacuated cells together with those near 'nests' of eggs, soon become discoloured and necrotic. The characteristic macroscopic brown flecks or discrete necrotic lesions then appear.

(vi) Survival and Spread: Survival occurs on infected corms or on tissue remaining from the previous crop. The nematode easily gets introduced into virgin land with banana soil and sets, usually brought from old infested plantations. Infected planting material is also the main means of spread.

(vii) Management Methods

(a) Physical Methods: The practice of peeling or paring of banana sets to remove necrotic tissue and dipping in hot water (53-55°C) for 20 min. is advocated.

(b) Cultural Methods:

Crop Rotation: Rice or green gram after banana suppressed the population of *H. multicinctus*.

Intercropping: Intercropping with sunnhemp in banana field was found effective in reducing *H. multicinctus* population by 38.4%, followed by marigold and cowpea which recorded a 29.0 and 22.3% reduction, respectively.

(c) Chemical Methods: In recent years, emphasis has been shifted to use of granular nematicides for the management of the spiral nematode (Table 2.5).

Table 2.5. Chemical control of *Helicotylenchus multicinctus* on banana using granular nematicides.

Nematicide/Dose	Method of application	Results
Phenamiphos - 1.2g a.i./m² in new plantations & 2.5g a.i./ m² in old plantations	At 4 months interval	---
Phenamiphos - 2g a.i./plant	---	Bunch weights increased
Aldicarb – 3-4g a.i./mat	At planting time & 4 months later	Gave earliest & highest yield
Carbofuran-42g/sucker	Applied uniformly over pared sucker dipped in clay slurry	---
Fensulfothion- 17g/sucker	Applied uniformly over pared sucker dipped in clay slurry	---
Ethoprophos - 4g a.i./ plant	At 4 months interval	Improved yields
Oxamyl – 3.6 kg a.i./ha	---	---

(d) Biological Methods

Antagonistic Bacteria: Soil application of *Pseudomonas fluorescens* at 20 g/plant (2.5 x 10⁸ cfu/g) was effective in increasing the plant growth parameters (pseudo stem height and girth, number of leaves, root length and weight) and in reducing the spiral nematode population in soil (58.87%) and roots (56.54%).

(e) Host Resistance: Banana cvs./hybrids viz., Neyvannan, Rasthali, Wather, Peyan, Kullan, Krishna Vazhai, Karpurvalli, Sirumalai, CO-1, Kunnan, Robusta, Gros Michel, Vennettu Kunnan, Hybrid-94, 100, 106 and 109 (all Matti x Tongat) and Hybrid-74 (Matti x Pisang Lilin) were found tolerant/resistant to *H. multicinctus* (Vadivelu *et al.*, 1987).

(f) Integrated Methods

Bioagents and Botanicals: Significant reduction in population of *Helicotylenchus multicinctus* was observed in banana plants treated with *P. fluorescens* at 2g/ *Glomus mosseae* at 25g along with press mud at 3 kg/plant. These treatments also enhanced the plant height, pseudo stem girth, number of leaves, leaf area and bunch weight (Jonathan and Cannayane, 2002).

2.1.1.3. The Lesion Nematode, Pratylenchus coffeae

(i) Economic Importance and Losses: P. coffeae has been reported to cause decline of banana plantations. The lesion nematode causes most damage to banana in association with *F. o. f.* sp. *cubense* (Vadivelu *et al.*, 1987). In India, crop loss caused by the root-lesion nematode in banana cv. Nendran was reported to be 45%.

(ii) Distribution: P. coffeae is reported from banana growing regions throughout the world almost equal to that of *R. similis*. In India, the nematode is known to occur on bananas in southern India, Gujarat, Orissa, Bihar and Assam.

(iii) Symptoms: The symptoms caused by *P. coffeae* are very similar to those caused by *R. similis*. Both *R. similis* and *P. coffeae* are associated with the blackhead toppling diseases of bananas.

(iv) Life Cycle: The life cycle of *P. coffeae* is similar to that of *R. similis* except that *P. coffeae* males penetrate and feed on the underground plant tissues and cause decay. The egg to egg cycle is completed in about 27 days at 26-32°C. The optimum temperature for reproduction is 29.5°C.

(v) Host-parasite Relationship: P. coffeae enter the roots and feeds on cortical cells resulting in the production of dark lesions in the cortex. The corm is similarly infected resulting in necrosis of the outer tissues.

(vi) Survival and Spread: The lesion nematode can survive for 6 months in the soil, in the absence of its host. Root lesion nematodes have been observed infesting the corm, so spread occurs through planting material.

(vii) Management Methods

(a) Cultural Methods

Fallowing: A fallow period of 6 months after destruction of all banana plants eliminated *P. coffeae.*

Intercropping: A significant reduction in *Pratylenchus* sp. population (85%) was observed in the banana field where *Tagetes erecta* was grown as intercrop. The yield of the plants significantly increased (12 kg/plant) when intercropped with *Tagetes* spp. compared with the untreated control (7 kg/plant) (Sundararaju *et al.*, 2002). Intercropping with sunnhemp in banana field was found effective in reducing *P. coffeae* population by 38.4%, followed by marigold and cowpea which recorded a 29.0 and 22.3% reduction, respectively. Rice or green gram after banana suppressed the population of *P. coffeae.*

Botanicals: A significant reduction in *P. coffeae* population and yield increase was recorded in plants that have received 50% N applied through neem cake (Sundararaju and Kumar, 2003). Among neem formulations evaluated for the management of *P. coffeae* and *M. incognita,* Econeem and Nimbicidine showed maximum efficacy in reducing the nematode population with increased plant growth and yield (15 kg/plant) (Sundararaju and Cannayane, 2002). A significant reduction in *P. coffeae* population and increased yield in plants treated with press mud (15 t/ha) was recorded. Application of press mud is economical and eco-friendly (Sundararaju *et al.,* 2002). Application of distillery sludge at 2.5 kg + vermicompost 1 kg + neem cake at 1 kg + poultry manure at 2.5 kg/plant at 3, 5, and 7 months after planting significantly reduced *P. coffeae* population. It was on par with distillery sludge at 2.5 kg + neem cake at 1 kg treatment. Neem cake at 200 g/plant recorded the maximum increase in plant growth parameters with significant reduction in lesion nematode (*P. coffeae*) population in banana cv. Nendran.

(b) Biological Methods

Antagonistic Fungi: Application of *Trichoderma viride* at 20 g/plant at the time of planting and repeated 3 months after planting is effective in controlling *P. coffeae* as well as reducing the incidence of Panama wilt in banana.

Application of *Paecilomyces lilacinus* at 10 g along with FYM at 500 g or neem cake at 250 g/plant effectively suppressed the lesion nematode population by 60 and 63%, respectively (Poornima *et al.,* 2004).

Antagonistic Bacteria: Soil application of *Pseudomonas fluorescens* at 20 g/plant (2.5 x 10^8 cfu/g) was effective in increasing the plant growth parameters (pseudo stem height and girth, number of leaves, root length and weight) and in reducing the lesion nematode population in soil (67.12 %) and roots (62.00%).

(c) Host Resistance

Banana cvs./hybrids viz., Kunnan, Pey Kunnan, Then Kunnan, Nattu Poovan, Pidi Monthan, Chirapunji, Singhlal, Sakkarachayam, Malai Kala, Manik Champa, Kartobium Tham, Vennettukunnan and Hybrid-74 (Matti x Pisang Lilin), H-21, H-55, H-59, H-65, H-84, H-109, H-110 were found tolerant/resistant to *P. coffeae* (Vadivelu *et al.,* 1987; Sundararaju *et al.,* 2004).

2.1.1.4. The Cyst Nematode, Heterodera oryzicola

The cyst nematode is an important nematode found on banana cv. Nendran in Kerala.

(i) Occurrence and Distribution: The infestation of a cyst nematode, *Heterodera oryzicola* on banana variety Nendran was observed for the first time in India at Banana Research Station, Kannara, Kerala during July 1983 (Charles and Venkitesan, 1984). These nematodes feed on fibrous roots and cause necrosis. In a subsequent survey, this nematode was recorded in several locations in Trichur, Ernakulam, Quilon and Calicut districts of Kerala especially on banana cultivated in paddy land. *H. oryzicola* also infects paddy and *Cynodon dactylon.* This creates practical difficulties in adoption of crop rotation with banana in rice fields. The infested plants are progressively debilitated leading to substantial yield losses.

H. oryzicola was also reported on banana cultivars Saldathi and Njalippnan from Goa (Koshy *et al.*, 1987).

(ii) Economic Importance and Losses: H. oryzicola has been reported to cause considerable damage to banana in certain pockets of Kerala. An initial inoculum level of 100 and 1000 viable cysts per plant could reduce bunch weight by 20.5 to 56.6 per cent. The nematode infection was observed to affect the quality of fruits by increasing acidity and reducing total sugars (Charles, 1989).

H. oryzicola attacks the minute tertiary feeder roots and not observed to attack the fleshy roots and rhizome portions. The affected feeder roots turn black in colour and detach easily (Charles, 1989).

(iii) Symptoms: The cyst nematode attacks thin lateral roots and the root weight is considerably reduced (62-87%). The damage caused to the root tissue may suppress the flow of food materials to the various parts of the plant resulting in general decline, stunting and premature defoliation.

(iv) Life Cycle and Histopathology: Life cycle of the cyst nematode on banana cv. Nendran is completed in 23 days from egg to egg stage. The second stage juveniles took to develop into adult males and white females with egg deposition within 12 and 17 days, respectively (Charles, 1989).

Histopathological studies revealed that the nematode attacked xylem vessals of the root with the head attached to the stele. The cortical cells and epidermis were ruptured at the infection site. Black discolouration was seen on the nearby cortical cells without any cellular change (Charles, 1989).

(v) Ecology, Population Fluctuation and Soil Types: Coastal sandy and sandy loam soils significantly favoured nematode activities affecting plant growth and nematode multiplication. A temperature of 25°C, flooded condition of the soil with a pH of 6.7 influenced higher percentage of larval emergence by 50th day (Charles, 1989).

(vi) Management Methods

(a) Physical Methods: The nematode can be controlled by treatment of suckers (by hot water or chemicals) before planting.

(b) Cultural Methods

Organic Amendments: Application of distillery sludge at 2.5 kg + vermicompost 1 kg + neem cake at 1 kg + poultry manure at 2.5 kg/plant at 3, 5, and 7 months after planting significantly reduced *H. oryzicola* population. It was on par with distillery sludge at 2.5 kg + neem cake at 1 kg treatment (Sundararaju *et al.*, 2002).

2.1.1.5. Root-knot Nematodes, Meloidogyne spp.

The root-knot nematodes, *M. incognita* and *M. javanica* attack bananas (Fig. 2.3.). In 1893, Cobb made one of the first associations of a root-knot species with bananas. Root-knot nematodes have not been reported to cause yield reductions in bananas, but may interact with other nematodes or soil pests. *M. incognita* caused 30.90% loss in fruit yield of banana (Jonathan and Rajendran, 2000).

(i) Symptoms: The infected plants were stunted, having small chlorotic leaves and galled roots (Fig. 2.3). Galling of the plant roots is most commonly found in areas previously planted with sugarcane. The galls vary in size and occur at the tips as well as in other areas along the root. Roots with galled tips cease to grow and sometimes develop secondary roots above the gall. Swollen female nematodes were found inside the galled and sometimes non-galled roots.

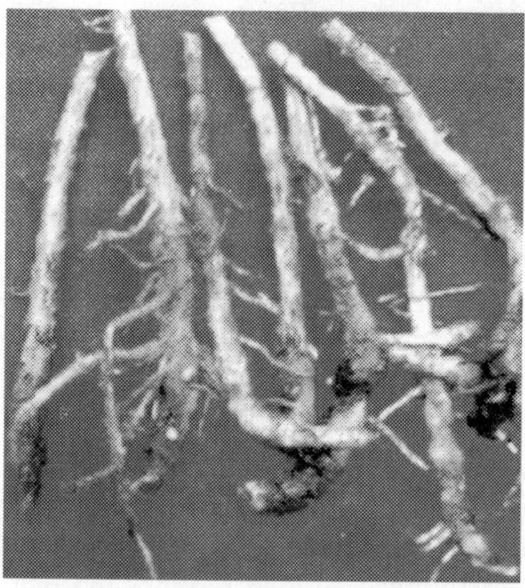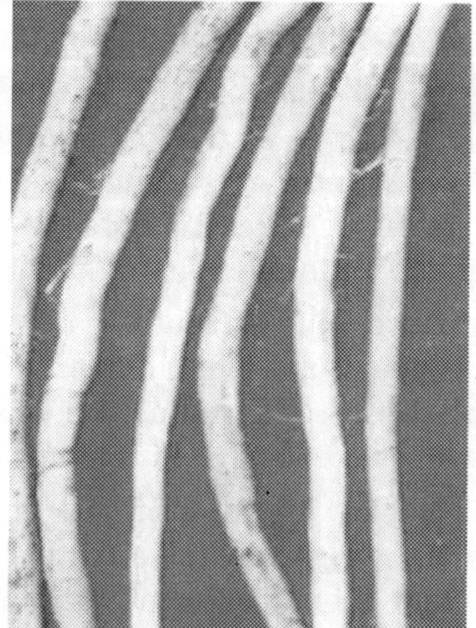

Fig. 2.3. Root-knot nematode on banana. Left – Infected; Right – Healthy. (Courtesy: NRC on Banana, Trichy).

(ii) Life Cycle: In thick, fleshy primary roots, egg-masses may not protrude outside the root surface and multiple cycles can be completed within the same root, depending on the longevity of this root and the severity of necrosis.

(iii) Survival and Spread: Root-knot nematodes survive on dicotyledonous plants, which are usually present in most soils in which bananas are growing. Survival and spread also occurs with the planting material on infected roots and corms.

(iv) Management Methods

(a) Physical Methods: The suckers can be pared to a depth of 1 cm and treated in hot water at 55°C for 20 min (Inomoto and Monteiro, 1989).

(b) Cultural Methods: Suckers should be selected from nematode-free mother plants. In wet lands, banana can be rotated with rice. Marigold can be grown in basins and incorporated into the soil around the plants.

Botanicals: Neem cake at 2 t/ha, FYM at 25 t/ha and press mud at 15 t/ha can be applied to encourage predacious nematodes and antagonistic fungi which in turn kill the nematodes. Application of distillery sludge at 2.5 kg + vermicompost 1 kg + neem cake at 1 kg + poultry manure at 2.5 kg/plant at 3, 5, and 7 months after planting significantly reduced *M. incognita* population. It was on par with distillery sludge at 2.5 kg + neem cake at 1 kg treatment.

(c) Biological Methods

Arbuscular Mycorrhizal Fungi: Prior colonization of banana suckers with *Glomus mosseae* was found to reduce the root and soil population of *M. incognita.*

Antagonistic Bacteria: Soil application of *Pasteuria penetrans* at 1 x 10^8 spores/g root powder mixed with 200 cc soil/plant and evenly distributed 45 days after planting recorded lowest final nematode population (473) at harvest compared to control (620). The yield was higher in plants treated with *P. penetrans* (3.5 kg/plant) compared to check (2.1 kg/plant) (Karuna *et al.*, 2001).

(d) Host Resistance: Banana cvs. Patkapuri, Pisang Jari Buaya, Poovan, Pedali Moongil, Kunnan, Pisang Seribu, Veneetu Mannan, Elakkibale, Palayankodan, Singhlal, Karpuravalli, Manik Champa, Boddida Bukisa, Sappumala Anamulu, Sabari, Hoobale, Mendi and Kothia were found resistant to *M. incognita* in Orissa (Ray and Parija, 1987). In Tamil Nadu, cvs./hybrids such as Chakia, CO 1, Pisang Lilin, H 74, H 94 and H 110 were found resistant to root-knot nematodes (Jonathan, 1994; Sundararaju *et al.*, 2004).

(e) Integrated Methods

Bioagents and Botanicals: Soil application of 2 kg FYM with *P. fluorescens* (with 1×10^9 spores/g) and *P. chlamydosporia* (with 1×10^6 spores/g) per plant at the time of planting and at an interval of 4 months significantly reduced the root-knot nematode by 76% compared to control.

2.1.1.6. *The Reniform Nematode, Rotylenchulus reniformis*

The reniform nematode, *R. reniformis* attack bananas.

(i) Economic Importance: In field plots where *R. reniformis* was the predominant parasitic nematode, yield reduction of 19.3 tonnes per ha was observed.

(ii) Symptoms: Nematodes feed on secondary or tertiary roots of bananas and destroy feeder roots. Necrotic lesions are produced in the area of the roots around the female head. The posterior portion of the female body and the egg mass are easily seen with a hand lens. High populations of *R. reniformis* cause severe necrosis and destruction of feeder roots.

(iii) Management Methods

(a) Cultural Methods: Crop rotation with Pangola grass gives effective control of the reniform nematode on bananas.

(b) Integrated Methods: Soil application of neem cake + *T. viride* + carbendazim was found effective in reducing the disease complex and in increasing the banana fruit yield (15.147 t/ha). This treatment also gave minimum lesion index (1.1) and root-knot index (1.0) as compared to control (4.0) with cost benefit ratio of 1:2.72 (Ravi *et al.*, 2001).

2.1.1.7. *Future Lines of Investigations*

- Intensive survey for *Radopholus similis* and *Heterodera oryzicola* in major banana growing regions.

- Crop loss assessment due to major nematode pests.

- Introduction of internal plant quarantine regulation for the burrowing nematode and disinfection of banana suckers at the sites of indigenous germplasm collections.

- Production of certified banana suckers.

- Research on biological control, crop rotation, cultural practices and sanitation methods.

- Studies on the use of mycorrhizae and other antogonistic fungi/bacteria.

- Studies on different races of *R. similis* in India.

- Role of nematodes in disease complex with fungi and bacteria.

- Developing resistant cultivars.

- Developing integrated management technologies using tolerant/ resistant cultivars, parasitic fungi and bacteria, minimal quantity of nematicides, organic amendments and vesicular arbuscular mycorrhizae should be the priority areas of research.

2.1.2. Citrus

Citrus fruits (*Citrus* spp.) rank third in area and production after mango and banana with an estimated production of 5.997 million tonnes from an area of 0.712 million hectares. There are many commercially grown *Citrus* spp. such as mandarins (*Citrus reticulata*), sweet orange (*C. sinensis*), acid lime (*C. aurantifolia*), lemon (*C. limon*), grape fruit (*C. paradisi*), pumello (*C. grandis*), etc. Although one or the other species of the citrus fruit is grown in almost all the states of India, but the major citrus growing states are Andhra Pradesh, Maharashtra, Karnataka, Punjab, Gujarat, Madhya Pradesh, Orissa and Manipur. Citrus is a highly sensitive crop and is attacked by number of insect pests, diseases and nematodes. These play major role towards the low productivity of citrus in India with national average of 8.4 tonnes per hectare, as compared to 25-30 tonnes per hectare in other citrus growing countries.

Nematodes are one of the important limiting factors in citrus production throughout India. Citrus crop being perennial in nature, harbour and encourage the build up of nematode population throughout the year. The first record of an association between a nematode and citrus appears to be that of Thirumala Rao (1956), who found *Meloidogyne* sp. parasitizing citrus roots in Andhra Pradesh. But it was not until the discovery in 1961 of *Tylenchulus semipenetrans*, on the roots of citrus trees in Uttar Pradesh (Siddiqi, 1961) that a nematode was found to cause a disease condition of citrus, later called "slow decline". At present 122 species within 57 genera of plant parasitic nematodes have been reported in association with citrus roots from many parts of the country. However, most of these nematodes are not known pathogens of citrus and their true relationship with their host plant still remains to be established.

Of the 122 species of plant parasitic nematodes reported in association with citrus roots from India, proof of pathogenicity to citrus is available for only eight species viz., *Tylenchulus semipenetrans* (Siddiqi, 1961), *Pratylenchus coffeae* (Siddiqi, 1964), *Hoplolaimus indicus* (Gupta and Atwal, 1972), *Meloidogyne africana* (Chitwood and Toung, 1960), *M. indica* (Whitehead, 1968), *M. javanica* (Mani, 1986), *Radopholus citrophilus* (Suit and DuCharme, 1953) and *Hemicycliophora arenaria* (Van Gundy, 1957).

Except for the citrus nematode (*T. semipenetrans*), other nematode species that are pathogenic on citrus have limited geographic distribution. This review will be devoted to these eight cases of nematode-citrus relationships that have been described in some detail.

2.1.2.1. The Citrus Nematode, *Tylenchulus semipenetrans*

Siddiqi (1961) reported the citrus nematode for the first time from India in a paper presented before the Indian Science Congress and observed that about 80 per cent of the citrus trees at Aligarh, Uttar Pradesh, were infested with this nematode. The nematode causes "slow decline" and is considered to be one of the factors responsible for die-back of citrus trees in India (Chona *et al.*, 1965).

(i) Economic Importance and Losses: Yield increases of 40 to 200 per cent have been obtained following the control of the nematode over a range of growing conditions (Swarup and Seshadri, 1974). *T. semipenetrans* was responsible for 69%, 29% and 19% loss in fruit yield of sweet orange (Baghel and Bhatti), lemon (Mukhopadhyaya and Suryanarayana, 1969) and sweet lime (Mukhopadhyaya and Dalal, 1971), respectively.

(ii) Occurrence and Distribution: The citrus nematode is known to be widely distributed in major citrus growing regions of the country. It has been reported from Uttar Pradesh (Siddiqi, 1961); Delhi, Punjab, Rajasthan, Maharashtra, West Bengal, Assam (Chona *et al.*, 1965); Himachal Pradesh (Mukhopadhyaya, 1970a), Sikkim (Edward and Rai, 1970); Orissa (Khuntia and Das, 1969); Haryana (Mukhopadhyaya and Suryanarayana, 1969); Bihar (Prasad and Chawla, 1965); Karnataka (Swamy *et al.*, 1973); Kerala (Nair, 1965); Tamil Nadu (Muthukrishnan and Sivakumar, 1973) and Andhra Pradesh (Krishnamurthyrao and Thammiraju, 1975).

(iii) Symptoms: 'Slow decline' of citrus is a diseased condition of trees with symptoms similar to those caused by drought and malnutrition. Affected trees exhibit reduced vigour (Fig. 2.4), chlorosis and falling of leaves, twig dieback and consequently, reduced fruit production (Prasad and Chawla, 1965). This decline of the tree is gradual and persists until the crop is so small that tree maintenance may become uneconomical.

As the nematode feeds on roots (Fig. 2.5) and reproduces, a large proportion of the feeder roots of citrus trees, particularly in the upper soil layers, is inactivated or destroyed, the uptake of water and minerals from the soil is reduced and the symptoms appear in the above ground tree parts. Heavily infested roots are darker in colour with branch rootlets shortened, swollen and irregular in appearance than in normal ones (Chona *et al.*, 1965).

Fig. 2.4. Sweet orange plants (9 month-old) infected with *Tylenchulus semipenetrans*.
Left – Healthy, Right – Infected.

Fig. 2.5. Citrus root infected with *Tylenchulus semipenetrans*.

Soil particles usually cling tightly, even after washing, to the gelatinous egg masses which cover the protruding part of the nematode body (Fig. 2.6). In heavily infested roots, the cortex separates readily from the vesicular stele.

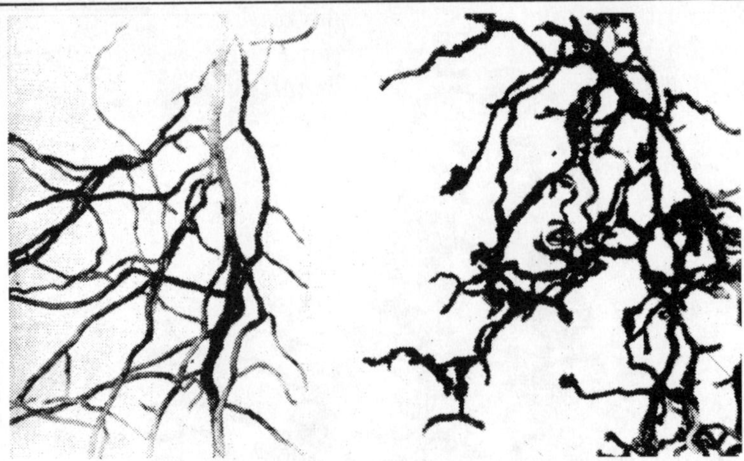

Fig. 2.6. Citrus root infected with *Tylenchulus semipenetrans*.
Left – Healthy, Right - Infected

(iv) Life Cycle: Van Gundy (1958) studied the life cycle in detail. Eggs hatched in 12 to 14 days at about 24°C and sex differentiation was possible at the second stage. The second stage male larvae developed to maturity within 7 days without feeding. The second stage female larvae required 14 days to locate the root, feed on the epidermal cells and molt. The life cycle from egg to egg required 6 to 8 weeks (Fig. 2.7).

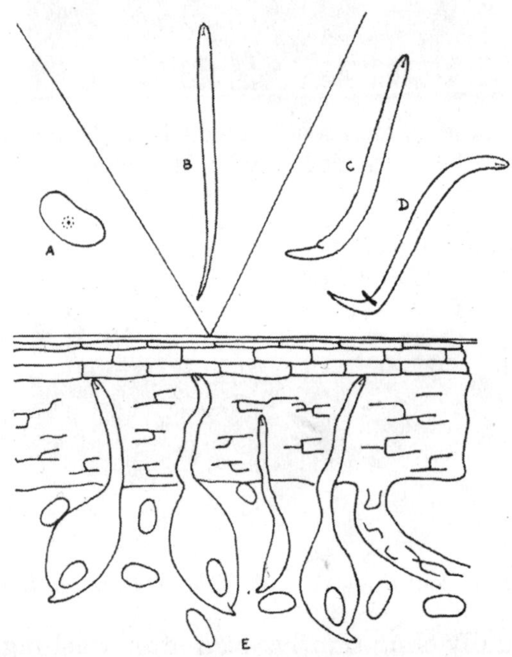

Fig. 2.7. Life cycle of *Tylenchulus semipenetrans*. A-Egg, B-Second stage larva, C-Young female, D-Male, E-Adult females attached to roots.

In experiments conducted at Israel, the minimum time required at 24°C for completion of life cycle was 14 weeks on *Poncirus trifoliata*, 10 weeks on *Ruta bracteosa* and 7 weeks on *Citrus aurantium* and *C. limettoides* (Cohn, 1966).

(v) Histopathology: Histopathological changes in the root tissue from the time of inoculation with second stage larvae to the establishment of feeding zone are described in citrus rootstocks viz., Karna Khatta (*Citrus karna*), Jamberi (*C. jambhiri*), Hill lemon (*C. pseudolimon*), Rangpur lime (*C. limonia*), Sweet lime (*C. limettioides*) (Misra and Edward, 1977) and Gajanimma (*C. pennivesiculata*) (Edward *et al.*, 1965).

(vi) Hosts: Mainly host range is confined to Rutaceae family. A total of 75 Rutaccous species mainly citrus and citrus hybrids are the hosts of this nematode. Other non Rutaceous hosts reported from India are grapes (Parvatha Reddy and Singh, 1978a), *Clerodendron inerme* (Nand *et al.*, 1994), *Jasminum sambac* (Bajaj *et al.*, 1988), *Garcina mangostana* (Chawla *et al.*, 1980) and loquats (Chona *et al.*, 1965).

(vii) Races/Biotypes: Based on different reports and studies on the infectivity of citrus nematode on different host's world over, four races of *T. semipenetrans* have been proposed by Inserra *et al.* (1980). (1) 'Poncirus Biotype' which reproduces on *Citrus* spp., *P. trifoliata* and their hybrids, grapes, but not on olive; (2) 'Citrus Biotype' which reproduces poorly on *P. trifoliata*, but infects *Citrus* spp., Carrizo and Troyer Citrange, Olive and Persimmon; (3) 'Mediterranean Biotype' which is close to citrus biotype, but does not reproduce on olive; (4) 'Grass Biotype' reproduces only on grass, *Andropogon rhizomatus*. Indian population of the nematode is considered related to Mediterranean biotype (Table 2.6).

Table 2.6. Biotypes of *Tylenchulus semipenetrans*

Host differential	Biotypes		
	Poncirus	Citrus	Mediterranean
Citrus sp.	+	+	+
Poncirus trifoliate	+	-	-
Grape	+	+	+
Olive	-	+	-
Persimmon	?	+	+
Distribution	California	California	Mediterranean region
	Japan	Italy	South Africa
	Israel	India?	India?

(viii) Economic Threshold Level: Citrus is a perennial crop for which to find out economic threshold level is slightly difficult. This also varies from field to field depending upon the agro practices followed. In India, economic threshold level has not been estimated under field conditions. However, it has been estimated in California that soil stages (juveniles/100 g soil) below 800 represents a non damaging population level. The orchards with more than 1600 juveniles/100 g soil may respond economically to nematicide treatments and at level more than 3600 during peak population growth period, treatment may improve yield substantially. The females per gram root are also used to define the damage level, with counts of <300, >700 and >1400 representing low, moderate and high ranges, respectively. In Florida, it was estimated that yield was not measurably reduced if populations were below 2000 juveniles/100cc soil during peak period of the soil population development (Duncan and Cohn, 1986).

(ix) Survival and Spread: T. semipenetrans is extremely sensitive to lack of moisture and high temperature. Most of the juveniles are killed if infected roots and soil are exposed to sun during summer. Under field conditions, the nematode survives mainly on the left over roots. In the absence of host, at 3 bars moisture stress, 9-27% of *T. semipenetrans* juveniles could survive after 450 days (Gaur and Sehgal, 1988). While at 30 bars of moisture stress at 40°C, only 5% of the populations could survive after 7 days (Sehgal *et al.*, 1990).

The dissemination of this nematode to newer areas is mainly through infected nursery plants. The irrigation water, particularly, flood irrigation is the prime source for its spread from plant to plant within the orchard. This can also spread through contaminated implements.

(x) Ecology and Population Fluctuation: The active population development period may vary from place to place depending upon agro climatic conditions. In Haryana, two peaks in October and April have been recorded on *C. sinensis* (Baghel and Bhatti, 1982). In Delhi, maximum populations have been recorded on *C. aurantifolia* in August (Prasad and Chawla, 1965). While contrary to these reports, one peak of nematode population has been recorded in Maharashtra in January-February on *C. reticulata* grafted on *C. jambhiri* (Singh, 1997a).

The distribution of *T. semipenetrans* in the rhizosphere of citrus trees is related with the active root zone. The maximum nematode populations are observed at a distance of 90 cm from tree trunk and up to 30 cm depth (Chhabra, 1978; Baghel and Bhatti, 1982), though, *T. semipenetrans* have been observed up to the depth of 180 cm (Chawla and Sharma, 1984). Sharma and Sharma (1977) reported maximum concentration of population at 30 cm distance from tree trunk in top 30 cm soil irrespective of plant age.

The population density is also related with the decline condition of the trees. Usually the population is more on the trees at initial stage of decline as compared to the population at advance stage of tree decline. This is mainly due to reduction in root mass in the advance stage of decline (Chona *et al.*, 1965; Bindra *et al.*, 1967; Mani, 1994; Singh, 1997c). The nutritional level of plant also influences the nematode population. The low levels of N and P result in higher population while higher dose of K favours low nematode population (Mangat and Sharma, 1981).

(xi) Interaction with Other Pathogens: The role of root-rot fungi in the disease syndrome caused by *T. semipenetrans* is significant. Misra and Edward (1977) reported severe root damage to citrus when the nematode was accompanied by secondary invaders like *Fusarium oxysporum* and *Macrophomina phaseoli*. The nematode helps weakly pathogenic fungus like *Fusarium* to cause secondary infection of citrus roots (Chhabra, 1972).

In the presence of *T. semipenetrans*, significantly more citrus plants become infected by *Fusarium oxysporum* and *F. solani* and the combined activity of the nematode and *F. solani* is thought to be a factor in the debilitation of citrus. *Fusarium* spp. seems to cause more damage in combination with the citrus nematode in cool, wet soils and with *Phytophthora* spp. in warm, wet soils.

(xii) Management Methods

(a) Physical Methods: The nematode can be controlled by dipping the bare roots of citrus seedlings in hot water at 45°C for 25 min or 46.7°C for 10 min (Bindra *et al.*, 1967). This treatment leaves no adverse effect on seedlings. During summer, exposing the soil to hot sun can control the nematodes substantially (Sehgal *et al.*, 1990).

(b) Cultural Methods: Populations of the citrus nematode in soil and roots can be reduced by removing the old feeder roots before the growth flush followed by application of FYM.

The citrus nematode is easily transported to new orchards through infested seedlings and budded plants which are usually carried 'balled' with infested soil. Therefore, nurseries should never be established on or near old citrus orchards and nursery soil should be sterilized before planting. Care should also be taken not to spread infection through tools, machinery and irrigation water used in infested groves.

Commencing from March, the bimonthly irrigation of citrus orchards with dung extract (prepared by mixing 2/3 cow dung + 1/3 water and kept for a period of two weeks with daily shaking; diluted with water in a ratio of 1: 10 before application) improved tree growth and markedly reduced the nematode population.

Intercropping: Interculture of acid lime with marigold and mustard markedly reduced the rate of multiplication of *T. semipenetrans* to 10.03 and 8.59, respectively (Mani, 1988b). Interculture of onion, garlic or marigold in the citrus orchards not only reduces the nematode population but also provides additional income.

Botanicals: Singh and Sitaramaiah (1973) reported that castor cake was highly effective in reducing 50% of the citrus nematode population. Mobin and Khan (1969) have also observed that neem, mahua, groundnut and mustard oil cakes were effective in reducing the nematode population in citrus rhizosphere. Application of neem cake at 20 kg per tree basin at an interval of 4 months can effectively reduce the nematode population of *T. semipenetrans* and *Meloidogyne javanica* (Mani and Murty, 1986).

Irrigation of citrus orchards with sewage water reduces the citrus nematode population to a very low level.

A number of plants possess nematicidal properties. Different plant parts and their extracts have been tested against *T. semipenetrans* by different workers. Aqueous and methanol garlic extracts were toxic to *T. semipenetrans in vitro* (Nath *et al.*, 1982). The extracts of leaves, stem and buds of *Datura stramonium, Ipomea carnea, Tagetes patula* and *Lowsonia alba* caused 50-100% mortality at 4 mg/ml or 1:5 dilution after 48 hrs (Kumari *et al.*, 1986). Similarly, water extract of *Calotropis procera* and *Nerium oleander* and essential oil of *Cymbopogon* grasses have been found toxic to *T. semipenetrans* (Sangwan *et al.*, 1985; Verma *et al.*, 1989). Castor (*Ricinus communis*), Aak (*Calotropis procera*) and Datura (*Datura strumonium*) chopped leaves applied at 20 and 40 g/kg soil in 9 sq. m. area around tree trunks can result in reduction of nematode population and increase in yield (Baghel, 1995).

(c) Chemical control

Nursery Bed Treatment: Soil application of aldicarb or carbofuran at 4 kg a.i. /ha was effective for the control of *T. semipenetrans* (Parvatha Reddy, 1988).

Seedling Bare Root-dip Treatment: The bare root-dip treatment of acid lime seedlings with monocrotophos or chlorpyriphos at 1000 ppm for 45 min. reduced the multiplication of *T. semipenetrans* by 90% (Mani, 1990a).

Main Field Treatment: In early seventies, fumigant nematicides such as DBCP, DD etc. tested against citrus nematode were very effective (Chhabra and Bindra, 1971; Mukhopadhyay and Dalal, 1971). Due to health and environmental reasons, these fumigants are banned now.

Other non fumigant compounds like carbofuran, ethoprophos, phorate, fensulfothion, dimethoate etc., have also been tested against citrus nematode under field conditions. Dichlofenthion at 45 l/ha reduced the nematode population by 80% (Mukhopadhyaya, 1970b). Mukhopadhyaya and Dalal (1971) achieved 97.6% control by ethoprophos when applied at 4 kg a.i./ha and 39.9% increase in yield of sweet lime during second year. Chhabra *et al.* (1977) found fensulfothion very effective at 3 kg a.i./ha in controlling the nematodes on pumello and increased the yield by 68-76%.

Significant reduction in the citrus nematode population (76 to 84 per cent) in sweet orange and increase in yield (61 to 244 per cent) was observed with soil application of aldicarb and carbofuran both at 6 kg a.i. per ha (Baghel and Bhatti, 1983b). The effectiveness of aldicarb and carbofuran was also demonstrated for the control of *T. semipenetrans* infecting acid lime (Parvatha Reddy, 1986) (Table 2.7).

Table 2.7. Effect of different nematicides on *Tylenchulus semipenetrans* infecting different *Citrus* spp.

Citrus spp.	Nematicide & Dosage	Type of application	Effectiveness
Acid lime	Aldicarb – 4 kg a.i./ha	Soil application	Reduced nema popn. by 75-80%
	Carbofuran - 4 kg a.i./ha	Soil application	Reduced nema popn. by 65-75%
Grape fruit	Aldicarb – 5.7 to 11.4 kg a.i./ha	Soil application	Reduced nema popn., improved fruit size, quality & fruit yield by 15.7 – 108.0%
	Carbofuran, Fensulfothion, Phorate - 5 & 10 kg a.i./ha	Soil application	Reduced nema popn. by 75-90% at lower conc. & by 90% at higher conc.
Naval orange	Carbofuran, Ethoprophos, Phenamiphos- 2.2 kg a.i./ha	Soil application	Reduced larval popn. in soil, increased fruit yield by 108-137%
Orange	Aldoxycarb, Carbofuran, Carbosulfan – 1.1 kg a.i./ha;	Through drip irrigation	Reduced female nemas on roots and increased fruit yield
	Phenamiphos – 10 ml/m^2	With irrigation water	Reduced larval popn., increased uptake of K, Ca and Cu, gave extra yield of 6.1-7.7 tonnes
Sweet lime	Ethoprophos - 4.0 kg a.i./ha	Soil application	Reduced nema popn. by 98%, checked popn. build-up for 16 months, increased fruit yield at 12 kg/tree

Sweet orange	Aldicarb – 4 kg a.i./ha	Soil application	Reduced nema popn. in soil
	Aldicarb – 6 kg a.i./ha	Soil application	Reduced nema popn. by 84%, increased fruit yield by 224% with cost: benefit ratio of 1:11
	Carbofuran - 4 kg a.i./ha	Soil application	Reduced nema popn. in soil
	Carbofuran - 6 kg a.i./ha	Soil application	Reduced nema popn. by 76%, increased fruit yield by 152% with cost: benefit ratio of 1:8
	Phorate – 6 kg a.i./ha	Soil application	Reduced nema popn. by 66%
	Oxamyl – 4 μg/ml	Foliar spray	Reduced nema popn. in pots

(d) Biological Methods

Antagonistic Bacteria: Application of *Pseudomonas fluorescens* at 20 g/tree three times in a year at 15 cm depth and 50 cm away from tree trunk helps in reducing the nematode population.

Pasteuria penetrans can be readily multiplied on the citrus nematode. The parasite seems to offer potential for the biocontrol of *T. semipenetrans* (Mani, 1988a) (Fig. 2.8).

In citrus, a monthly rate of 1.1 kg a.i. per ha of avermectins (*Streptomyces avermitilis*) for 7 months gave maximum increase in yield and reduction of *T. semipenetrans* population (Garabedian and Van Gundy, 1983).

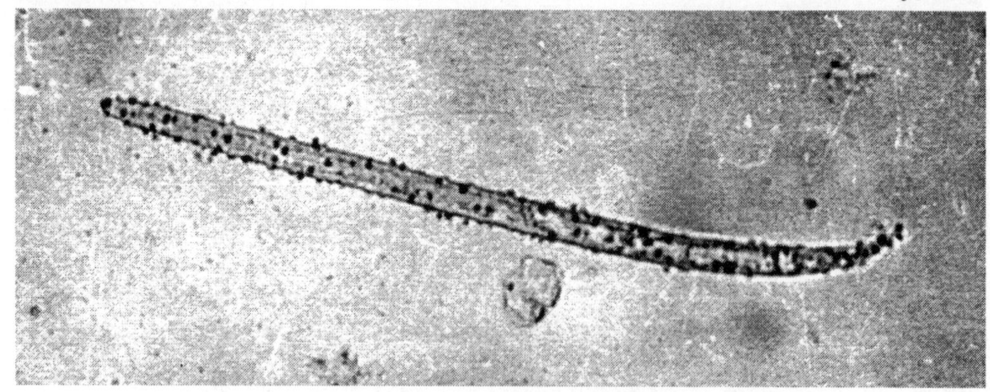

Fig. 2.8. Citrus nematode larva infected with *Pasteuria penetrans*.

Antagonistic Fungi: Several nematophagous fungi such as *Arthrobotrys cladodes*, *Dactylaria* sp., *Monacrosporium gephyrophagum*, *Laginidium* sp. etc., can trap the *T. semipenetrans* and could be exploited for nematode control (Gowda *et al.*, 1982). Endoparasitic fungus, *Paecilomyces lilacinus* is a potential biocontrol agent for controlling citrus nematode (Mani *et al.*, 1989).

Since the parasitic fungus, *P. lilacinus* has many attributes of a successful biological control agent, experiments were conducted in nematode infested citrus grove to determine its efficacy in controlling *T. semipenetrans*. It was observed that fruit diameter of the orange cv. 'Valencia' was significantly affected by application of 3 nematicides and *P. lilacinus*. The fruit diameter from fungus treated tree was greater than that from trees treated with temik, vydate and mocap. Similarly, the number of nematodes in roots and soil around citrus trees inoculated with *P. lilacinus* was significantly lower than that in the non-treated or the nematicide treated trees (Jatala, 1986). Application of *P. lilacinus* in the form of spore suspension checked the multiplication of *T. semipenetrans* by 71.7 to 97.8% (Mani and Murthy, 1989) (Fig. 2.9).

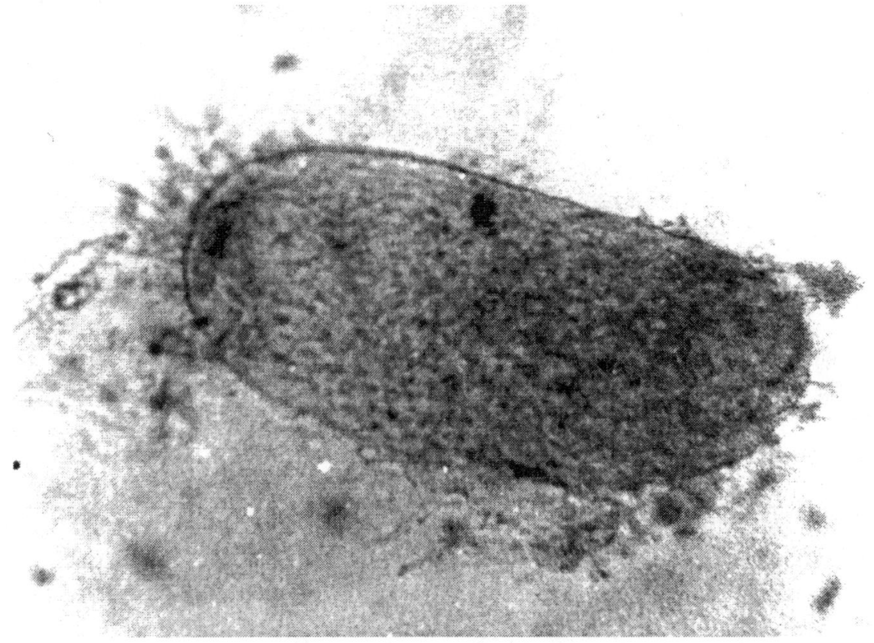

Fig. 2.9. Citrus nematode egg infected with *Paecilomyces lilacinus*.

Arbuscular mycorrhizal fungi: Arbuscular mycorrhizal fungus, *Glomus mosseae* when added to the soil with *T. semipenetrans* in *Citrus jambhiri,* limited the nematode development and partially neutralized the adverse effect of the nematode (Fig. 2.10) (Baghel *et al.*, 1990).

(e) Host Resistance: Citrus rootstocks which are resistant or highly tolerant to the citrus nematode would offer practical and effective means of control. Trifoliate orange (*Poncirus trifoliata*), Troyer Citrange and Swingle citrumello and its hybrids (Citrumello) were highly resistant to the citrus nematode (Parvatha Reddy and Singh, 1978; Chhabra and Bindra, 1974; Parvatha Reddy and Agarwal, 1987; Mani and Reddy, 1986). The citrus rootstock hybrids evolved at the Indian Institute of Horticultural Research, Bangalore, by crossing Rangpur lime (*Citrus limonia*) with *P. trifoliata* viz., CRH-3, CRH-5 and CRH-41 were highly resistant to the citrus nematode (Parvatha Reddy *et al.*, 1987).

Fig. 2.10. Management of *Tylenchulus semipenetrans* using *Glomus mosseae*.
A – *G. mosseae*, B – Control, C – *T. semipenetrans* + *G. mosseae*, D - *T. semipenetrans*.

Inter-generic hybrids were evolved, involving the commercial rootstocks Rough lemon, Rangpur lime, Cleopatra mandarin and Orlando tangelo, as seed parents and *Poncirus trifoliata* (the citrus nematode resistant compatible donor) as pollen parent to overcome the citrus nematode problem. Twenty-five promising hybrids and the 5 parents were screened against *T. semipenetrans* by artificial inoculation under glasshouse conditions. Nine hybrids: CRH-57, 47, 10, 35, 17, 5, 41, 3 and 12 were resistant while 2 hybrids, CRH-8 and 24 were susceptible (Prasad *et al.*, 1998, 1999). The remaining 14 hybrids were moderately resistant. Among the parents, *Poncirus trifoliata* was resistant while all the others were highly susceptible.

Resistance, based on the number of larvae and adult females on the feeder roots, was evaluated in 57 selections belonging to 4 genera (*Citrus, Fortunella, Poncirus* and *Severinia*) and 24 species/hybrids. All *Severinia disticha* and *P. trifoliata* strains and hybrids except 2 were rated resistant. Pummelo (*C. maxima*) X *P. trifoliata* and Smooth Flat Seville (*C. aurantium*)

X Swingle citrumelo (*C. paradisi* X *P. trifoliate*) exhibited a moderately resistant reaction. It is suggested that Swingle citrumello and *P. trifoliata* strains should be used as citrus rootstock parents. Rangpur lime *(C. limonia)*, grapefruit *(C. paradisi)* and *C. pectinofera* were moderately resistant. Of 13 Rangpur lime (*Citrus limonia*) and rough lemon (*C. jambhiri*) strains tested, none were found to be resistant to *T. semipenetrans*. Rough lemon (Poona), Rough lemon (Madhya Pradesh), Rough lemon 14, Rough lemon 8779 and Rough lemon 8782 were moderately resistant, while all the others were susceptible.

(f) Integrated Methods

Using Bioagents and Botanicals: Management of the citrus nematode based on the application of neem cake and castor cake both at 10 kg per plant along with a nematophagous fungus, *Paecilomyces lilacinus* at 250 g (grown on paddy seeds) per plant 3 times in a year at 15 cm depth and 50 cm away from the trunk was found to be extremely effective in reducing the citrus nematode population with a consequent increase in the growth of acid lime trees. The above treatment also gave highest parasitization of egg masses and eggs of *T. semipenetrans* and increased spore density of *P. lilacinus* in soil (Parvatha Reddy *et al.*, 1991a, 1993).

Trichoderma harzianum in combination with neem oilcake was effective in increasing the growth of acid lime trees and reducing the citrus nematode population both in soil and roots. The parasitization of citrus nematode females with *T. harzianum* increased in the presence of oil cakes (Parvatha Reddy *et al.*, 1996b).

Incorporation *Verticillium lecanii* with neem cake facilitated the effective management of *T. semipenetrans* on acid lime (Parvatha Reddy *et al.*, 1996a).

Integration of neem cake with *Pseudomonas fluorescens* gave maximum reduction in citrus nematode population both in soil and roots and increased plant growth of acid lime (Parvatha Reddy *et al.*, 2000).

Using Arbuscular Mycorrhizal Fungi and Botanicals: Inoculation of endomycorrhiza, *Glomus fasciculatum* in the soils amended with neem cake was found effective for the management of *T. semipenetrans* on acid lime. This strategy can help in combating the menace of citrus nematode at nursery stage and also provide highly mycorrhizal seedlings of acid lime for transplanting in the main field for the management of *T. semipenetrans* under field conditions (Parvatha Reddy *et al.*, 1995).

Using Two Bioagents: Application of bacterial bioagent, *Pasteuria penetrans* (at 2 x 10^9 spores/plant) and fungal bioagent, *P. lilacinus* (at 50 g/plant with 4 x 10^7 spores/g) was effective in reducing the *T. semipenetrans* population and in increasing the parasitization of larvae by *P. penetrans* and eggs by *P. lilacinus* (Parvatha Reddy and Nagesh, 2000).

Using Nematicides and Botanicals: Neem cake at 1 kg/plant combined with carbofuran at 2 kg/ha reduced 47.1% nematode population and increased the yield by 41.1% (Baghel, 1995).

Using Bioagents and Chemicals: Integration of *P. lilacinus* (at 4 g/plant) with carbofuran (at 30 mg a.i./plant) was found effective in increasing the plant growth parameters of acid lime and in reducing the citrus nematode population both in soil and roots (Parvatha Reddy *et al.*, 1996c).

2.1.2.2. The Lesion Nematode, Pratylenchus coffeae

(i) Economic Importance and Losses: P. coffeae appears to be the most pathogenic species and is responsible for the "Citrus slump disease" in Florida (Fig. 2.11). The nematode root rot of citrus in Uttar Pradesh is principally caused by *P. coffeae* (Siddiqi, 1964). This nematode constitutes the dominant factor in the nematode complex of citrus in Aligarh. O'Bannon and Tomerlin (1973) reported 49 to 80 per cent growth reduction in young trees up to 4 years depending upon different rootstocks.

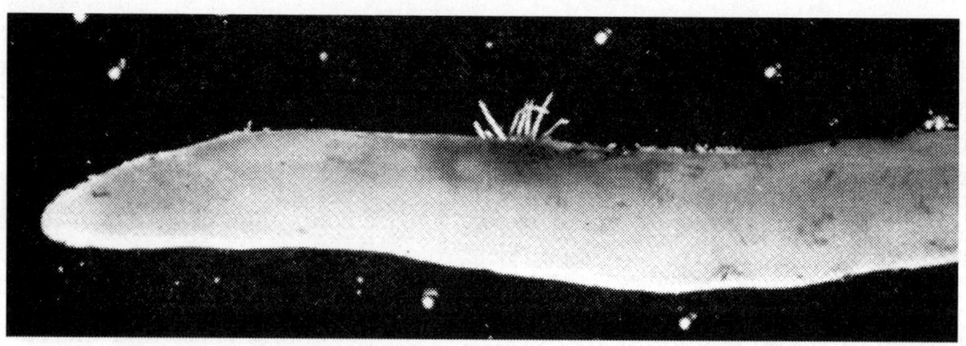

Fig. 2.11. Rough lemon roots infected with *Pratylenchus coffeae*.

(ii) Distribution: P. coffeae is a pest of citrus in India, Japan and the U.S.A. In India, it has been reported from Uttar Pradesh, Punjab and South Western India on citrus (Sethi & Swarup, 1971; D'Souza *et al.*, 1970).

(iii) Symptoms: P. coffeae infected citrus plants exhibit poor growth, general dieback and produce undersized fruits.

(iv) Hosts: Citrus limon, C. sinensis, C. reticulata and banana have been found to be hosts of *P. coffeae* in Uttar Pradesh (Siddiqi, 1964). The nematode is transferable from banana to citrus plants and from lemon to orange plants and *vice versa.*

(v) Life Cycle: P. coffeae is a migratory endoparasite of the root cortex where it feeds and multiplies. The favoured place of entrance is slightly back the elongating zone in the piliferous region. A colony or 'nest' is established showing brownish lesions on the root. When decay sets in due to secondary invasion by bacteria and fungi, the nematodes quit and migrate through the soil to attack fresh roots. The first molt takes place within the egg and three molts occur outside. *P. coffeae* requires about 45 to 48 days to complete the entire life cycle. Of this time, 15 to 17 days are necessary for eggs to hatch, 15 to 16 days are spent as larvae, and it takes 15 days from maturity to egg production.

(vi) Histopathology: Adults and larvae of *P. coffeae* are found in the cortical tissues of the feeder roots where a tiny, brownish-black lesion is formed which gradually expand to girdle the rootlets. When the nematode invaded a root-tip, the meristem often was destroyed and lateral root initiation usually occurred near the destroyed root tip. Inoculation experiments have shown that the young plants of *C. limon* and *C. sinensis* show poor growth when infected with *P. coffeae* (Siddiqi, 1964). A marked reduction in the top as well as root weight of the infected seedlings was noticed. *C. limon* seedlings inoculated with 20 *P. coffeae* had peak populations at 36 weeks with 10,000 nematodes per g root extracted; shoot growth was reduced by 22 per cent. *P. coffeae* extensively invades the cortex of feeder roots, causing cavities and cell necrosis.

(vii) Survival and Spread: P. coffeae survived best in excised infected roots of *C. jambhiri* stored in soils with initial moisture content near field capacity and at temperature ranging from 10-32°C and were infective after 4 months. The nematodes did not survive extended storage at temperatures over 38°C. The optimum temperature for reproduction was 29.5°C.

(viii) Management Methods

(a) Cultural Methods: Sanitation is important to reduce damage by *P. coffeae*. Citrus nurseries in Florida, South Africa, Brazil and elsewhere have either mandatory or voluntary programmes to certify that young trees are free of *P. coffeae*.

(b) Chemical methods: Infestation in citrus seedlings of *P. coffeae* was eliminated by treating with fensulfothion and fenamiphos at 4.4 kg a.i. per ha and the stem diameter of the treated seedlings increased by 44 and 46 per cent, respectively after 9 months. The population of *P. coffeae* can be effectively reduced by applying aldicarb or carbofuran at 4 kg a.i. per ha (Baghel and Bhatti, 1983a).

(c) Host Resistance: Out of 125 selections of citrus, most of the selections were considered highly susceptible supporting population more than 1000 per gram root. Only four selections of *Microcitrus australis* x *M. australasica* hybrid and Rubiboux 70-A5, Trifoliate orange (*Poncirus trifoliata*) had nematode populations of less than 100/g root. They were considered to show some resistance to *P. coffeae* because of low numbers.

2.1.2.3. Root-knot Nematodes, Meloidogyne javanica, M. indica

In India, Thirumala Rao (1956) reported that *Meloidogyne* sp. caused considerable damage to citrus in Andhra Pradesh when a susceptible crop like tobacco or okra is grown as an intercrop. This is the first report of nematode damage to citrus in India.

Chitwood and Toung (1960) observed the root-knot nematode resembling *M. africana* infecting *C. sinensis* from Delhi and proved its pathogenicity to citrus. Whitehead (1968) reported *M. indica* on citrus from India (Fig. 2.12).

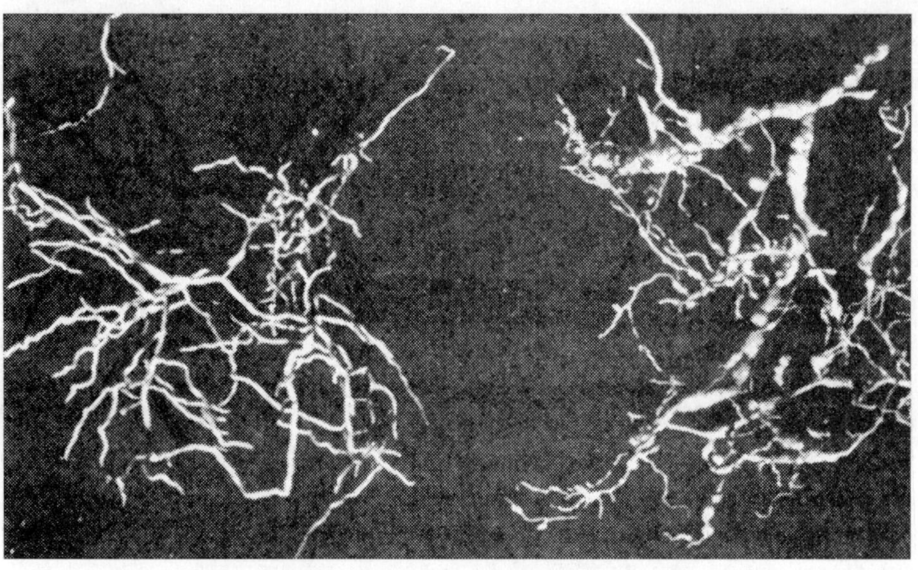

Fig. 2.12. Acid lime roots infected with *Meloidogyne indica*.
Left – Healthy roots; Right – Infected roots.

Mani (1986) reported that *M. javanica* caused severe crop loss of acid lime in Andhra Pradesh and it was pathogenic to acid lime and sweet orange (Fig. 2.13). It is confined to only coastal Andhra Pradesh.

(i) Distribution: The root-knot nematode on citrus has been reported from Andhra Pradesh, Gujarat and Delhi.

(ii) Symptoms: M. javanica infected trees are poor in vigour, unthrifty in appearance and show severe stunted growth. They fail to flower and produce fruits even after several years of planting. The roots have conspicuous galls on pioneer and fibrous roots. In advanced stage, large cavities can be

observed in place of galls. Egg masses can be seen as thin films spread over the root surface. Nematode infestation gets aggravated if vegetables like okra, brinjal, cucurbits, tomato and tobacco are grown as intercrops in orchards or as rotational crops in nurseries (Mani, 1986).

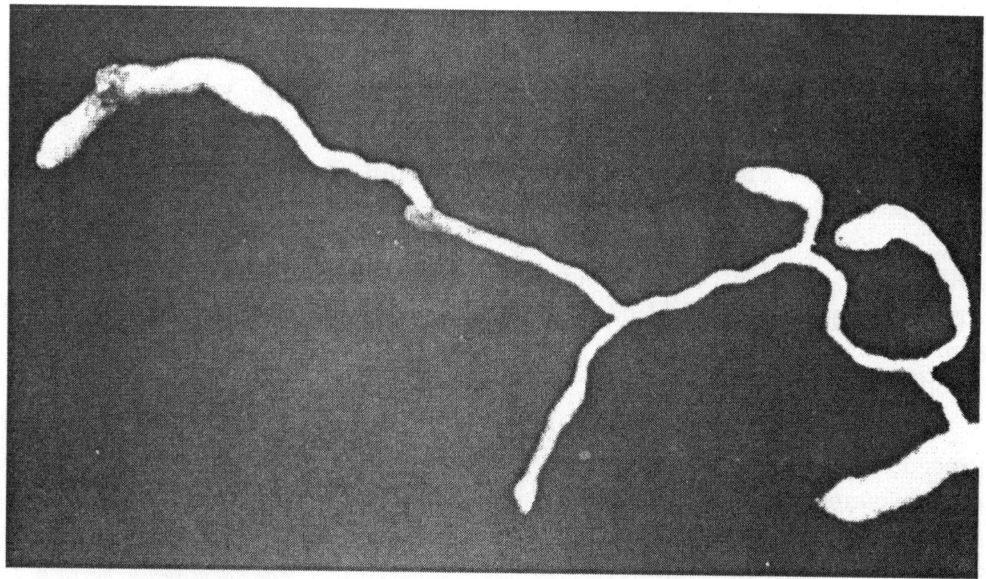

Fig. 2.13. Tomato roots infected with *Meloidogyne javanica.*

(iii) Hosts: M. javanica has been reported on *C. aurantifolia* and *C. sinensis* (Mani, 1986), whereas *M. indica* infects *C. aurantium, C. sinensis* (Whitehead, 1968) and *C. aurantifolia* (Parvatha Reddy *et al.*, 1981).

(iv) Management Methods

(a) Physical Methods: Soil solarization is one of the effective ways of suppressing nematode populations and can be employed mostly during hot weather days. The nursery area should be ploughed well and leveled after breaking the clods. The prepared field may be covered with polythene sheet of 100 gauge thickness for about 7-10 days before sowing.

(b) Cultural Methods: At least two to three deep ploughings should be given during bright sunny days at an interval of 10-15 days.

It is essential to select root-knot nematode-free planting material to prevent the introduction of nematode inoculum into orchards. It is preferable to avoid growing vegetables as intercrops or as rotational crops in citrus orchards and nurseries. Cereals and millets can be grown in nursery area according to grower's choice.

Intercropping: Growing sunnhemp in tree basin reduced the population of root-knot nematodes. *Crotalaria* roots exert a toxic effect on *Meloidogyne* population. *Meloidogyne* larvae freely entered roots of *Crotalaria* but failed to survive. The possibility of growing *Crotalaria* in tree basin for the control of root-knot nematodes infecting citrus needs further investigation.

Crop Rotation: Once the life of the citrus orchard comes to an end, it is highly preferable to grow cereals and millets for at least 2-3 seasons before initiating new orchard in the same field. This will help in reducing the populations of *Meloidogyne* species and *T. semipenetrans* which are common in citrus orchards.

Botanicals: Field experiments carried out in lime orchards indicated that application of neem cake at 20 kg per tree basin reduced the population of *M. javanica* by 25 per cent three months after application. Repeated applications once in four months significantly reduced the root-knot nematode population. FYM or compost at 20-25 t/ha or oil cakes, particularly neem or karanj cakes at 2-3 t/ha may be incorporated in soil before sowing or transplanting.

(c) Chemical Methods: Aldicarb and carbofuran at 4 kg a.i. per ha reduced the larval population of *M. javanica* on acid lime in soil by 77.8 and 62.8 per cent, respectively. Chemicals should be applied in two split doses at 4 – 6 months interval at the time of formation of new roots to protect them against nematode infection.

(d) Biological Methods: A fungal parasite, *Paecilomyces lilacinus* consistently and effectively controlled the population of *Meloidogyne* spp.

(e) Host Resistance: It was observed that *C. reticulata* was not infected by *M. javanica*. Sour orange, Lime, Rough lemon, Grape fruit, Troyer citrange, Cleopatra mandarin and Trifoliate orange were highly resistant to *M. incognita* (Race 1 and 3) as well as *M. javanica*.

The response of *Citrus* spp. and related rootstocks to a population of *Meloidogyne javanica* was evaluated by some researchers in a screen-house experiment. Palestine and Rangpur lime, Rough lemon, Sour orange, Sexton and Thentriton tangelos and Volkamer lemon were not infected by *M. javanica*. Galls and tip swellings were observed on the roots of *Poncirus trifoliata* and Troyer citrange. There was no evidence of nematode development. Symptoms induced by the nematode were stelar division, syncytia formation in the vascular tissues and necrotic cells.

2.1.2.4. The Burrowing Nematode, *Radopholus citrophilus*

Perhaps the most spectacular disease with which the burrowing nematode, *R. citrophilus* has been associated is that of **'spreading decline'** of citrus in Florida, USA. Suit and DuCharme (1953) have demonstrated that this nematode causes spreading decline in citrus trees in Florida, USA. This nematode is of great economic importance to citrus in Florida and causes reduction in top and root growth. Although attempts have been made to eliminate this pest, several thousand hectares currently under cultivation still remain infested. Spreading decline has been detected only in those citrus groves located on the sandy ridge zone of central Florida. It is responsible for 40-80% loss in fruit yield.

(i) Symptoms: Trees with heavily infected roots wilt more readily than healthy trees during periods of water stress. The trees appear under-nourished and branches show twig die-back. The leaves are small and sparse. The size of the fruit is small and overall yields are less. Diseased trees are non-thrifty, lack vigour and produce fewer fruits. Affected trees have abundant dead twigs and branches.

R. citrophilus parasitizes the tips of new feeder roots soon after they are formed, resulting in stubby, somewhat swollen root tips. At the nematode infection sites, brown to black lesions are formed and some times cankers develop. Extensive feeder root damage takes place 50 cm below ground level in affected trees with only 20 to 30% feeder roots at depths of 25 to 75 cm. Feeder roots are virtually absent at depths below 75 cm in infested orchards.

(ii) Life Cycle: The burrowing nematode is a migratory endoparasitic nematode. Juveniles and adults penetrate roots in the region of elongation. The nematodes feed on and kill cells forming cavities and tunnels in the cortex and stele of roots. Eggs are laid randomly within the roots. Once inside, the nematodes may spend their entire lives within the root. The life cycle takes about 18-20 days at 24-26°C. The nematode may survive in soil without feeding for about 6 months.

(iii) Histopathology: Females and larvae enter growing feeder roots near their tips in the region of cell elongation and root hair production. The nematode upon penetration, feeds on the cortical parenchyma cells and gradually burrows towards the stele, creating tunnels and cavities in the cortex and stele. Large number of nematodes accumulate in the phloem-cambium ring region. This part of the root is often completely destroyed leaving a cavity filled with nematodes. Cell reactions involve hypertrophy and when the nematode penetrates pericycle, hyperplasia and tumor formation occur. Wound gum accumulates next to the stele in the invaded area, imparting tan to amber colour in the older portions of the lesions (DuCharme, 1959).

(iv) Biotypes: The nematode appears to have at least 2 biotypes: Biotype 1, which reproduces poorly on Milam lemon roots, and Biotype 2, which reproduces well and significantly damages Milam Lemon and other resistant rootstocks.

(v) Survival and Spread: Large root fragments that remain buried in soil after tree removal may help support populations during fallow. The nematode is spread in contaminated rootstock, machinery, subsoil water and it migrates rapidly along developing root systems. The infested area gradually expands in all directions regardless of rows or direction of cultivation. The measurable spread (about 15 meters per year) is characteristic of the disease. The nematode spread is aided by the flow of water in soil.

(vi) Management Methods

(a) Regulatory Methods: Since the burrowing nematode that attacks citrus is presently confined to Florida, all other citrus growing areas have stringent quarantines and special regulations to prevent the introduction of *R. citrophilus* via affected planting material.

Sanitation is one of the most important methods to manage the burrowing nematode. The major management tactics consist of 'nematode-free nursery stock' certification programme. Citrus nursery certification programme in Florida, USA, results in added crop value of more than US$ 17 million per year by slowing the spread of *R. citrophilus*.

(b) Cultural Methods: Barriers on a small scale are being used to confine *R. citrophilus* within limited areas of infestation (Poucher *et al.*, 1967). These man-made barriers surrounding the infested area consist of a strip of land 3 to 8 meter wide, free of citrus trees, and treated with a nematicide at 6 months intervals to keep the soil in the barrier zone free of roots as well as burrowing nematodes. The surface of the soil is also treated with herbicides to keep the barrier free of all plant growth. A biological barrier composed of a band of resistant citrus trees or a non-host of *R. citrophilus* has been suggested to block the migration of nematodes from infested areas (DuCharme and Suit, 1956; Ford, 1967).

Attempts are being made to eradicate *R. citrophilus* from infested groves of Lakeland fine silt in Florida by the "pull and treat" method (Poucher *et al.*, 1967). This procedure involves destruction of all trees in an area overlapping the infested site by at least 16 meters in all directions and removal of as many roots as possible from the soil. The cleaned soil is treated with DD at 650 liters/ha, and the area is then kept free of all vegetation for 2 years. After the period of fallow, the site is replanted preferably with trees grafted on a tolerant rootstock.

(c) Chemical Methods: Various chemicals tried for the management of *R. citrophilus* on citrus as bare root dip treatment is presented in Table 2.8.

Table 2.8. Effect of various chemicals as bare root dip for the management of *R. citrophilus* on citrus

Chemical	Dosage	Duration & type of treatment	Effectiveness
Fensulfothion	1000 ppm	30-60 min. dip	Eliminated infection
Phenamiphos	250-1000 ppm	30-60 min. dip	Eliminated infection
Ethoprophos	1000 ppm	30-60 min. dip	Eliminated infection
Thionazin	1000 ppm	30-60 min. dip	Eliminated infection

Various chemicals tried for the management of *R. citrophilus* on citrus under field conditions is presented in Table 2.9.

Table 2.9. Effect of various chemicals for the control of *R. citrophilus* on citrus under field conditions

Chemical	Dosage	Duration & type of treatment	Effectiveness
DBCP	40 litres/ha	Soil appln. thrice a year (March, May & April)	Controlled infection
EDB	150 litres/ha	Pre-plant soil treatment	Controlled infection
Fensulfothion	100-400 ppm	Sub-surface soil drench	Eradicated infection in Rough lemon seedlings
Oxamyl	11.4 kg a.i./ha	As foliar spray 6 times a year at 6 weeks interval	Reduced nematode population
Phenamiphos	11.2 kg a.i./ha	Split application	Reduced nematode population
Ethoprophos	100-400 ppm	Sub-surface soil drench	Eradicated infection in Rough lemon seedlings
	7-25 kg/ha	3 field applications	Increased twig growth
Thionazin	15 kg/ha in 400 litres of water	3 soil drenches around trees + post treat. drench	Increased feeder roots
	100-400 ppm	Sub-surface soil drench	Eradicated infection in Rough lemon seedlings

(d) Biological Methods: Application of AMF, *Glomus etunicatum* in nursery soil as well as on transplantation increased the plant vigour and reduced the damage caused by *R. citrophilus* on citrus (O'Bannon and Nemec, 1979).

(e) Host Resistance: Citrus rootstocks such as Milam, Ridge Pineapple, Carrizo Citrange and Algerian Navel were reported to be resistant to the burrowing nematode (Ford and Feder, 1964).

2.1.2.5. The Lance Nematode, *Hoplolaimus indicus*

Khan *et al.* (1964) reported *H. indicus* from the soil around the roots of *C. sinensis, C. aurantifolia* and *C. reticulata* from India. This nematode is only known from India where it is widely distributed. It causes damage to citrus in Punjab.

(i) Occurrence and Distribution: H. indicus has been reported on citrus from Punjab (Gupta and Gupta, 1966), Delhi, Rajasthan (Sharma *et al.*, 1969) and Uttar Pradesh (Rashid *et al.*, 1973).

(ii) Symptoms: H. indicus at 1,000 numbers per 500 g of soil caused some growth reduction of citrus, but a marked reduction occurred at or above an initial population level of 2,000 (1 male; 4 females) (Gupta and Atwal, 1972). None of rough lemon, karna khatta, sweet orange, mandarin, lemon or trifoliate orange was completely resistant to nematode attack at population levels above 4,000 per 500 g of the soil. Lemon proved somewhat more tolerant than other species. The greatest nematode populations were found in roots and soil of trees 10 to 20 years old and in an intermediate stage of decline; smallest populations were associated with trees one to five years old at initial stages of decline and 10 to 20 years old in the final stages of decline.

(iii) Hosts: H. indicus is reported on sugarcane, banana, pea, guava, tomato, rice, brinjal, finger millet, amaranthus, black gram, chillies, cowpea, cucumber, groundnut, okra, maize, mustard, Para grass, cauliflower, harayali grass, mango, peach, wheat, bajra, sorghum and cabbage besides citrus. In citrus, it has been reported on *C. jambhiri, C. karna, C. reticulata, C. sinensis, C. limon* and *Poncirus trifoliata* (Khan *et al.*, 1964).

(iv) Life Cycle: The first molt occurs within the egg and the development outside the egg consists of 3 larval stages and the adult, with the usual 3 molts. Sex differentiation is indicated early in the second molt by the presence, in the female only, of 4 specialized ventral chord nuclei coinciding the position with that of the vagina. The optimum conditions for maximum population growth are a temperature of 30°C, a soil pH of 7.0, sandy loam soil (10-20% clay) and 16% moisture content.

2.1.2.6. The Sheath Nematode, *Hemicycliophora arenaria*

H. arenaria was reported on *Citrus jambhiri* from sandy desert region of California, USA (Van Gundy, 1957). *H. arenaria* is considered important on *Citrus* species only. Growth reduction of citrus ranged from 12% at 25°C to 37% at 30°C.

(i) Symptoms: H. arenaria feeds ectoparasitically near root tips of citrus. Root elongation ceases soon after juveniles insert their stylets into cells near root cap. Galls are formed in the meristem region or the zone of elongation by

the hyperplasia of pericycle leading to enlarged cortex, some cells being multinucleate. Such galls are always at root tips, and thus can be distinguished from those caused by root-knot nematodes (Fig. 2.14). Lateral root initials are stimulated near root tip, and if feeding occurs in the zone of elongation, abnormally long root hair often develops.

Fig. 2.14. *Hemicycliophora arenaria* on rough lemon roots (Terminal galls). Left – Infected; Right – Healthy. (Courtesy: Union Carbide Agril. Products Co. Inc., 1986).

(ii) Life Cycle: H. arenaria reproduces by parthenogenesis. At 28-30°C, female laid on an average 6.2 eggs. A gelatinous substance on eggs causes them to adhere to soil particles. The time from egg to egg was 15-18 days at 30°C (Van Gundy, 1959). The nematode multiplied best in soil containing 90% sand, 5% silt and 5% clay. Adult females can survive in water without feeding for 4-5 weeks.

(iii) Histopathology: Histochemical studies by Mc Elroy and Van Gundy (1968) revealed an accumulation of protein in the food cells during the first 6 hours and enlargement of nuclei in the adjacent cells. Afterwards, the food cells which are hypertrophied with distorted nuclei, are injested. As feeding continues the contents of neighbouring cells are removed and cell walls are thickened. Eventually the food cells are crushed and pushed to the root surface by the new meristem formed by the pericycle, which formed the new food cells.

(iv) Management Methods

(a) Physical Methods: Dipping of bare rooted nursery plants in hot water maintained at 46°C for 10 minutes eliminated nematode infestation (Van Gundy and Mc Elroy, 1969).

(b) Host Resistance: Sweet and sour orange, trifoliate orange and citrange are non hosts to the sheath nematode.

2.1.2.7. Future Lines of Investigation

* The population threshold level of *T. semipenetrans* on different varieties under different soil types need to be worked out for effective control measures.

* *T. semipenetrans* is economically important and widely distributed throughout India, but how much economic losses are being caused by this nematode in India need to be worked out.

* The dissemination of nematodes to newer areas is mainly through nursery plants. There is need to develop methodologies for obtaining nematode free seedlings through nematode management at nursery level.

* With the emphasis on drip irrigation system to save the water in citrus plantations, there is need to work out etiology of nematodes under changed irrigation system and also work out the efficient use of nematicides through low volume irrigation system for controlling nematodes.

* The biological control of citrus nematode has not been explored well. Intensive research is needed on this aspect.

* Except citrus nematode, no other nematodes on citrus have been given due attention in India. Systematic survey of parasitic nematodes associated with citrus plants, their densities and pathogenic potential need to be worked out.

* Very often other soil microorganisms are involved with infection of the nematode causing more damage to the plants. *Fusarium* sp., a weak pathogen is reported to cause more damage in presence of citrus nematode, but there is also need to study the associative effects of citrus nematode with *Phytophthora* sp. which is the cause of very severe problem of root rot in citrus.

* There is need to integrate different methods of nematode control and find out their workability under field conditions.

2.1.3. Papaya

Papaya (*Carica papaya*) is one of the most important fruit crops valued for its rich source of vitamins and nutrient content. It is native to tropical America and its place of origin is said to be in Southern Mexico and Costa Rica. It is believed to have been introduced into India during the 16th century. It is grown both in tropical and sub-tropical countries of the world like Brazil, Australia, USA, Taiwan, Puerto Rico, Peru, Pakistan, Bangladesh, Malaysia, India and various parts of Central and South Africa. The ripe papaya fruits are used for table purpose, raw fruits are cooked and used as vegetable. Immature fruits are used for the extraction of proteolytic enzymes papain, which is mainly used as a drug in digestive disorders and in the treatment of ulcers and diphtheria. It has also varied industrial uses viz., in tanning industry, degumming of silk,etc.

India is the largest producer of papaya in the world. It is cultivated on a commercial scale in Andhra Pradesh, Maharashtra, Gujarat, Karnataka, West Bengal, Assam, Kerala and Tamil Nadu. Papaya is grown in 73, 000 ha with annual production of 2.568 million tonnes. Productivity of papaya is the highest (35.2 t/ha) among the fruit crops, which has attracted the growers for its commercial cultivation resulting in the spectacular increase in area and production in last few decades. In the last decade, phenomenal increase in area and production is recorded from Southern states owing to demand of fruits for industrial use.

On an average, 1000 g of ripe papaya fruit contains 2500 I.U. of Vitamin A, 8 I.U. of Vitamin B, 33 Sherman units of Vitamin B2 and 70 mg of Vitamin C. Besides, it contains papain, a proteolytic enzyme which acts as papsin and is useful in digestive disorders.

Nematodes are one of the important limiting factors in the production of papaya throughout the country. Papaya being perennial in nature, harbours and encourages the build up of nematode population round the year. The first record of an association between a nematode and papaya appears to be that of Prasad (1960) who reported *Meloidogyne* spp. parasitizing papaya roots at New Delhi. At present, 32 nematode species within 20 genera have been found associated with papaya roots from various parts of the country (Khan and Khan, 1998). However, most of these nematodes are not known pathogens of papaya and their true relationship with the host plant is yet to be established.

Root-knot nematodes (*Meloidogyne incognita* and *M. javanica*), the reniform nematode (*Rotylenchulus reniformis*) and the dagger nematodes (*Xiphinema basiri*) have been identified as proven pathogens on papaya.

2.1.3.1. Root-knot Nematodes, *Meloidogyne* spp.

M. incognita and *M. javanica* have been reported to be the major nematode pests of papaya in India.

(i) Economic Importance and Losses: Ponte (1980) and Taylor *et al.* (1982) reported 10 to 20 per cent reduction in papaya fruit yield due to root-knot nematodes.

(ii) Distribution: The root-knot nematodes on papaya have been recorded from Delhi, Bihar, Haryana, Karnataka, Madhya Pradesh, Maharashtra, Orissa and Uttar Pradesh.

(iii) Symptoms: General symptoms visible in field include poor growth, yellowing of foliage, dropping of leaves, reduction in leaf production, weak vigour and premature dropping of fruits. Papaya orchards infected with *Meloidogyne* spp. show patches of poor growth with many plants missing in the rows. Roots exhibit typical below-ground symptoms *i.e.* galls of varying sizes. The lateral branching of roots is limited. In heavily infected old roots, adjacent galls join together and form large galls. In mild infestation, root tips become swollen and root growth inhibition is distinctly seen (Fig. 2.15).

(iv) Life Cycle: Khan and Saxena (1993) observed that *M. incognita* race 1 takes 30 days to complete the life cycle on papaya cv. Honey Dew in the months of August-September with temperature ranging from 28 to 35°C. Similarly, Baghel (1992) reported that *M. javanica* takes 32 days to complete the life cycle on papaya cv. Pusa Nanha, 30 days on Washington, 26 days on Solo II and 21 days on Honey Dew.

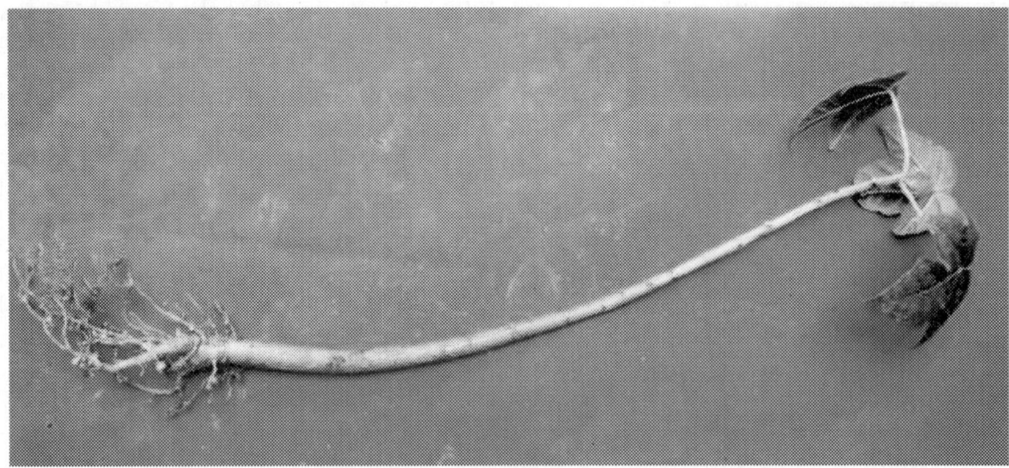

Fig. 2.15. Papaya plant infected with *Meloidogyne incognita*.

(v) Histopathology: Infection with *Meloidogyne* sp. caused localization of total lipids.

(vi) Management Methods

(a) Physical Methods: Hot water treatment of *M. incognita* infected papaya seedling roots at 50°C for 10 min. gave satisfactory control provided the main field was not infested.

(b) Cultural Methods: Neem cake at 2.5 t/ha gave reduction in root-knot nematode (*M. incognita*) population in both soil and roots and increased yield by 260% (10.4 kg/plant compared to 5.5 kg/plant in control) (Nayak, 1986).

(c) Chemical methods

Nursery treatment: Application of carbofuran and phenamiphos each at 2 kg a.i./ha, a day or two before seeding in nursery, is effective in the management of root-knot nematodes (36.2 and 43.4%, respectively) and getting higher healthy transplantable seedlings of papaya (12.9 and 17.3%, respectively). Highest germination (94%) was obtained with carbofuran at 4 kg a.i. per ha. Maximum plant height was recorded in fensulfothion at 2 kg a.i. per ha.

Main field treatment: Carbofuran and aldicarb at 2 kg a.i. per ha were effective against nematodes and gave highest yield.

Neem cake at 2.5 tons per ha and carbofuran at 6 kg a.i. per ha gave reduction in root-knot nematode population in both soil and roots. Highest yield of 11 kg per plant was recorded in carbofuran followed by neem cake (10.4 kg), whereas control plants yielded 5.5 kg (Nayak, 1986).

Nayak (1986) reported that aldicarb followed by carbofuran and ethoprophos all at 6 kg a.i. per ha were effective in controlling *M. incognita* on papaya.

(d) Host Resistance: The papaya cv. Pusa 22-3 was resistant, while Thailand, Waimanalo, Mukund Farm, Pusa 1-15 and Pusa 1-45 were moderately resistant (Parvatha Reddy *et al.*, 1988). Nayak (1986) reported that cvs. CO-2 and CO-3 were found to inhibit the development and reproduction of *M. incognita*. Khan *et al.* (1992) found that cv. CO-2 was resistant, while Peradeneya, Selection No. 7 and Washington were moderately resistant against *M. incognita* Race 1. Pusa Delicious, Pusa Dwarf and Pusa Giant were reported to be resistant to *M. incognita* (Laqman Khan, 2001).

(e) Integrated Methods: Nursery bed treatment with *T. harzianum* and *P. lilacinus* each at 5 or 10 g/kg soil resulted in production of highly vigorous papaya seedlings whose roots were colonized with both the bioagents. There was significant reduction in root galling in the combination treatments (2.9 to 3.2) compared to control (8.9) (Rao and Naik, 2003).

Simultaneous application of *P. lilacinus* and *T. harzianum* both at 1 g per plant gave maximum increase in plant growth parameters and highest reduction in reproduction factor and root galling in papaya (Khan, 1991).

2.1.3.2. The Reniform Nematode, Rotylenchulus reniformis

Prasad *et al.* (1964) first reported this nematode on papaya from Northern India. *R. reniformis* caused 28% loss in fruit yield of papaya (Rajendran and Naganathan, 1977b).

(i) Distribution: The reniform nematode has been found in the states of Gujarat, Madhya Pradesh, Orissa, Punjab, Tamil Nadu and Uttar Pradesh (Khan and Khan, 1998).

(ii) Symptoms: Nematode affected plants are stunted, foliage is slightly yellow in colour with reduced leaf size. Under severe nematode infection, roots show discolouration with dark brown lesions which are prone to attack by soil pathogens. Egg masses on roots are covered with soil particles.

(iii) Histopathology: Sivakumar and Seshadri (1972) studied histopathological changes in papaya due to the infection of *R. reniformis*. The infected tissue at the site of feeding showed hypertrophy, hyperplasia, thickening of cell walls, granular protoplasm and enlarged nuclei and nucleoli. *R. reniformis* was mainly a phloem feeder.

(vi) Ecology: The incidence of *R. reniformis* is higher in wetter soils.

(v) Management Methods

(a) Chemical methods: Foliar application of phenamiphos and oxamyl on papaya was effective for the management of reniform nematode (Ayala *et al.*, 1971).

Rajendran and Naganathan (1981) reported that carbofuran at 2 kg a.i. per ha gave significant reduction in soil population of *R. reniformis* and increased papaya fruit yield by 38.4 per cent over control.

(b) Host Resistance: Papaya cultivars CO-1, CO-2, CO-3, CO-4, CO-5, Thailand and Giant were tolerant/resistant to *R. reniformis*. CO-2 and CO-3 were found to inhibit the development and reproduction of *M. incognita* (Nayak, 1986). Patel *et al.* (1989) reported that papaya cvs. Solo and Washington were resistant, while Coorg Honey Dew was moderately resistant. Papaya cv. Mammoth was found resistant, while CO-1, Peradeneya, Sel.7 and Washington were moderately resistant (Khan and Khan, 1998).

2.1.3.3. The Dagger nematode, Xiphinema basiri

(i) Economic Importance and Losses: The pathogenic capability of *X. basiri* on papaya has been established by Sundarababu and Muthukrishnan

(1990). Inoculation of 2000 nematodes per 2 kg soil caused significant reduction in plant height (30.4%) and weight (38.6%) on papaya.

(ii) Symptoms: The nematode feeds on the roots of papaya and induced stubbiness of roots, thickness at root tips resulting in galling and fish hook formation (Sundarababu and Muthukrishnan, 1990).

(iii) Host Range: X. basiri is associated with banana, citrus, fig, guava, litchi, mango, mulberry, plum, rose, tomato, okra and castor in India.

2.1.3.4. Root-knot, *Meloidogne incognita* and Wilt, *Fusarium solani* Disease Complex

(i) Management Methods

(a) Integrated Methods: Khan *et al.* (1997) reported that application of both the bioagents (*P. lilacinus* and *T. harzianum*) limited the damage caused by the disease complex and gave a 35% increase in plant growth compared to individual applications.

2.1.4. Pineapple

Pineapple (*Ananas comosus*) is grown in 81,000 ha with annual production of 1.229 million tonnes. There has been phenomenal increase in production and productivity (15.1 t/ha) due to adoption of improved production technologies especially in Karnataka, West Bengal and Kerala. However, productivity is still low in North Eastern States (8-10 t/ha) owing to traditional methods of cultivation compared to West Bengal (24.88 t/ha). The major pineapple producing states include West Bengal, Assam, Bihar, Tripura, Kerala, Meghalaya, Manipur and Karnataka.

At present, 26 species belonging to 19 genera of plant parasitic nematodes have been reported in association with pineapple roots from India. The most common nematode found to be detrimental to crop production are *Meloidogyne* species and *Rotylenchulus reniformis*.

2.1.4.1. The Root-knot Nematode, *Meloidogyne incognita, M. javanica*

(i) Economic Importance and Losses: M. incognita has been found on pineapple plants in various parts of the country. Losses for the first crop were greater than 40% due to the attack of root-knot nematodes.

(ii) Symptoms: The root growth is greatly slowed down. The nematode also reduces plant weight, leaf weight and fruit weight by 47.5%. The root infested simultaneously by several larvae stops the growth and may produce many secondary roots.

(iii) Life Cycle: Second stage juveniles penetrate roots in the meristematic region of the root tip and become sedentary after 2-3 days. Development through subsequent molts leads to vermiform adult males and saccate, sedentary females. Reproduction is by mitotic parthenogenesis and female nematodes produce eggs contained in a gelatinous matrix.

(iv) Survival and Spread: Eggs contained in galled tissue can survive 20 days at relative humidity of 90%. Juveniles of *M. javanica* can survive in desiccated soil without a host for 20-24 weeks. *M. javanica* can survive, although at low levels, as long as 2 years in fallow field soil. The nematode spread over long distances in soil adhering to workers feet, implements and equipment that is moved from field to field. The nematode is also spread by infected planting material.

(v) Ecology: Population of *M. incognita* showed two seasonal peaks, once during April-May and another in November. Soil temperature range of 21.2°C-27.5°C and soil moisture range 14.0-18.6% were found to favour the population development.

(vi) Management Methods

(a) Cultural Methods: Yield increases have been observed when planting was done in a field, which was previously had pigeon pea, owing to decrease in infestation by *M. incognita*. Planting of pineapple after pangola grass reduced the population of *M. incognita, Criconemoides* spp., *Helicotylenchus* spp. and *R. reniformis* and increased the fruit yield in the planted crop (23.5 MT/acre as against 15.1 MT/acre in control) and first ratoon crop (18.4 MT/acre as against 9.3 MT/acre in control) (Ayala *et al.*, 1967).

The addition of chopped pineapple material to soil at 125-375 tonnes/ha significantly reduced galling caused by root-knot nematode.

(b) Chemical Methods: Dipping crowns in solutions of oxamyl at 2,400-4,800 ppm, or solutions of phenamiphos at 300 ppm provides a good protection against *M. incognita*.

(c) Biological Methods: Addition of *Dactylella ellipsospora* reduced plant injury caused by the root-knot nematode.

(d) Host Resistance: *Ananas ananasoides* and three of its hybrids are highly resistant to attacks by *M. incognita*.

2.1.4.2. The Reniform Nematode, Rotylenchulus reniformis

The reniform nematode is a major problem in pineapple throughout the world. Heavy infestations combined with moisture stress can result in complete ratoon failures.

(i) Economic Importance and Losses: The reniform nematode can reduce the pineapple plant populations by more than 83% in the fields.

(ii) Symptoms: Leaves of infected plants are less erect than those of healthy plants, are reddish in colour and show poor growth. Heavy infestations may result in plant collapse and death. Reniform nematode infection inhibits secondary root formation and root systems are poorly developed. Improper management of reniform populations typically leads to ratoon crop failures.

(iii) Life Cycle: Egg hatch is stimulated by root exudates of certain host plants, and second stage juveniles leave the egg and move into the soil. Once in the soil they undergo 3 molts without feeding, yielding adult males and pre-adult females. Females enter the root system, initiate a feeding site, become sedentary and develop into swollen, mature egg-producing females. Males do not feed.

(iv) Survival and Spread: The reniform nematode survives extended periods (6 months to 2 years) without a host. The nematode survives fallow periods in the egg stage or as anhydrobiotic juvenile stages, depending on soil moisture.

(v) Management Methods

(a) Cultural Methods: Rotation of pineapple with French marigold (*Tagetes patula*), Rhodes grass (*Chloris guyana*), sunnhemp (*Crotalaria juncea*), *Desmodium unicatum*, sugarcane or pangola grass were effective in controlling *R. reniformis*. Sugarcane is generally considered a non-host for *R. reniformis*, so planting pineapple into sugarcane soils may decrease problems with the nematode, provided that weed hosts are not present.

Irrigated fallow would reduce reniform nematode population as nematode prefers low moisture regimes in pineapple crop.

(b) Chemical Methods: Application of carbofuran at 2.25 kg a.i. per ha, oxamyl at 1 kg a.i. per ha and phenamiphos at 4.5 kg a.i. per ha were effective in controlling *R. reniformis* and in increasing fruit yields. Foliar application of phenamiphos at 600-2400 ppm gave effective control of reniform nematode. Foliar application of oxamyl may be as effective as phenamiphos if applied at twice the rate.

(c) Host Resistance: Pineapple varieties/species such as Venezolana, *Ananas ananasoides* and several of its hybrids are highly resistant to *R. reniformis* and do not allow the initial infestation to be maintained.

(d) Integrated Methods: A series of chemical and non-chemical methods have been suggested for integrated management of the reniform nematode:

- Avoiding monoculture.

- Increasing soil pH above 6.0 which reduces nematode population.

- Avoiding continuous application of ammonium sulphate as nematode prefers acid environment.

- Irrigated fallowing would reduce nematode population as nematode prefers low moisture regimes in pineapple crop.

- Application of fenamiphos at 0.56 to 3.36 kg a.i./ha or oxamyl spray at 1.12 to 4.48 kg a.i./ha helps in reducing the nematode population.

2.1.5. Grapevine

Grapevine (*Vitis vinifera*) is one of the important commercial crops grown in different climatic zones of the world for table delicacies and wine preparation. It is fairly good source of minerals like calcium, phosphorus and iron and vitamins like B_1 and B_2. In India, it is grown in tropical states like Maharashtra, Karnataka, Tamil Nadu, Andhra Pradesh and Punjab. Grapevine is grown in 60,000 ha with annual production of 1.546 million tonnes. In grapes, India has recorded the highest productivity (25.7 t/ha) in the world.

Currently, about 20,000 tonnes of fresh grapes are exported, accounting for just 2% of the total production. Out of this, about 12,000 tonnes are exported to the Middle East markets and the rest to U.K, Europe and few South-east Asian markets. Among these markets, U.K. is the most paying besides being reliable. Middle East markets are not dependable and are also less paying.

Several nematode species are known to be associated with grapes but the major ones belong to the genera *Meloidogyne, Pratylenchus* and *Xiphinema*. Other nematodes found associated with this crop are *Tylenchulus semipenetrans, Rotylenchulus reniformis, Paratrichodorus minor* and *Longidorus attentuatus*. Besides, several new species of nematodes have been recorded on grapes from different parts of India.

2.1.5.1. Root-knot Nematodes, Meloidogyne spp.

Two species of root-knot nematodes, *M. javanica* and *M. incognita* are recognized as the major pests of grapes causing economic damage. In India, *M. javanica* is most prevalent in northern part of country (Baghel *et al.*, 1980) and *M. incognita* in southern part (Darekar and Patil, 1985). *M. incognita* is also reported from some parts of Haryana. *M. incognita* was responsible for 55% loss in fruit yield of grapes (Rajagopalan and Naganathan, 1977b), while *M. javanica* caused 53% loss in yield (Baghel and Bhatti, 1983b).

(i) Symptoms: The root-knot nematode infestation is not manifested by any typical above-ground symptoms. Patches of poorly branched vines with scant foliage, pale and small leaves, and poor bearing are the indications of root-knot nematode damage. In young plants, premature decline, weak vegetative growth are commonly associated with nematode attack. The visibly unthrifty growth is generally attributed to moisture stress, low fertility, nutritional deficiency and other adverse conditions. However, the confirmation of nematode attack is possible by assaying soil and root samples. The root system shows typical localized swellings particularly on feeder roots and young secondary roots and females may be found on internodal trunk just below the ground level. *Meloidogyne incognita* has been reported to stimulate the production of many new fine rootlets above the site of nematode infection resulting in "hairy root" condition. Depending upon the variety of grape, *M. javanica* forms galls of varying size and shape and distorts the normal appearance of roots.

(ii) Histopathology: Studies conducted to investigate histological changes in root-knot nematode resistant grape (*Vitis rotudifolia*), due to *M. incognita* revealed that juveniles entered the roots but were not able to induce nurse cell formation. The only host plant reaction observed was the dissolution of middle lamellae in some cells surrounding nematode head. Investigations on the impact of *M. incognita* revealed that the physiological efficiency of vine is reduced. Fewer nematodes become established and fewer eggs were produced in roots of nematode resistant vine than in susceptible vine. The production was more on resistant vines as less energy reserve was required against nematode infection, reproduction and to repair the damage caused by nematodes. *M. incognita* did not affect the concentration of reducing sugars at nematode feeding sites on French Colombard (susceptible) and Thompson Seedless (moderately resistant) cultivars whereas non-reducing sugars increased in susceptible and decreased in moderately resistant cultivar, indicating that there was more translocation of photosynthates to the feeding sites of susceptible cultivar than that of resistant cultivar. This explains why nematode caused more damage to susceptible cultivars and fewer eggs are produced in resistant cultivar.

(iii) Management Methods

(a) Physical Methods: Hot water treatment of lightly root-knot infected rooted cuttings at 50°C for 10 min or at 48°C for 30 min before planting has been used to disinfect planting material.

(b) Cultural Methods: Use of nematode-free planting material is one of the most important cultural practices adopted to avoid nematode infection.

Intercropping: *Tagetes patula* used as intercrop, reduced 39-44% nematode population and significantly increased vine yield. Under field conditions, 27.0 to 38.7% soil population and 30.1 to 35.7% root population of nematode on grapevine was reduced when intercropped with marigold, asparagus or sunnhemp. Marigold was found most effective (Baghel and Gupta, 1986). Intercropping with marigold, asparagus or sunnhemp gave effective control and resulted in increased fruit yield. Cultivation of onion or garlic in the basin area brings additional remuneration besides reducing the nematode population.

(c) Chemical methods: Single application of aldicarb at 5 kg a.i./ha improved both mean yield (42.2%) and mean cane mass. Aldicarb at 5 kg a.i./ha or oxamyl at 10 kg/ha applied in two equal splits to control nematodes in vineyard gave significant yield improvement of 58% and 97%, respectively and reduced nematode population (Loubser and Klerk, 1985). Aldicarb at 2.5 kg /ha applied once annually at bud breaking stage to control nematode in vine, gave best results in terms of nematode reduction and crop yield (Loubser, 1985). Application of carbofuran, benfurocarb or phorate at 6 kg a.i./ha, in *M. javanica* infested vineyard gave 63-68% nematode reduction and improved yield (120-133%). Carbofuran at 13 g/sq m at bud breaking stage is recommended for controlling *M. javanica* (Anon, 1989b).

(d) Biological Methods

Antagonistic Bacteria: Sundarababu *et al.* (1999) observed that commercial formulation of *Pseudomonas fluorescens* reduced root-knot nematode population effectively and enhanced yield in grapevine. Soil application of talc formulation of *P. fluorescens* containing 15×10^8 cfu/g at 4 g/vine around root-knot infested grapevine at 15 cm depth in the basin at the time of pruning (during July) significantly reduced root galling due to *M. incognita* (39%), number of egg masses (250%) and increased fruit yield (166%) (Shanthi *et al.*, 1998). Natural suppression of *M. incognita* by *Pasteuria penetrans* has been reported on grapevine in Tamil Nadu (Mani, 1996).

Antagonistic Fungi: *Paecilomyces lilacinus* when applied at 20 g/sq.m./vine reduces *M. incognita* infestation by 33% on grapevine.

Arbuscular Mycorrhizal Fungi: Endomycorrhizal fungus, *Glomus mosseae* adversely affected *M. javanica* on grapevine cv. Perlette.

(e) Host Resistance: Use of nematode resistant root-stocks has an added advantage over other methods of control. The cost factor is reduced to minimum. In grapevine, Dogridge (*Vitis champini)*, Salt creek (*V. champini*), 1613 (*V. solanis* x Othello), Harmony (1613 x Dogridge), St. George (*V. rupestris*), A x G1 (Aroman x Ganzin1), are well known nematode resistant root-stocks. Lake Emerald and Tompa are recently recognized as resistant root-stocks (Mortensten, 1985). Banquabad, Cardinal, Early Muscat, Joazbeli, Loose Perlette and Reisling (all commercial varieties) have been recognized as resistant to *M. javanica* (Anon, 1989b). Black Champa, Dogridge, 1613, Salt Creek, Cardinal, Banquabad were recognized as resistant to *M. incognita*. Deshmukh *et al.* (2004) reported that grape varieties Black Champa, Dogridge and Degrasset resistant, while Pandhri Sahebi and Champanel moderately resistant to *M. incognita*.

(f) Integrated Methods

Botanicals and Chemicals: Application of neem cake (200 g/plant) + carbofuran (10 g/plant) reduced the nematode populations in the soil and roots. Combination of garlic and carbofuran at 0.1 g a.i./sq.m. reduced nematode population (44.2-47.6%) and increased yield (88.3-105.6%) over control with cost: benefit ratio of 1: 4 to 1: 5 (Anon, 1993b).

Two Bioagents: The combined application of bioagents *P. lilacinus* and *P. penetrans* was most effective in increasing the fruit yield of grapevine besides checking *M. incognita* with cost: benefit ratio of 1: 3.3 (Sivakumar and Vadivelu, 1999).

Bioagents and Botanicals: Application of neem cake at 200 g and *Paecilomyces lilacinus* at 50 g/vine helps in improving vine growth and yield. Further a technology was developed to enrich farm yard manure (FYM) by applying nematode bio-control agents such as *Trichoderma harzianum* and *Paecilomyces lilacinus*. To achieve this, one kilogram of neem based formulation of this bio-agent was applied to 1 tonne of farm yard manure. One hundred kgs of neem cake was applied to the farm yard manure to enhance the rate of enrichment by bio-agents. The farm yard manure was kept moist for 15 days under shade with thorough mixing of farm yard manure at 5 day interval. This enriched FYM was applied to the field at the rate of 2 kgs per plant for 5-6 times at an interval of 2 months. This integrated method reduced the nematode problems significantly and improved the yield levels of grapes.

Cultural and Chemicals: Combined application of neem cake and nematicide at reduced doses, along with intercropping with onion and garlic is highly profitable.

2.1.5.2. The Reniform Nematode, Rotylenchulus reniformis

The reniform nematode is probably the most widespread and has been reported on grapes in the Philippines, India, Egypt and Algeria.

(i) Management Methods

(a) Host Resistance: Grapevine cvs. Dakshi, Joazbeli and Mukchilani are reported to be resistant to the reniform nematode.

2.1.5.3. *The Dagger Nematode, Xiphinema index*

Dagger nematodes are ectoparasites and feed on succulent tissues of young roots. *X. index, X. diversicudatum* and *X. americanum* have been found pathogenic on grapes.

X. index inoculated at 500 nematode per vine altered auxin relationship and reduced vine growth. Concomitant inoculation of grape cv. Thompson Seedless with 500 *X. index* and *P. vulnus* caused greater stunting of vine shoot and root than did inoculation of either nematodes alone. Individually *P. vulnus* caused greater stunting than *X. index*.

(i) Symptoms: Dagger nematodes feeding on grape roots result in terminal swelling, cessation of root elongation and distortion due to malformation of rootlets (Fig. 2.16).

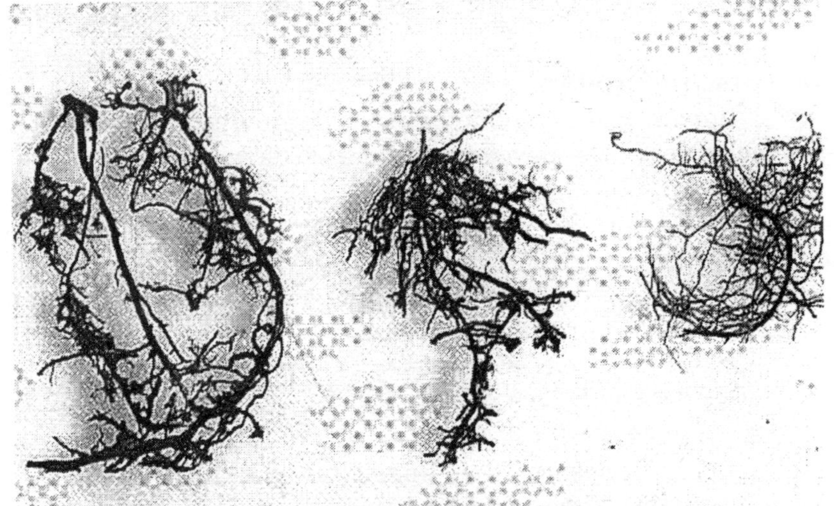

Fig. 2.16. Dagger nematode (*Xiphinema index*) damage to grapevine roots.
Left – Infected, Right – Healthy.
(Courtesy: Union Carbide Agril. Products Co. Inc., 1986).

Discolouration, decay of roots and death of growing points of feeder roots occur at later stage.

(ii) Histopathology: Epidermal and outer cortical cells collapse at feeding site and show necrosis. Multinucleate enlarged cells have been found beneath the layer of necrotic cells. *X. index* rarely preferred feeding on root tip.

(iii) Interaction with Other Pathogens: *X. index* transmits the fan leaf yellow mosaic vein banding virus (GVFV-GVYMV) disease of grape (Fig. 2.17). Leftover roots of dead and old plants in vineyards act as source of viral inoculum of dagger nematode year after year. *X. americanum* has been found transmitting tobacco mosaic virus, grape yellow vine virus, peach rosette mosaic virus and tomato ring spot virus.

Fig. 2.17. Grapevine infected with fanleaf virus disease.
(Courtesy: Union Carbide Agril. Products Co. Inc., 1986).

(iv) Management Methods

(a) Cultural Methods: In France, to save grapevines from attack by *Xiphinema index* and subsequently infection by grapevine fan leaf virus, a 7-year rotation with cereal and alfalfa crop has been recommended (Lamberti, 1981).

2.1.5.4. *Root-lesion Nematodes, Pratylenchus spp.*

Pratylenchus vulnus commonly found on grape is economically most important among lesion nematodes and its occurrence and severity is more in heavier soils.

Simultaneous inoculation of *P. vulnus* and *Xiphinema index* at 1000 and 500 nematodes, respectively per vine (cv. Thompson Seedless) enhanced the stunting of shoot and root as compared to individual inoculation of either nematode alone.

(i) Symptoms: Plants infected with *P. vulnus* show loss of vigour and reduction in fruit production. Infected young vines remain very weak, often fail to establish root system and eventually die. Below-ground symptoms on roots distinctly shows lesions, which are initially brown and later turn black. In severe infection, black lesions combine and girdle the roots.

(ii) Histopathology: Infected plants have generally reduced root system. Nematode penetration has been found restricted to few layers of cortical cells only. Adult female and eggs were not found in meristematic and vascular tissues. Inoculation of 1000 *P. vulnus* per vine has reported to be pathogenic and resulted in suppression of root and shoot growth. *P. vulnus* infection resulted in reduction in potassium and zinc uptake, which was measured in leaf blade and petiole of vines.

2.1.5.5. Root-knot, *Meloidogyne incognita* and Wilt, *Fusarium moniliformae* Disease Complex

(i) Management Methods

(a) Biological Methods: Soil application of *Pseudomonas fluorescens* at 100 g and FYM at 20 kg/vine gave effective management of disease complex and improved the plant stand by reducing the final soil nematode population (56.9%), root gall index (1.8) and % disease incidence (15.67%). This treatment also increased the number and weight of fruit bunches (17.83 and 155.40%, respectively) and fruit quality (more TSS – 13.53 Brix, TSS: acid ratio – 14.87, lower acidity – 0.91%) (Senthil Kumar and Rajendran, 2004).

2.1.6. Mulberry

Sericulture industry is dependent on Mulberry (*Morus* spp.) cultivation as the leaves form the main food of silkworms. Under the existing conditions, the yield is 3 to 3.5 tonnes/ha which can be stepped up to 30 tonnes by adopting improved cultivation aspects. Mulberry varieties are also grown for fruits. They bear small white, black or red fruits which are sweet and can be eaten. The fruits are a delicacy in Kashmir and Afghanistan. The fruits can be dried and powdered for feeding poultry and hogs.

The root-knot nematodes are the major limiting factors in the successful production of mulberry.

2.1.6.1. Root-knot Nematodes, Meloidogyne spp.

Govindaiah *et al.* (1991) reported 11.8% loss in herbage yield of mulberry due to *M. incognita*.

(i) Management Methods

(a) Physical Methods: Hot water treatment is effective with mulberry at 50°C for 15 min. Hot water immersion must be followed up with a cold water treatment.

(b) Cultural Methods: Intercropping of mulberry with marigold at wider spacing of 30 to 45 cm controlled *M. incognita*, increased leaf moisture and leaf yield of mulberry (Govindaiah *et al.*, 1990).

(c) Integrated Methods: Combined inoculation of AMF and *Pseudomonas fluorescens* had positive effect on root-knot nematode control on mulberry (Kumutha, 2001).

2.1.7. Passion Fruit

Passion fruit (*Passiflora edulis*) is native of Brazil. In India, it grows wild in the Nilgiris, Wynad, Kodaikanal, Shevroys, Kodagu and Malabar. Recently, it is being cultivated in some areas of Himachal Pradesh, Nagaland and Mizoram. The juice of passion fruit with an excellent flavour, is quite delicious, nutritious and liked for its blending quality. It is extensively used in confectionery, in preparation of cakes, pies and ice-cream. A rich source of vitamin A, it also contains fair amount of sodium, magnesium, sulphur and chlorides.

Although a number of plant parasitic nematodes are reported associated with passion fruit, only root-knot and reniform nematodes are known to cause economic damage.

2.1.7.1. The Root-knot Nematode, Meloidogyne incognita

The root-knot nematodes are the major limiting factors in the successful production of passion fruit.

(i) Interaction with Other Pathogens: The purple passion fruit is more susceptible to the disease complex involving the root-knot nematode and *Fusarium* wilt fungus.

(ii) Management Methods

(a) Host Resistance: Passion fruit cvs. Yellow and Kaveri were found highly tolerant to *M. incognita*.

Use of rootstock such as *Passiflora caerulea*, which is tolerant to root-knot nematodes, has been suggested.

2.1.8. Cherry

Cherry (*Prunus avium*) is a delicious fruit rich in protein, sugars and minerals. It has more calorific value than apple. Due to higher return, cherry is gaining popularity in temperate regions of the country (Kashmir, Himachal Pradesh and hills of Uttar Pradesh).

Root-knot and lesion nematodes are the major limiting factors in the successful production of cherry.

2.1.8.1. Root-knot Nematodes, Meloidogyne spp.

(i) Management Methods

(a) Physical Methods: Hot water treatment of cherry seedlings at 50-52°C for 5-10 min eliminates the nematode infection.

(b) Host Resistance: Cherry rootstocks – Mazzard, Stockton Morello, English Morello and Montmorency were entirely free of root-knot galls.

2.1.8.2. The Lesion Nematode, Pratylenchus vulnus, P. penetrans

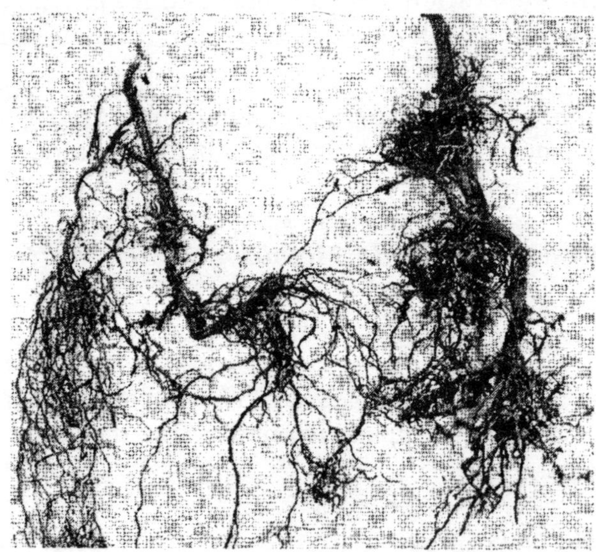

Fig. 2.18. Cherry roots infected with *Pratylenchus vulnus*
(Courtesy, Thorne, 1961).

(i) Management Methods

(a) Host Resistance: Cherry replants on Mahaleb rootstock grew well and *P. vulnus* and *P. penetrans* caused no root lesions.

2.2. TEMPERATE FRUIT CROPS

2.2.1. Peach

Peach (*Prunus persica*) is a temperate fruit rich in proteins, sugar, minerals and vitamins. In India, peach is grown in Himachal Pradesh, Punjab, Haryana, Delhi, west Uttar Pradesh, Jammu and Khasmir and Assam.

Root-knot, lesion and ring nematodes have been recognized as pathogenic causing economic damage to peach crop.

2.2.1.1. *Root-knot Nematodes, Meloidogyne spp.*

Of the many nematodes associated with peach trees, the root-knot nematode was the first known to cause damage in the form of knotted, swollen roots, reduced root and top growth and leaf chlorosis.

(i) Distribution: Infestation of *Meloidogyne* spp. has been reported from Delhi, Punjab, Haryana and Himachal Pradesh. *M. javanica* and *M. incognita* occur commonly in warmer peach growing areas, whereas *M. hapla* is restricted to colder regions only.

(ii) Symptoms: Above-ground symptoms on peach plants infected with *Meloidogyne* spp. include – poorly developed foliage, lack of vigour, die-back of twigs and yield reduction. In periods of drought, infected trees wilt more quickly than non- infected ones. The peach grove infested with root-knot nematode is usually composed of unevenly sized trees which is due to uneven distribution of nematodes and their population variation within one field.

The root system of nematode infested trees has poorly developed secondary and tertiary branches bearing galls of varying sizes. When many juveniles penetrate a root, root tip is devitalized and elongation of the root ceases resulting in the production of many fine tertiary branches forming brush-like appearance. Nematode damage to roots create 'suit effect' in which plant photosynthates are diverted downwards, making the nutrients available for root repair, thus impairing the tree from desired energy for the production of fruit resulting into yield loss.

(iii) Histopathology: The larvae of *M. javanica* readily penetrate susceptible peach rootstock (Okinawa) and develop into mature females. The production of giant cells was detected as early as fourth day after nematode penetration. Giant cell formation takes place in pericycle area, a few cells away from nematode head. The cytoplasm of affected cells become dense, followed by enlargement of nuclei and nucleoli. The affected cells undergo fast multiplication of nuclei without cytoplasmic division, thus forming syncytium. In most cases giant cells are located within vascular tissues.

(iv) Interaction with other Pathogens: Simultaneous occurrence of *Agrobacterium tumefaciens* with *M. javanica* on Lovell peach increased the incidence of crown gall disease. Not only there was significant higher incidence of crown galls but also 27.5% increase in nematode galls.

(v) Management Methods

(a) Physical Methods: Hot water treatment of peach seedling roots at 50-52°C for 5-10 min eliminates the nematode infection.

(b) Cultural Methods: Nematode-free planting stock is essential for transplanting in field.

(c) Biological Methods: In simultaneous inoculations of *Glomus mosseae* and *M. javanica* on peach cv. Floridasun, the fungus negated the growth suppressive effect of nematode and reduced the multiplication of nematodes.

(d) Host Resistance: Nemagaurd, Nemared, Okinawa and Rancho are the most popular root-stocks resistant to *M. incognita, M. javanica* and *M. arenaria* (Sharpe *et al.*, 1983). Yunnan, Shalil and Bokhara have been found resistant to *M. incognita* but susceptible to *M. javanica.* Peach cvs. Chinaflat, Bidbillos Early, Japanese Double and Elberta showed no gall formation in screening tests against *M. incognita* (Verma, 1987).

2.2.1.2. The Ring Nematode, Macroposthonia xenoplax

It is ectoparasitic and economically important nematode which is one of the factors in bacterial canker and peach tree short life (PTSL) disease. The estimated losses that occurred in peach orchards suffering from PTSL in Georgia were 30-70% after five years (Wehunt *et al.*, 1980). *M. xenoplax* was responsible for 33% loss in fruit yield of peach (Anon, 1990a).

(i) Distribution: M. xenoplax was first observed on peach grown in Northern India. It is distributed in Himachal Pradesh and Haryana.

(ii) Symptoms: Parasitism of *M. xenoplax* causes chlorosis, leaf drop, stunting and loss of vigour on top of peach trees. Peach roots showed lesions, pits and fewer feeder roots. The roots are predisposed to bacterial canker caused by secondary infection of *Pseudomonas syringae.*

(iii) Life Cycle: According to Seshadri (1964), eggs are deposited singly, 8-15 in number per female in 2-3 days egg laying period. Total life cycle was completed in 25-34 days at 22-26°C in sandy soil at pH 7.0 and 15.5% moisture. It takes 11-13 days for egg stage, 3-5 days for second stage, 4-7 days for third stage and 5-6 days for fourth stage development and preoviposition period of adult. Several generations are completed in a year.

(iv) Interaction with Other Pathogens: Parasitization by *M. xenoplax* predisposes peach trees to bacterial canker caused by *Pseudomonas syringae* (Lownsbery *et al.*, 1977). *M. xenoplax* has been strongly implicated in the PTSL condition in some parts of USA. PTSL disease syndrome is characterized by a sudden collapse of the new growth above soil line and death of trees in late winter or early spring shortly after foliation (Taylor *et al.*, 1970).

(v) Management Methods

(a) Physical Methods: A 37-100% reduction in population of *M. xenoplax* and other nematodes has been achieved by soil solarization in South Africa (Barberecheck and Broembsen, 1986).

(b) Chemical Methods: Lownsbery (1959) found that preplant fumigation with D-D at 350 litres/ha improved the growth of peach replants in orchards where *M. xenoplax* was the dominant nematode.

(c) Biological Methods: *Hirsutella rhossiliensis* reduced nematode population by 50%. Rate of nematode decay was much faster for parasitized second or third-stage juveniles (Jaffee *et al.*, 1988).

(d) Host Resistance: Lorell rootstock of peach has been recorded as most tolerant to *M. xenoplax* (Rom *et al.*, 1985).

2.2.1.3. The Lesion Nematode, Pratylenchus vulnus, P. penetrans

P. penetrans was responsible for 'peach replant problem' in Canada.

(i) Symptoms: Above-ground symptoms include tree decline with general nutrition deficiency symptoms and reduced growth. Under-ground symptoms are root lesions or discolouration of roots usually reddish brown later turning dark and black. The track of lesions on root becomes visible as the nematodes feed and migrate from one place to another. Rapid deterioration of feeder roots and reduced vigour are symptoms caused by *P. vulnus*. The nematode poses threat to replanting of peach.

(ii) Life Cycle: The nematode enters roots and moves inter and intracellularly and feed on cortical cells, the nematode larvae and adults both have capacity to enter or leave the roots. Adult females lay eggs singly within the root system or in the soil. Hatching begins to take place immediately after egg laying. The second, third and fourth stages feed on roots. In unfavourable conditions, nematodes leave the roots and migrate to new favourable roots for penetration and feeding.

(iii) Management Methods

(a) Physical Methods: Hot water treatment of peach seedlings at 52°C for 15 min eliminates *P. penetrans* infection.

(b) Chemical Methods: Best control of *P. vulnus* in peach can be obtained by combining pre-planting treatment with D-D at 150-400 litres/ha and post-planting treatment with DBCP at 35 litres/ha annually to the standing crop.

(c) Biological Methods: Pinochet *et al.* (1995) found that precolonization of peach roots with *Glomus mosseae* was effective in reducing the *P. vulnus* population.

2.2.2. Strawberry

Strawberry (*Fragaria vesca*) is an attractive, luscious, tasty and nutritious fruit with a distinct and pleasant aroma and delicate flavour. It is rich in vitamin C and iron and is mainly consumed as fresh. Jam and syrup are also prepared from strawberry. It is cultivated in tropical and sub-tropical areas round the year. It is commercially cultivated in Himachal Pradesh, Uttar Pradesh, Maharashtra, West Bengal, Nilgiri Hills, Delhi, Haryana, Punjab and Rajasthan.

Bud and leaf, lesion and root-knot nematodes have been recognized as pathogenic causing economic damage to strawberry crop.

2.2.2.1. *The Bud and Leaf Nematode, Aphelenchoides fragariae*

Strawberry dwarf is caused by at least three species of *Aphelenchoides*. *A. fragariae* causes spring dwarf, a cool weather disease of strawberry. The summer dwarf is caused later in the season by *A. besseyi*. Susceptible varieties showed 60 to 82% reduction in yield after 2 years of infestation.

(i) Symptoms: Spring dwarf (caused by *A. fragariae*) symptoms appear in the early spring as soon as plant growth starts. Bud and leaf development is retarded and plants fail to develop normal foliage. The crown is a compact mass of crinkled, twisted, distorted leaves with short twisted petioles. When the apical bud is killed, numerous adventitious buds develop and give rise to multiple-crowned plants (Fig. 2.19). Generally fruit buds are killed. The petioles of certain cultivars develop a prominent red colouration. Hence, the disease is called "red plant".

Summer dwarf has very similar symptoms, but is caused later in the season by *A. besseyi*. The severely stunted plant has short petioles and elongated, crinkled and asymmetrical leaflets. Leaf margins usually curl upwards in the young leaves and downward in the older ones. As is true with spring dwarf, the main bud is killed, resulting in multiple crowns.

(ii) Life Cycle: The life cycle is completed in 10-11 days at 18°C. The eggs hatch in 4 days and the larvae mature in 6-7 days. About 32 eggs are laid by a female. Occasionally the mother is killed when eggs hatch in the uterus.

Fig. 2.19. Strawberry plant infected with *Aphelenchoides fragariae* showing "spring dwarf" symptom (Courtesy: Jensen).

(iii) Interaction with other Pathogens: A. fragariae and/or *A. ritzemabosi* occasionally interacts with a bacterium, *Rhodococcus fascians*, causing "cauliflower disease". Both nematode and bacterium must be present to produce the disease. Diseased plants are stunted with deformed stems, leaves and flowers which in advanced stages resemble tiny cauliflowers.

The disease manifests in the form of malformation of the inflorescence, shortening and thickening of the aerial parts of strawberry plants. *A. fragariae* and *A. ritzemabosi* act as vectors of *R. fascians* and play an obligatory role in the cauliflower disease complex on strawberry. Strawberry plants infected with both *A. ritzemabosi* and certain strains of *R. fascians* resemble small cauliflowers. In extreme cases plants are stunted and flowers are deformed. The petals fail to develop or become small and greenish. The sepal becomes much enlarged. Stamens and receptacles are malformed so that fruits do not develop. Auxiliary buds are continually produced in the crowns. Therefore, both the organisms actively contribute to the cauliflower syndrome. The bacterium attaches to the cuticle of the nematodes and is transported to the crown tissue of strawberry plants.

(iv) Management Methods

(a) Regulatory Methods: Selecting nematode-free propagation material is of paramount importance to reduce the impact of bud and leaf nematode. Whenever possible, certified stocks should be used.

(b) Physical Methods: Hot water treatment of strawberry runners at 50°C for 5 min or 48°C for 15 min or 46°C for 20 min eliminates the nematode infection.

(c) Cultural Methods: Uprooting and destroying infected plants, and propagation through healthy planting stock are recommended for the control of nematodes.

Regenerating plants from clean, dormant, excised axillary buds is effective for eliminating *A. fragariae* from strawberry.

(d) Chemical Methods: Foliar sprays with phenamiphos at 4-6 ml/3.8 litres of water gave good control of *A. fragariae*. Chemicals used for disinfestation of strawberry plants include parathion as a foliar spray and soil application of aldicarb in granular form.

(e) Host Resistance: In Poland, the strawberry varieties "George Soltwedel", "Regina" and "Talizman" are comparatively resistant as are "Saksonka" and "Festival'naya" in USSR.

2.2.2.2. The Lesion Nematode, Pratylenchus penetrans

(i) Symptoms: Infested plants are stunted, pale green, have reduced yields and eventually die. Root symptoms appear as distinct brown lesions in early or light infestation. The entire root system eventually becomes black and necrotic as populations increase and secondary organisms invade (Fig. 2.20).

Fig. 2.20. The lesion nematode (*Pratylenchus penetrans*) on strawberry. A-Left, infected; Right, healthy. B-Lesion nematodes inside the root. C- Lesions on the root.

(ii) Management Methods

(a) Physical Methods: Hot water treatment of dormant strawberry plants at 54.4°C for 1 min or 52.8°C for 3 min or 51°C for 7.5 min or 49.4°C for 17.5 min eliminates the nematode infection.

(b) Integrated Methods: When the soil is heavily infested with *P. penetrans*, intelligent combination of nematicides and crop rotation is desirable:

- Soil application of nematicides reduce nematode population to a very low level.

- An advisable cropping sequence could then be: (a) strawberry which is susceptible but does not cause a large increase of the residual nematode population, (b) potato which is moderately susceptible and causing moderate nematode reproduction, (c) oats or barley which are not susceptible but causing nematode reproduction, and (d) beet which is not susceptible and suppressing nematode reproduction, after which moderately susceptible crops can be grown again successfully.

2.2.2.3. Root-knot Nematodes, Meloidogyne spp.

(i) Management Methods

(a) Host Resistance: Strawberry cvs. Tiago, Chandler and Torrey were found resistant to root-knot nematodes (Laqman Khan, 2001).

(b) Integrated Methods: Application of neem cake in combination with carbofuran applied at the time of planting, increased the yield by 35.2% over control (Laqman Khan, 2001).

2.2.2.4. The Root-knot Nematode, Meloidogyne hapla and Wilt, Verticillium dahliae Disease Complex

(i) Management Methods

(a) Chemical Methods: Soil application of ethylene dibromide (EDB) exhibited dual efficacy against *M. hapla - V. dahliae* disease complex in strawberry (Meagher and Jenkins, 1970).

2.2.3. Apple

Apple (*Malus pumila*) is the most important temperate fruit of the north-western Himalayan region. It is predominantly grown in Jammu and Kashmir, Himachal Pradesh and hills of Uttarakhand, accounting for about 90% of the total production. Its cultivation has also been extended to

Arunachal Pradesh, Sikkim, Nagaland, and Meghalaya in north-eastern region and Nilgiri hills in Tamil Nadu. Apple is grown in an area of 2,31,000 ha with annual production of 1.739 million tonnes.

The lesion nematode is the major limiting factor in successful production of apple crop.

2.2.3.1. The Lesion Nematode, Pratylenchus vulnus

(i) Management Methods

(a) Physical Methods: Hot water treatment of apple rootstocks at 50°C for 15 min or at 47°C for 30 min eliminates the nematode infection. The material should first be stored in a cold place (at 1 to 3.3°C), the treatment conducted in January, and the plants subsequently restored in a cold place until planting. Immersion in hot water should be followed immediately by a cold bath.

2.3. ARID ZONE FRUIT CROPS

2.3.1. Pomegranate

Pomegranate (*Punica granatum*) is an ancient and favourable table fruit in tropical and sub-tropical regions of the world. It is commercially grown for its sweet-acidic fruits which provide a cool refreshing juice and is valued for its medicinal properties. Its popularity is also due to the ornamental nature of the plant especially when bearing bright red flowers throughout the year. It is commercially cultivated in Maharashtra and in small scale in Karnataka, Andhra Pradesh, Gujarat and Tamil Nadu.

In India, it is considered as a crop of the arid and semi-arid regions because it can be grown even under conditions of severe drought and frost. Pomegranate is grown in an area of 1,13,000 ha with annual production of 0.793 million tonnes. The fruit rind, juice, leaf and roots are used in the preparation of various Ayurvedic medicines. The fruit juice contains glucose and fructose. The juice and seed contain large quantities of tannin and agolic acid which are essential in curing several diseases.

The root-knot nematode is an important limiting factor in successful cultivation of pomegranate crop.

2.3.1.1. The root-knot Nematode, Meloidogyne incognita

The root-knot nematodes were responsible for 24.64 to 27.45% loss in fruit yield of pomegranate (Singh *et al.*, 2003).

(i) Symptoms: Heavy root galling and visible damage to pomegranate trees in young orchards under irrigation is frequently encountered (Fig. 2.21).

Fig. 2.21. Pomegranate roots infected with *Meloidogyne incognita*.

(ii) Management Methods

(a) Cultural Methods: Application of neem cake at 3 kg/tree was effective in reducing the nematode population (57.83%) and in increasing the fruit yield (33.64%) over control.

(b) Chemical Methods: Soil application of phenamiphos gave good control of root-knot nematodes, provided protection to roots for 60 days against nematode invasion and improved fruit yields (Siddiqui and Khan, 1986).

Table 2.10. Biological control of plant parasitic nematodes in fruit crops.

Fruit crop	Nematode	*Effective Bioagents*	Reference
Banana	*M. incognita*	*P. lilacinus*	Devarajan & Rajendran, 2002
		G. fasciculatum	Prakash Babu, 2000
			Suma *et al.*, 2002
			Vidya & Reddy, 1998
			Umesh *et al.*, 1988

	R. similis	*P. lilacinus*	Devarajan & Rajendran, 2001
		T. viride	Sosamma *et al.*, 1994
		G. fasciculatum	Shanthi, 2003 Prakash Babu *et al.*, 2002 Prakash Babu, 2000 Channabasappa *et al.*, 1995
		P. penetrans	Tabassum *et al.*, 2002
	M. incognita & *R. similis*	*G. mosseae*	Vidya & Reddy, 1998 Parvatha Reddy *et al.*, 1997
Citrus	*T. semipenetrans*	*P. lilacinus*	Manzoor *et al.*, 2002
		P. fluorescens	Shanthi *et al.*, 1999 Parvatha Reddy *et al.*, 2000
		P. penetrans	Parvatha Reddy & Nagesh, 2000; Mani, 1988a
Grapevine	*M. incognita*	*P. fluorescens*	Shanthi *et al.*, 1999
		B. subtilis	Mani *et al.*, 1999
		P. penetrans	Stirling & White, 1982

Chapter 3

VEGETABLE CROPS

More than 40 vegetable crops belonging to Solanaceous, Cucurbitaceous, Leguminous, Cruciferous, root crops and leafy vegetables are grown in Indian tropical, subtropical and temperate regions. Important vegetable crops grown in the country are onion, tomato, brinjal, peas, beans, okra, chilli, cabbage, cauliflower, pumpkin, bottle gourd, cucumber, musk melon, water melon, radish, carrot, palak and methi. West Bengal, Uttar Pradesh, Bihar, Orissa, Tamil Nadu, Gujarat, Karnataka, Maharashtra, Andhra Pradesh, Jharkhand and Haryana are the major vegetable-growing states.

India is the second largest producer of vegetables next only to China contributing 14.40% of the total world production. Vegetables occupy an area of 6.756 million ha with a production of 101.434 million tonnes (Table 3.1). India produces 13.6% of the world onions. In vegetables, India occupies the first position in the production of okra, second in cabbage, cauliflower, onion, tomato, brinjal, peas and third in potato. The per capita availability of vegetables has increased from 76 g in 1951 to 256 g in 2001 as against 280 g/day recommended by ICMR for a balanced diet. Increase in production is mainly due to increase in productivity. Productivity is still lower in India (15 t/ha) than the world average productivity (15.7 t/ha). There is scope for increasing the productivity of vegetables in India further. The vegetable production target for 2020 has been fixed at 150 million tonnes.

Table 3.1. Area, production and productivity of vegetable crops in India (2004-05) (Chamber, 2006)

Vegetable crops	Area ('000 ha)	Production ('000 tonnes)	Productivity (tonnes/ha)
Potato	1542.3	29188.6	18.9
Tomato	497.6	8637.7	17.4
Brinjal	530.3	8703.8	16.4
Chilli	834.3	847.8	1.0
Cabbage	290.3	6147.7	21.2

Cauliflower	238.2	4507.9	18.9
Okra	358.3	3524.9	9.8
Onion	593.9	7515.4	12.7
Garlic	107.1	436.1	4.1
Peas	276.7	1971.8	7.1
Total	6756.0	101434.0	15.0

Potato, tomato, brinjal, chilli, cabbage, cauliflower, onion, peas, gourds, okra and melons are now grown on commercial scale with increasing demand, better output, change in production system and development of suitable cvs. These crops have high potential for foreign exchange earning. Today the emphasis on vegetable breeding is on value addition, disease resistance and better quality.

3.1. SOLANACEOUS VEGETABLE CROPS

3.1.1. Potato

Potato (*Solanum tuberosum*) is one of the important vegetable crops grown in India. It is being cultivated in 1.542 million hectares producing 29.118 million tonnes of potatoes with an average yield of 18.9 tonnes per hectare. It thrives best in cool climate. Therefore it is a summer crop in the hills and a winter crop in the plains. The major potato growing states include Uttar Pradesh, West Bengal, Bihar, Punjab, Gujarat, Madhya Pradesh and Karnataka.

Cyst, root-knot, dry rot and lesion nematodes have been recognized as the major nematode pests of potato.

3.1.1.1. Root-knot Nematodes, Meloidogyne spp.

Thirumalachar (1951) noticed root-knot galls on potato tubers from Simla during 1950 for the first time and since then it has been recorded from all the potato growing regions of the country. The root-knot nematodes, *Meloidogyne* species are by far the most important nematode pests of potatoes in India. Severe infestation of root-knot nematodes leading to crop failure have been noticed in Mahasu district of Himachal Pradesh and Hassan in Karnataka where farmers did not get even as much as planted by them.

An initial level of 2 root-knot larvae per gram of soil resulted in 100% tuber infection with an overall yield reduction of 42.5% (Prasad, 1989).

(*i*) *Distribution: M. incognita* has been recorded on potato in Himachal Pradesh, Uttar Pradesh, Assam and Karnataka; *M. javanica* in Himachal

Pradesh, Uttar Pradesh, Punjab, Haryana, Delhi, Bihar and Maharashtra; and *M. hapla* in Gulmarg (Jammu and Kashmir) and Shimla (Himachal Pradesh).

(ii) Symptoms: Plants affected by root-knot nematodes are generally stunted, slightly yellow and may wilt during hot weather. The diagnostic symptoms of a *Meloidogyne* attack on potatoes are the galls found on both roots and tubers. Tuber surface becomes uneven and warty because of numerous blisters like galls (Fig. 3.1). When an infested tuber is transversely cut, the white glistening swollen females of the size of pin head can be easily seen embedded in the potato tissue. Such tubers are invariably small, unfit for marketing, rot quickly and cause considerable yield losses. Galling on tubers may render them unsaleable.

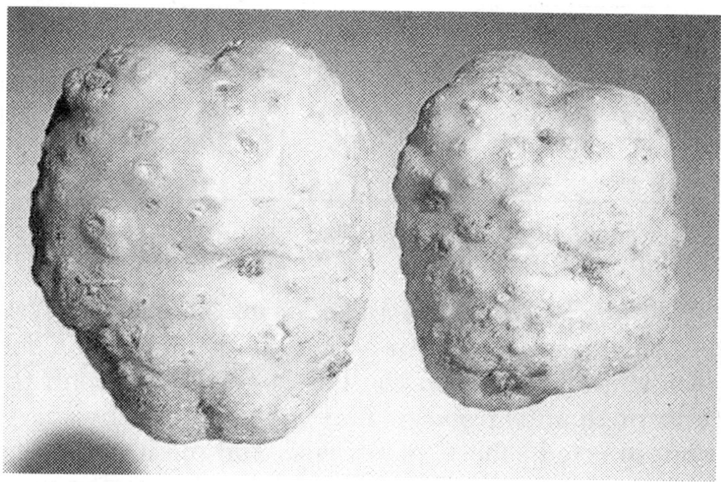

Fig. 3.1. Potato tubers with blisters incited by *Meloidogyne incognita* (Courtesy: Sasser, 1971).

(iii) Life Cycle: There are up to five generations on the susceptible host under favourable conditions. Both roots and tubers are infected, however, the first generation occurs mainly on the root systems, while the succeeding generations attack tubers.

(iv) Interaction with Other Pathogens: Meloidogyne species are associated with the fungi such as *Rhizoctonia solani* and *Verticillium* species and the bacterium, *Ralstonia solanacearum* in forming disease complexes.

Jatala *et al.* (1990) reported the interactions between *R. solanacearum* and nematodes, especially *Meloidogyne* spp. in potato in relation to breeding for combined resistance.

(v) Ecology: Root-knot damage to potatoes is usually associated with light soil. Aggravated attack by *Meloidogyne* species on potatoes under irrigation is usually ascribed to spread of the nematodes in irrigation water. *M. hapla*

requires cool climate, whereas *M. incognita* and *M. javanica* require higher temperatures for their development and reproduction.

(vi) Races: Four races in *M. incognita* and 2 races in *M. arenaria* have been identified based on the North Carolina Differential Host Test which employed 5 differential hosts for delineating races (Table 3.2).

Table 3.2. Races of *Meloidogyne incognita* and *M. arenaria*

Meloidogyne species/Races	Cotton Deltapine 61	Tobacco NC 95	Pepper Calif. Wonder	Watermelon Chart. Grey	Peanut Florunner	Tomato Rutgers
M. incognita 1	-	-	+	+	-	+
2	-	+	+	+	-	+
3	+	-	+	+	-	+
4	+	+	+	+	-	+
M. arenaria 1	-	+	-	+	-	+
2	-	+	+	-	+	+

In India, all 4 races of *M. incognita* have been reported and the race 1 is dominant (Sharma and Gill, 1992).

(vii) Survival and Spread: Since *Meloidogyne* spp. attack a large number of plant species, their population can be maintained on weeds and volunteer crops. They over winter usually in the form of eggs, although the ability of juveniles to go through anhydrobiosis may contribute to the survival of some *Meloidogyne* spp. Infected tubers, plant parts and planting material, as well as movement of infested soil by farm machinery and irrigation water are the main avenues of disseminating *Meloidogyne* spp.

(viii) Management Methods

(a) Physical Methods

Heat Treatment: Incubation of potato tubers at 45°C for 48 hours has been shown to kill about 98.9 per cent *M. incognita* without affecting tuber viability (Nirula and Bassi, 1965). Hot water treatment of potato tubers at 46-47.5°C for 120 min gave effective control of root-knot nematode infection.

Rabbing: In small holdings, burning of trash in the field before taking up tuber planting not only helps in reducing the nematode population but also enriches the soil.

(b) Cultural Methods: Tubers from root-knot infested potato crops should not be used as seed, as even outwardly clean looking seed may be infested. The movement of soil and water from the infested fields should be avoided. The field should be kept free from weeds, since root-knot nematodes have a wide host range and most of the weeds help in the build up of nematodes. Thus, clean cultivation (free from weeds which act as hosts) reduces nematode population to a great extent.

Deep ploughing and drying of soil in the summer months facilitates the drying and death of infective larvae thereby reducing initial inoculum in the soil. Dry fallow with 2 or 3 deep ploughings during the hot summer months gives excellent control of root-knot nematodes.

Intercropping: Potatoes grown with white mustard in a pot of infested soil were less heavily attacked by root-knot nematodes than potatoes growing alone. Potato root diffusate was ineffective in the presence of leachings from the roots of mustard seedlings. Mustard oil increased the yield of potatoes by reducing the severity of nematode attack. The active principle involved in mustard is allyl isothiocyanate which is toxic to the nematodes. The inter-planting of marigold with potato has resulted in a 7% increase in tuber yield.

Intercropping of onion and maize reduce galling by *M. incognita* on potato roots. Potatoes intercropped with maize have been found to be extremely vigorous with well developed root systems.

Time of Planting: In long season, warm climate regions, potatoes may be planted during the winter months and harvested before injury occurs in the spring without visible infection. Frequently potatoes are planted in January-February and harvested in April-May before the second generation of nematodes develop and attack the tubers. Planting of potato during third or fourth week of March in Shimla Hills would reduce both root and tuber infestation of *M. incognita* without affecting the yield (Ahuja, 1983; Prasad *et al.*, 1983). The yields were maximum which was concomitant with the lowest tuber infestation and lowest larval population in the soil at harvest. Late planting of autumn crop and early planting of spring crop in North Western Plains; while in hills, early planting of summer crop in fourth week of March is ideal. At Jalandhar, early planting of spring crop in first week of January and late planting of autumn crop in second and third week of October reduced the root-knot infection in potato.

Crop Rotation: In North Western Hills, rotation of potato with French marigold , leafy vegetables, maize and wheat helped in minimizing the nematode damage and increasing the productivity of the crop (Raj and Nirula, 1969). Crop rotation with non-host crops like maize, wheat, barley, beans, etc. reduces nematode population. Growing of trap crop like French marigold, *Tagetes patula* in alternate rows with potato reduced the root-knot nematode larval population in soil, root galling and tuber infestation while the yields increased up to 123% over control. Similarly, growing *Cassia* sp. and incorpo-ration of saw dust at 2.6 t/ha also helped in reducing tuber infestations. Potato crop can be followed by peanuts without any damage to peanuts and nematode populations will be reduced.

Botanicals: The oil cakes (neem, groundnut, mustard and castor) were effective in reducing parasitic nematode population (*Hoplolaimus, Tylenchorhynchus, Meloidogyne* and *Helicotylenchus* species) and in increasing yields of potato. Application of eucalyptus leaf waste from oil distillation plants at 2.5 t/ha helps in reducing the nematode population.

(c) Chemical Methods: Application of carbofuran at 3 kg a.i./ha in heavily infested fields and 2 kg a.i./ha in moderately infested fields in two equal splits, first at planting and second at the time of earthing up, is recommended. Application of carbofuran at 3 kg a.i./ha or aldicarb at 2 kg a.i./ha or ethoprop at 10 kg a.i./ha in two equal splits, first at planting and second at the time of earthing up, increased the yields by 18 to 23% while accounting for 80% reduction in tuber infestation.

(d) Biological Methods: The bacterium, *Pasteuria penetrans* has offered possibilities of biocontrol. Arbuscular mycorrhizal fungi such as *Glomus fasciculatum* and *G. mosseae* have shown promise in reducing the root-knot nematode infestation.

(e) Host Resistance: Potato cv. Kufri Dewa was found resistant to root-knot nematodes.

(f) Integrated Methods: Application of neem cake/FYM/compost enriched with *Trichoderma harzianum*/*Paecilomyces lilacinus* gave effective control of root-knot nematodes.

Following 2 years crop rotational sequence of maize-wheat-potato-wheat coupled with summer fallow after 2 or 3 deep ploughings in North Western Plains reduced root-knot damage significantly.

3.1.1.2. *Cyst Nematodes, Globodera rostochiensis, G. pallida*

The potato cyst nematode is established as one of the major crop protection problems of the world. The ability of this nematode to build up to damageable levels in a short span of 5-6 years, substantial yield reductions in the crop, lack of inexpensive nematicides for soil treatment capable of providing adequate level of control under field conditions, the relative ease with which the cysts are dispersed with soil adhering to the seed tubers and the long persistence of eggs within the cyst in the absence of the host makes this nematode as probably the most important pest problem on potatoes.

The avoidable yield loss in susceptible potato due to cyst nematodes was 99.5 to 99.8% in summer and autumn crops. Total failure of the crop has been reported under severe infestation conditions. An average loss of about 9% of global potato is accounted to the cyst nematodes amounting to about 45 million tonnes (Prasad, 1989).

(i) Distribution: In India, it was first reported by Jones (1961) from Nilgiri Hills, Tamil Nadu. It was also reported from Kodai Hills, Tamil Nadu, and Munar Hills of Kerala (Ramana and Mohandas, 1998). Out of 9,000 ha under potato, 3,000 ha are infected by this nematode in Nilgiris. In Kodai Hills, about 200 ha are infected (Thangaraju, 1983). Tomato and brinjal are also attacked by this nematode.

(ii) Symptoms: The disease usually occurs in patches. The infested plants exhibit typical symptoms of patchy growth of weak and stunted plants. The patches increase in time with continuous potato cropping. Under conditions of severe infestation, the plant growth is stunted and wilting occurs during hot part of the day. Plants show tufting of leaves at the top as the outer leaves turn yellow and die. Root system is smothered, secondary roots are induced at the collar region and the plants can be easily pulled out. Tubers formed are less in number and reduced in size. In extreme cases, tuber formation is arrested. The total photosynthesis per plant is also significantly reduced as a result of reduced leaf area and this is reflected in reduced potato yield.

(iii) Life Cycle: G. rostochiensis is amphimictic, with sessile, globose females and motile, vermiform males preceded by four larval stages. The second stage larva is the resistant, dormant and infective stage remaining inside the egg within the cyst formed by tanning of the female's cuticle. The cysts remain in the soil after death of the host, each containing about 500 eggs. After hatching, the second stage larva enters the root of a host plant, just behind the root tip or at the site of a new lateral root and moves through the root away from the tip before starting to feed on a group of pericycle, cortex or endodermis cells which are modified to form a large syncytial transfer cells on which the nematode feeds throughout the rest of its development to the adult stage. Sex is determined by nutrition, a poor food supply resulting in the formation of a male, a good supply, a female. Thus with heavy infections males predominate, the reverse with light infections.

The adult females swells greatly, rupturing the root cortex and protrudes from the root, exposing the vulval region while the head and neck remain inserted in the root (Fig. 3.2). As they develop, males become coiled within the earlier larval cuticles and on maturity they rupture them, then leave the root and are attracted to females by a secretion. Both sexes mate several times. Males do not feed but remain active for up to 10 days, females normally mate soon after rupturing the root cortex and within 50 days of larval invasion.

(iv) Host Range: All known varieties of potato, tomato and brinjal are severely attacked by the cyst nematode. A number of other Solanaceae are recorded as hosts, but the majority of hosts are species and varieties of tuberous *Solanum*. The only susceptible plant outside the family is snap dragon, *Antirrhinum majus*.

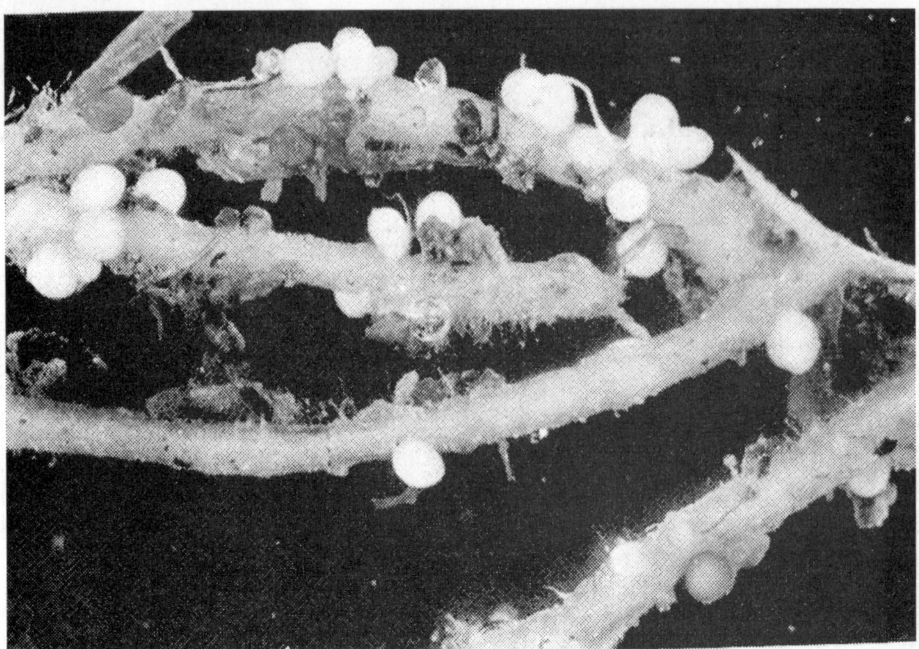

Fig. 3.2. Potato roots infected with *Globodera rostochiensis* (Courtesy: Sasser, 1971).

(v) Interaction with Other Pathogens: The potato cyst nematode and the black-scurf fungus *Rhizoctonia solani* together caused much greater reduction in yield than did the organisms separately. The cyst nematode also interacts with *Verticillium dahliae*, enhancing the severity of the wilt caused by the fungus. The leaf symptoms appeared earlier and the tuber yields were depressed in the presence of the nematode.

The potato cyst nematodes, *Globodera rostochiensis* and *G. pallida* in association with the wilt bacterium, *R. solanacearum* causes wilt complex in potatoes.

(vi) Ecology: Well-drained, aerated sands, silts and peat soils with a moisture content of 50-75% water capacity are suitable for survival and movement of the free-living stages of the potato cyst nematode. A minimum soil temperature of 10°C is necessary for larval activity.

(vii) Spread and Survival: The cysts can be spread by the movement of soil through seed material for a very long distance. They can also be spread through field tools, implements, seedlings grown in infested soil for transplanting elsewhere, labour's feet, etc. from one field to another. The cysts can remain viable in the soil without a host for 30 years.

(viii) Physiological Races: At present, five pathotypes in *G. rostochiensis* and four pathotypes in *G. pallida* are differentiated using the following differentials (Table 3.3).

Table 3.3. Pathotypes of *Globodera rostochiensis* and *G. pallida*

Differential host	Nematode pathotypes*				
	1	2	3	4	5
	Globodera rostochiensis				
Solanum tuberosum	+	+	+	+	+
ex *andigena* CPC 1673	-	+	+	-	+
ex *kurtzianum* 60.21.19	-	-	+	+	+
ex *vernei* 58.1642/4	-	-	-	+	+
ex *vernei* 62.33/3	-	-	-	-	+
ex *vernei* 65.346/19	-	-	-	-	-
	Globodera pallida				
Solanum tuberosum	+	+	+	+	
ex *multidissectum* P 55/7	-	+	+	+	
ex *vernei* 62.33/3	-	-	+	+	
ex *andigena* x *multidissectum* D 47/1	-	-	-	+	
ex *andigena* x *multidissectum* D 47/2	-	-	-	-	

*(+) indicates ability to produce females, (-)indicates inability to produce females

In India, potato cyst nematodes are restricted to Nilgiri and Palani hills in South India, where 5 pathotypes – Ro 1, Ro 2 of *G. rostochiensis*, and Pa 1, Pa 2 and Pa 3 of *G. pallida* have been detected (Amalraj and Rao, 1992).

(ix) Management Methods

(a) Regulatory Methods: In India, there is a Quarantine Act against the cyst nematode of potato (*G. rostochiensis*) in Nilgiris. Infected potatoes are not allowed to be transported from Nilgiris to other parts of India for seed and table purposes in order to prevent the spread of potato cyst nematode from Tamil Nadu to other states and Union Territories.

Potato seed pieces free of the cyst nematode can be produced commercially by seed certification. Use of certified planting material is fundamental to the prevention of cyst nematodes spreading into uninfested regions.

(b) Physical Methods: Fassuliotis and Sparrow (1955) demonstrated that irradiation of potato tubers with x-rays inhibited sprouting and also development of *G. rostochiensis*. Cysts of this nematode exposed to 20,000 r contained only brown and dead eggs, while at 40,000 r the eggs lost their contents completely.

Careful washing of soil adhering to potato tubers can considerably reduce the risk of spreading cyst nematodes. Commercial potato tuber washing apparatus has been developed.

(c) Cultural Methods: The potato cyst nematode (*G. rostochiensis*) can be eliminated by selecting nematode-free planting material.

Early Maturing Varieties: Early maturing varieties of potato like Irish Cobbler usually suffers less than do late maturing varieties. These varieties start growth in spring at soil temperature too low for much nematode activity and may be harvested before many of the nematodes can reproduce.

Crop Rotation: Continuous cropping with potatoes increases the cyst population of *G. rostochiensis*. More than three year rotations with wheat, strawberry, cabbage, cauliflower, peas, maize and beans reduces the nematode population to a safe level. Growing non-host crops like radish, garlic, beet root, French bean, cruciferous vegetables, turnip or green manuring crops bring down the cyst nematode population by more than 50%. A crop rotation pattern (using potato-French bean-peas-potato), wherein potato is grown once after 3-4 other crops, decreased the nematode population by 98 to 99% and increased yields of potato by 90%. In Nilgiris, crop rotation with cabbage and carrot gave effective control of cyst nematodes. Growing of vegetables and resistant potato in a long term rotation trial brought down the cyst propagule population up to 92% and was effectively used for nematode management at Nilgiris.

Botanicals: Incorporation of eucalyptus litter at 10 t/ha at the time of preparation of land reduces the population of the potato cyst nematodes. Application of eucalyptus distillation waste at 2.5 t/ha helps to reduce the cyst nematode population in soil.

(d) Chemical Methods: Application of carbofuran at 3 kg a.i./ha or aldicarb at 2 kg a.i./ha or ethoprophos at 10 kg a.i./ha was found to reduce the infestation of root-knot and cyst nematodes in potato crop. The yield increase ranged from 18 to 27%. The efficacy of these treatments further increased when applied in equal split doses.

Application of aldicarb at 2 kg a.i./ha as spot treatment was the most economical dose which ensured reasonable level of nematode control while increasing yields (14 to 219% in summer and 15 to 115% in autumn), followed by carbofuran (18 to 206% in summer and 10 to 115% in autumn) and phenamiphos (15 to 105% in summer and 10 to 90% in autumn). The least build up of cyst nematode population was noticed in aldicarb and carbofuran (Rf=0.5 to 1.2) followed by phenamiphos (Rf=1.5 to 1.9) and control (Rf=4.7 to 9.2).

(e) Biological Methods: Application of *Pseudomonas fluorescens* at 10 g/m^2 gives good control (Mani *et al.*, 1998). Application of neem cake at 1 t/ha or FYM at 20 t/ha or compost at 10 t/ha enriched with *Trichoderma harzianum / Paecilomyces lilacinus* gave effective control of cyst nematodes.

(f) Host Resistance: In the United States, both USDA and Cornell University scientists have evolved two commercial potato varieties – Peconic and Wauseon – which are resistant to the cyst nematode.

In India, Potato variety Kufri Swarna developed by the Central Potato Research Institute Substation at Ooty is resistant to both the species of cyst nematode and the late blight disease. The variety is also draught resistant and has shorter dormancy which is advantageous to Nilgiris region where potato is taken up mainly as a rain fed crop almost throughout the year and thus covers about 40% of potato area (Anon, 1985).

(g) Integrated Methods: The combination of disease escape (by planting early maturing varieties) and the use of nematicides gave good control of the potato cyst nematode. Hygiene in the form of seed certification, combined with crop rotation in seed growing areas is effective. Use of a resistant variety followed by a nematicide would kill 99% of the nematode population. The potato cyst nematode problem in Nilgiris is being managed by chemical treatments, crop rotations and utilizing the available sources of resistance (Kufri Swarna). The most effective management combines crop rotation, use of nematicides and resistant varieties to keep the nematode at an economically acceptable level (Jones, 1969) (Table 3.4).

Table 3.4. Integrated management of the potato cyst nematode

S.N.	Management method(s)	Resulting kill popn. (% initial popn)	Kill (%)	Popn. after growing & harvesting a susc. cv., calculated at 2 assumed nema multpln. rates (% initial popn)	
				30 x	70 x
1	4 years without potato	3	97	90	‹ 100
2	1 year with resistant Potato	20	80	› 100	› 100
3	Nematicide treatment (Fumigant)	25	75	› 100	› 100
4	1 and 2	0.6	99.4	18	42
5	1 and 3	0.75	99.25	22.5	52.5
6	2 and 3	5	95	› 100	› 100
7	All 3 methods	0.15	99.85	4.5	10.5

Better control of *G. rostochiensis* by *P. penetrans* was possible with soil solarization combined with application of FYM or karanj cake which resulted in greater kill of nematodes (Sitaramaiah and Naidu, 2003).

A well known integration of methods can control *G. rostochiensis* in potato in a 4-year rotation combining soil fumigation (causes 30% reduction in nematode population), cultivation of non-host (50% reduction), cultivation of resistant potato variety (30% reduction), cultivation of non-host (50% reduction) and then cultivation of susceptible variety of potato (Oostenbrink, 1972).

3.1.1.3. The Potato Rot Nematode, *Ditylenchus destructor*

D. destructor is the most important pest of potato tubers and is responsible for dry rot of tubers. High yield losses occur in the areas where climatological conditions favour establishment of the potato rot nematodes. The effect of nematodes will manifest itself at harvest or storage when infected tubers will rot.

(i) Symptoms: D. destructor enters potato tubers through lenticels and initially causes small white mealy spots just below the surface that are only visible if the skin is removed. Infested areas enlarge and coalesce and light brown lesions, consisting of dry granular tissue, may be visible beneath the skin. As the infestation progresses, the tissues dry and shrink and the skin becomes cracked and papery (Fig. 3.3). Internal tissues gradually darken and there are often secondary invasions of fungi, bacteria, mites, etc. If stored in moist conditions, a general rot may ensue and spread to neighbouring tubers.

(ii) Life Cycle: The life cycle is very similar to *D. dipsaci. D. destructor* does not seem to have a resistant stage, as it does not form 'nematode wool' or readily withstand desiccation but apparently survives in soil on alternative weed and fungal hosts. Thorne (1961) suggests that it may over winter as eggs.

(iii) Survival and Spread: D. destructor has a wide host range, can survive on weeds and on a wide range of soil-inhabiting fungi. It can also survive on infected tubers left in the field. Spread occurs by introduction of infected tubers and in soil adhering to seed pieces. Irrigation water and cultivation by infested farm tools and machinery are other sources of inoculum dissemination.

(iv) Management Methods

(a) Cultural Methods: The use of healthy seeds, destruction of infected plant parts in the field, control of weed hosts, late planting when soil moisture is inadequate for optimum invasion and early harvest reduces losses. Two to three years rotation of potato with beans, corn and fallow kept free of weeds gave a very marked reduction in nematode populations. Satisfactory control of *D. destructor* can also be obtained by crop rotation with small grains, vetch and lupine provided potatoes are grown once in 3-4 years. Rotation of potatoes with sugar beet and other non-host crops can reduce nematode populations.

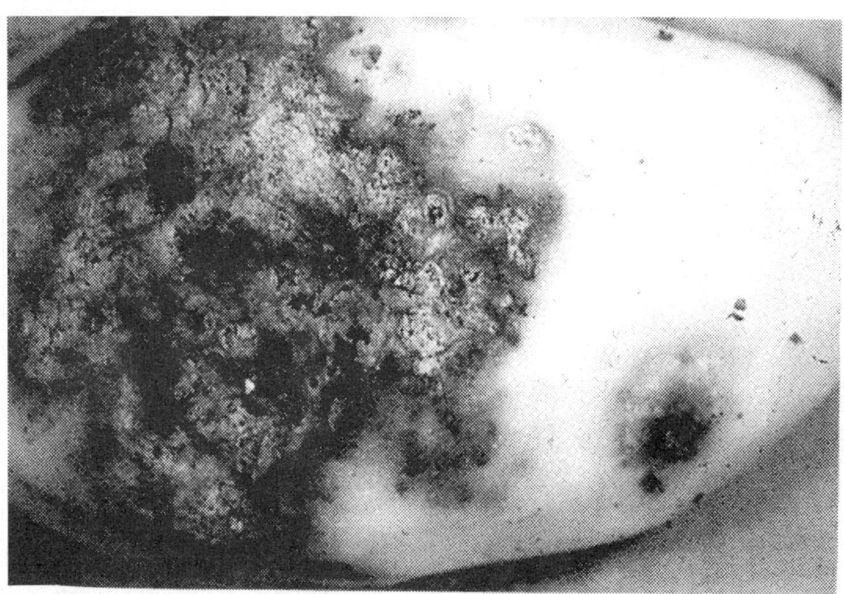

Fig. 3.3. Dry rot of potato tuber incited by *Ditylenchus destructor*
(Courtesy: Sasser, 1971).

The application of ammonium nitrate and ammonium sulphate (at 144 and 118 kg/ha, respectively) while planting potato reduced infection from 7% to 0.2-0.5%, and raised the yield from 10.6 t/ha to 16.3 and 25.2 t/ha, respectively.

(b) Chemical Methods: Darling (1959) found that heavily infested land treated with ethylene dibromide at 35 litres/ha in June followed by 17.5 litres/ha in August, gave clean potato crops in the following two years.

(c) Integrated Methods: The control of potato rot nematode was achieved by the combination of disease escape, hygiene, in the form of seed certification and crop rotation as follows (Winslow and Willis, 1972):

- Healthy seed, planted late, harvested early and stored as cool and dry as possible.

- Proper rotation of potatoes with non-host crops and growing of potatoes not more frequently than once in 3 or 4 years.

- Field hygiene in the form of removal of old infested tubers and weed control.

3.1.1.4. *The Lesion Nematode, Pratylenchus penetrans*

(i) Symptoms: High populations of lesion nematodes cause areas of poor growth; plants are less vigorous, turn yellow and cease to grow. Damage is often caused by direct feeding and, usually, only cortical tissues are affected. Large nematode populations cause extensive lesion formation and cortex destruction of unsuberized feeder roots (Mai *et al.*, 1981). Tubers are often attacked and small lesions are formed on the surface.

(ii) Survival and Spread: Infected tubers are sources of nematode inoculum and aid in survival of the nematodes.

(iii) Management Methods

(a) Physical Methods: Hot water treatment of infected tubers at 50°C for 45 to 60 min. aid in reducing nematode spread.

(b) Cultural Methods: In areas infested by *P. penetrans*, beet should precede the susceptible potato in the crop rotation system. The cultivation of legumes should be avoided since nematodes breed well in them.

(c) Host Resistance: Potato cv. Russet Burbank is tolerant, while Peconic appears to be less susceptible to the lesion nematode.

3.1.1.5. Other Nematodes, *Quinisulcius capitatus* and *Helicotylenchus dihystera*

The pathogenicity of the stunt nematode, *Quinisulcius capitatus* frequently occurring in hill tracts was established on potato cv. 'Kufri Jyoti' at Simla (Prasad and Sharma, 1985). The nematode build up ranged from 5 to 8 times at initial inoculum levels of 10 to 1000 at 45 days. Besides, the plant growth parameters were also reduced which affected the tuber production. The tuber reduction ranged from 14 to 29% and was negatively correlated with nematode build up.

The spiral nematode, *Helicotylenchus dihystera* was also pathogenic to potato accounting for 9-27% reduction in yield in 90 days with 2-4 times nematode build up (Prasad, 1990).

3.1.1.6. Root-knot Nematode, *Meloidogyne* spp. and Wilt, *Verticillium dahliae* Disease Complex

The root-knot nematodes often interact with *V. dahliae* in development of disease complex in potato.

(i) Management Methods

(a) Cultural Methods: In fields infested with root-knot nematodes and *V. dahliae,* Scholte (1990) found that a five-year rotation of maize-sugar beet-barley-barley-potato had significantly higher potato yields than continuous potato.

3.1.1.7. The Lesion Nematode, *Pratylenchus penetrans* and Wilt, *Verticillium dahliae* Disease Complex

(i) Management Methods

(a) Chemical Methods: Methylisothiocyanate has been successfully used for soil fumigation to reduce the populations of soil-borne fungus, *V. dahliae* and alleviate the effects of disease complex involving wilt-inducing fungus (*V. dahliae*) and the lesion nematode (*P. penetrans*) on potato (Rowe *et al.*, 1987).

3.1.8. The Cyst Nematode, *Globodera rostochiensis* and Wilt, *Verticillium dahliae* Disease Complex

The association of fungi like *Rhizoctonia solani* and *V. dahliae* with *G. rostochiensis* has been reported on potato indicating that the cyst nematode is involved in the disease complex.

(i) Management Methods

(a) Chemical Methods: Soil fumigation with methyl bromide resulted in yield increases of 68% for potato fields infested with *Globodera* spp. and *V. dahliae* (Hide and Corbett, 1974).

In *V. dahliae-G. rostochiensis* disease complex on potato, a mixture of methyl bromide and chloropicrin was found to have comparatively more fungicidal and nematicidal efficacy (Hide and Corbett, 1974).

3.1.2. Tomato

Tomato (*Lycopersicon esculentum*) is the most important and remunerative vegetable crop in India. It is being cultivated in 0.497 million hectares producing 8.637 million tonnes of tomatoes with an average yield of 17.4 tonnes per hectare. Uttar Pradesh, Maharashtra, Karnataka, Bihar and Orissa, are major tomato-growing states in India. A rich source of minerals, vitamins and organic acids, tomato fruits provides 3-4% total solids, 15-30 mg/100 g ascorbic acid, 7.5-10.0 mg/100 ml titratable acidity and 20-50 mg/100 g fruit weight of lycopene. Tomato fruits contain 0.4% protein, 2.5% carbohydrates, 1.5 mg iron and 2 mg vitamin C in 100g of fresh weight. Fruits are good for people suffering from constipation, jaundice and indigestion.

Root-knot, reniform and stubby root nematodes are the major limiting factors in the successful production of tomato crop.

3.1.2.1. Root-knot Nematodes, Meloidogyne spp.

Lal and Ansari (1960) reported *M. arenaria* for the first time on tomato from Bihar, India (Fig. 3.4). *M. incognita* was responsible for 30.57 to 46.92% loss in fruit yield of tomato (Bhatti and Jain, 1977; Reddy, 1985; Darekar and Mahse, 1988), while *M. javanica* caused 77.5% loss in yield (Anon, 1993).

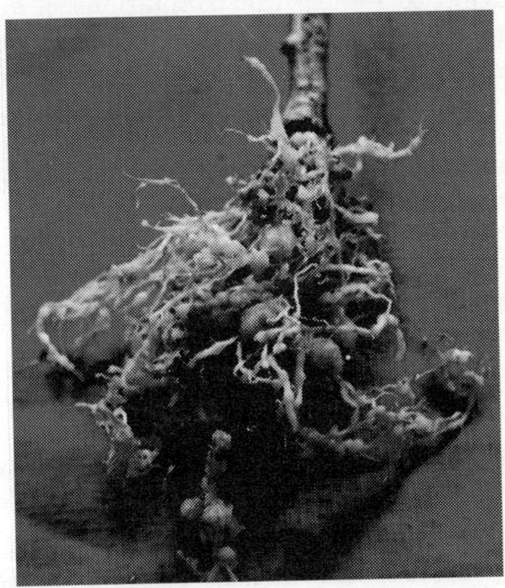

Fig. 3.4. Heavy galling of tomato roots with *Meloidogyne incognita*.

(i) Interaction with Other Pathogens: M. javanica increased the extent of damage by pre and post-emergence phases of damping off caused initially by *R. solani* and *Pythium debaryanum* in tomato (Nath *et al.*, 1984). Pani and Das (1972) have reported the association of root-knot nematode with bacterial wilt of tomato. The growth of tomato plants was significantly reduced in simultaneous inoculations with *M. incognita* and tobacco mosaic virus or when nematode preceded the virus inoculation (Goswami and Chenulu, 1974).

(ii) Management Methods

(a) Physical Methods

Heat Treatment: The most important nematode pests controlled by greenhouse steaming are cyst nematode of potato attacking tomatoes and root-knot nematodes on tomatoes.

Rabbing: Rabbing of root-knot infested tomato nursery with bajra husk or paddy husk or saw dust at 7 kg/sq.m. a week prior to seeding is recommended for economic and effective management of root-knot disease (ICBR 1: 5.12).

Soil Solarization: Soil solarization of nursery beds with 100 gauge LLDPE clear plastic film during summer (in between April 15th and June 15th) for 15 days is recommended (Fig. 3.5). It increased the number of transplantable seedlings by 615% giving net profit of Rs. 7,661 from 1000 sq.m. (ICBR 1: 5.65) and decreased the root-knot disease and weeds by 66% and 93%, respectively (Table 3.5).

Fig. 3.5. Soil solarization for the management of root-knot nematodes in tomato Nursery.

Table 3.5. Effect of soil solarization of nursery beds on root galling and number of transplantable seedlings of tomato in 1000 sq. m.

Treatment	No. of seedlings	Weight of 100 seedlings	Root-knot index	Net profit	Cost: Benefit Ratio
With plastic	2,60,400	36 g	1.37	Rs. 7,661	1:5.65
Without plastic	36,400	26 g	3.97	---	---
% Increase (+)/decrease (-) over control	(+) 615	(+) 38	(-) 66	---	---

Red plastic mulch suppresses root-knot nematode damage in tomatoes by diverting resources away from the roots (and nematodes) and into foliage and fruits (Adams, 1997).

(b) Cultural Methods: Several cultural methods like selection of nematode-free nursery sites, destruction of infested roots after crop harvest, crop rotation with non-hosts (maize, wheat, sorghum) or antagonistic crops (marigold, mustard, sesame), flooding, fallowing, deep summer ploughing and harrowing either alone or in combination proved to be reasonably effective and economical to check multiplication of root-knot nematodes on tomato.

Jain and Bhatti (1987) reported that planting of nematode-free seedlings in an infested field which received three summer ploughings increased the yield of tomato by 55%.

Decrease in *M. incognita* population on tomato occurs when the field was left fallow or following crop rotation with marigold, spinach and bottle gourd (Khan *et al.*, 1975).

Summer Ploughing: Ploughing not only leads to disturbance and instability in nematode community but also causes their mortality by exposing them to solar heat and desiccation. Three summer ploughings each at 10 days interval during June (average atmospheric temperature ranging between 40-46°C) at Hisar (Haryana) led to 96.5 % reduction in *M. javanica* population, while fallowing itself during the same period registered 44.5% reduction (Jain and Bhatti, 1987). Additional use of plastic sheets for covering soil either in nursery beds or in field further enhanced nematode reduction. Such an approach also helps in reducing the intensity of weeds, fungi and bacteria in the soil.

Crop Rotation: To control *M. incognita* and *M. javanica*, a tomato crop can be followed by peanut without risk of damage to the peanuts. While the peanuts are growing, the nematodes can not reproduce. Instead many of the larvae in the soil die or become non-infective because of starvation and the attacks of predators, fungi and diseases. If the population is reduced sufficiently, tomatoes can be grown again after peanuts without serious injury (Taylor and Sasser, 1978). Tomato rotated with cotton-wheat-cotton resulted in low gall index due to *M. javanica* (root-knot index 1-2) and gave maximum yield (38% increase over continuous tomato) (Hashmi and Hashmi, 1990).

Sesamum-tomato sequence produced the least reproduction rate (0.47), least root-knot index (1.33) and highest yield (86 q/ha) of tomato which was at par with niger-tomato sequence that had reproduction rate of 0.56, root-knot index of 1.66 and highest yield of 75.5 q/ha of tomato (Sahoo *et al.*, 2004a).

In cropping sequence studies for the management of *M. javanica* on tomato, nematode population decreased to low levels when carrot, capsicum and onion were grown (Kanwar and Bhatti, 1992). The population of *M. incognita* was reduced considerably under groundnut-mustard-tomato cropping sequence (Sharma *et al.*, 1980).

Tomato nursery beds previously planted with trap crop (marigold) gave maximum reduction in root-knot nematode population in soil which also increased the germination of tomato seeds and production of more healthy (nematode-free) seedlings (Rangaswamy *et al.*, 1999).

Inclusion of rice, maize, groundnut and *Stylosanthes* in rotation sequences controlled root-knot nematodes and increased yields of tomato. In crop rotation trials, it was found that a crop of groundnut or strawberry grown prior to tomato in soil heavily infested with root-knot nematodes greatly increased yields of tomato. Rotation of tomato with leek (*Allium porrus*), onion, groundnut and sunnhemp reduced the population of root-knot nematodes.

Intercropping: Intercropping with marigold or mustard with tomato reduced the damage of root-knot and reniform nematodes. The root-knot development on tomato was low when interplanted with *Tagetes erecta*. The population of *Tylenchorhynchus, Helicotylenchus, Hoplolaimus, Rotylenchulus* and *Pratylenchus* was also markedly reduced (Alam *et al.*,1977; Khan *et al.*, 1971a). α-Terthienyl is the active principle in *Tagetes* spp. which is toxic to these nematodes.

Marigold intercropped with tomato at 1: 4 and 1: 6 ratios and mustard at 1: 2 ratio was found to be effective in reducing root galls, egg masses and nematode population at harvest and gave cost: benefit ratio of 1: 8.36, 1: 7.88 and 1: 3.31, respectively (Rangaswamy *et al.*, 1999).

Interculture of onion with tomato reduced gall index due to *M. javanica* significantly and improved tomato fruit yield. Intercropping with 2 rows of onion with one row of tomato was the most effective treatment in reducing the nematode infection and population and enhancing tomato yield (Ram and Gupta, 2001).

In pots having *Zinnia elegans* with tomato, the population of *M. incognita* and *R. reniformis* declined. The root-knot index on tomato was reduced to 1.08 compared to 3.75 in control. The nematode reproduction factor of *M. incognita* and *R. reniformis* was reduced to 0.88 and 0.80, respectively compared to 3.86 and 2.43, respectively in control (Tiyagi *et al.*, 1986).

Cultivation of tomato with castor results in considerable reduction in number of root-knot galls on tomato (Hackney and Dickerson, 1975).

Influence of Fertilizers: Increased levels of potash have significantly reduced the number of galls by *M. javanica* in tomato (Gupta and Mukhopadhyaya, 1971).

The growth of tomato increased (35%) and root galling (root- knot index 1 as against 4 in control) due to *M. incognita* , nematode population in soil (80% reduction) and egg mass production reduced (egg mass index 1 as against 4 in control) in soil amended with 20% fly ash (Haq *et al.*, 1985). The increase in plant growth may be ascribed to increased nutrient availability (N, K, Ca, Mg, Na, B, SO_4) and the reduction in nematode population might be due to toxic compounds (polycyclic aromatic hydrocarbons, dibenzofuran and dibenzo-p-dioxin mixtures) present in fly ash. Row application of fly ash at 0.6 kg/sq. m. enhanced the yield of tomato by 90.4% (Khan and Ghadipur, 2004).

Botanicals: Soil application of neem and subabul leaves to nursery beds at 0.5 t/ha gave better seedling growth and reduced root galling in tomato (Jain *et al.*, 1988).

Singh and Sitaramaiah (1966) applied finely divided oil cakes to root-knot infested soil and noted a reduction in disease incidence in tomato. The oil cakes (neem, groundnut, mustard and castor) were effective in reducing parasitic nematode population (*Hoplolaimus, Tylenchorhynchus, Meloidogyne* and *Helicotylenchus* species) and in increasing yields of tomato.

Application of neem cake at 15 g/spot or 100 g/m furrow, three weeks prior to transplanting of tomato has been reported to give 36-450% enhanced yield with corresponding reduction in gall index to 2.3 as against 4.3 in control (Anon, 1992).

Best results in respect of root-knot reduction due to *M. javanica* and increase in yield of tomato were obtained by amending the soil with saw dust at 2.5 tonnes per ha 3 weeks before planting and then applying N through urea at 120 kg per ha (Singh and Sitaramaiah, 1971). Rice hull ash at 2.5 tonnes per ha increased the yield of tomato by 133 to 317 per cent and reduced root-knot incidence by 46 to 100 per cent (Sen and Dasgupta, 1981). Application of poultry manure at 2.0, 2.5 and 3.0 tonnes/ha 15 days prior to tomato seeding is recommended for effective management of root-knot disease and production of more transplantable seedlings in tomato nursery giving ICBR of 1: 4.54, 1: 5.63 and 1: 6.68, respectively.

Green manuring with naffatia/besharmi or neem leaves (ICBR 1: 2.87) or water hyacinth (ICBR 1: 2.527) or calotropis (ICBR 1:2.26) or congress grass (ICBR 1: 2.09) each at 3 kg/sq.m. 15 days prior to seeding is recommended for economic and effective management of root-knot disease in tomato nursery. Incorporation of green materials of naffatia and neem leaves each at 2 kg/1.44 sq.m. (ICBR 1: 15.7) or congress grass and neem leaves each at 2 kg/1.44 sq.m. (ICBR 1: 15.6) or calotropis, naffatia, congress grass and neem leaves each at 1 kg/1.44 sq.m. (ICBR 1: 14.7) 15 to 20 days prior to tomato seeding gave effective management of root-knot nematodes and production of more healthy seedlings in tomato nursery during kharif season.

(c) Chemical Methods

Nursery bed treatment: Nursery bed treatment with aldicarb at 4g a.i./m^2 and carbofuran at 2g a.i./m^2 were effective in increasing seedling growth and reducing root-knot nematode population of tomato (Jain and Bhatti, 1983; Ramakrishnan and Balasubramanian, 1981). In the main field, the above treatments were also effective and increased the fruit yields (88.2 and 55.9% increase in yield of tomato, respectively).

Bare Root-dip Treatment: Carbofuran and oxamyl at 1,000 ppm for 15 to 30 min. (Parvatha Reddy and Singh, 1979); phenamiphos at 250 to 750 ppm for 30 min. (Thakur and Patel, 1985); dimethoate at 500 ppm for 6 hr, phosphamidon and dichlofenthion both at 1000 ppm for 8 hr (Jain and Bhatti, 1978) and thionazin at 500 ppm for 15 min (Reddy and Seshadri, 1975) gave effective control of root-knot nematodes on tomato when used as bare-root dips.

Main Field Treatment: Aldicarb, carbofuran, ethoprophos and phenamiphos each at 1 to 2 kg a.i. per ha were found effective in reducing the root-knot nematode population and in increasing fruit yields of tomato (Ahuja, 1983).

Dichlofenthion was found effective against *M. incognita* in tomato (Alam *et al.,* 1973, Saxena *et al.,* 1974).

(d) Biological methods

Antagonistic Bacteria: Two bacteria *Bacillus licheniformis* and *Pseudomonas mindocina* and 2 fungi *Acrophialophora fusispora* and *Aspergillus flavus* were evaluated for their efficacy as bio-control agents individually and in various combinations against *M. incognita* infecting tomato. Individually, *B. licheniformis* gave best control although when all 4 agents were used, nematode multiplication was reduced by 82.06% (Siddiqui and Husain, 1991).

Application of *Bacillus thuringiensis* causes 95% mortality of *M. javanica* juveniles due to β-exotoxin production and suppression of gall formation, egg mass production and nematode population in soil. Field application of a nematicidal *B. thuringiensis* strain to tomato in Puerto Rico reduced galling in roots due to *M. incognita* and increased the yield significantly (Zuckermann *et al.,* 1993).

Root material containing *Pasteuria penetrans* spores at 212-600 mg powder/kg soil, when broadcast in soil, showed reduction in galling on tomato roots and final nematode population. Tomato yield increased by 20% over untreated check, when *P. penetrans* was introduced to the nursery soil at 1x10^5 spores/g soil. The persistence of *P. penetrans* in the field soil was evident since about 50% of the final juvenile population was encumbered with the bacterial spores (Walia and Dalal, 1994). *P. penetrans* is highly resistant to heat, desiccation and has ability to survive for more than 2 years in soil, qualifies as the most potential biocontrol agent against root-knot nematodes infesting vegetable crops.

Application of *P. fluorescens* at 10 g/m² in nursery beds gave good control of root-knot nematodes (Shanthi and Sivakumar, 1995).

Avermectins B_1 and B_{2a} (applied to soil through drip irrigation systems) at rates ranging from 0.093 to 0.34 kg a.i./ha applied as a single dose or 0.24 kg a.i./ha applied as three doses each at 0.08 kg on tomatoes against *M. incognita* were as effective as oxamyl and aldicarb at 3.36 kg a.i./ha (Garabedian and Van Gundy, 1983). The aqueous solution of avermectins (250 ml of 0.001%/m²) significantly reduced root galls of tomato seedlings raised in root-knot infested nursery beds which yielded robust and healthy seedlings with no root infection (Parvatha Reddy and Nagesh, 2002). In a root-knot infested micro-plot study, avermectins at 0.015 kg/ha effectively controlled *M. incognita* on tomato and increased yield by 11 and 8% compared to the untreated and carbofuran treated plots, respectively. (Parvatha Reddy and Nagesh, 2002). Seedling bare root-dip in avermectin 100% gave highest reduction in root-knot index (1.3), soil nematode population (132.8), number of females/g root (12.8), egg masses/g root (5.4) and eggs/egg mass (125.2) in tomato (Jayakumar *et al.*, 2005).

Antagonistic Fungi: Application of 'Royal 350' (*Arthrobotrys irregularis* cultured on oat seed medium) at 140 g/m² a month before transplantation of tomato resulted in good protection against root-knot nematodes. Among the egg parasites, efficacy of *Paecilomyces lilacinus* (commercially formulated as 'Biocon' in Philippines) has been found to be comparatively higher in suppressing the population of *Meloidogyne* spp. and *R. reniformis* on tomato. Bare root-dip treatment of tomato seedlings in a spore suspension of *P. lilacinus* at 4×10^5 spores/ml reduces *M. incognita* infestation in roots by 50% and increases the growth of seedlings by 20%. Application of organic amendments increases the effectiveness of *P. lilacinus* against *M. incognita*.

Arbuscular Mycorrhizal Fungi: Arbuscular mycorrhizal fungus (AMF), *Glomus fasciculatum* was effective against root-knot nematodes and *G. mosseae* against reniform nematodes on tomato. Organic amendments (oil cakes, calotropis leaves) in combination with AMF enhance the colonization of AMF on tomato roots which further increased plant growth and reduced gall index. Sundaram and Arangarasan (1996) observed that inoculation of *G. fasciculatum* at 10 g/plant was effective in increasing the uptake of nutrients and yield and reducing root galling (10) compared to 135.3 in control, followed by *G. mosseae, Gigaspora margarita* and *Acaulospora laevis*. Application of *G. fasciculatum* (Shimoga banana isolate) at 50 g/sq.m. (200 spores/g of soil) gave minimum number of galls per root system (39.2) at harvest and recorded maximum yield (26.47 t/ha) with an increased cost: benefit ratio (1: 3.34) and decreased nematode population (Kiran Kumar *et al.*, 2004).

(e) Host Resistance: Tomato cvs. All Round, Anahu, Anahu R, Atkinson, Beefeater, Beefmaster, Better Boy, Big Seven, Cavalier, Eurocross, Nematox, SL-120, NTR-1, SL-12, Patriot, VFN-8, VFN Bush, Piersol, Radiant, Nemared, Ronita, Bresch, Healani, Kewalo, Campbell-25, Punuui, Arka Vardan (Fig. 3.6), Pelican, Hawaii-7746, Hawaii-7747 and Hisar Lalit have been reported to be resistant to root-knot nematodes.

Fig. 3.6. F1 Tomato hybrid 'Arka Vardan' resistant to root-knot nematodes and yield potential of 70 t/ha.

In a field with resistant tomato (Hisar Lalit) grown for a year, the yield was 27% higher than susceptible variety (HS-101) (Kanwar and Bhatti, 1990).

(f) Integrated Methods

Using Bio-agents and Botanicals: In nursery, integration of *Pasteuria penetrans* (at 28×10^4 spores/m^2), *P. lilacinus* (at 10 g/ m^2 with 19×10^9 spores/g) and neem cake (at 0.5 kg/m^2) gave maximum increase in plant growth and number of seedlings/bed (Fig. 3.7). Parasitization of *M. incognita* females was highest when neem cake was integrated with *P. penetrans*, while parasitization of eggs was highest when neem cake was integrated with *P.*

lilacinus. In field, planting of tomato seedlings (raised in nursery beds amended with neem cake + *P. penetrans*) in pits incorporated with *P. lilacinus* (at 0.5 g/plant) gave least root galling and nematode multiplication rate and increased fruit weight and yield of tomato (Parvatha Reddy *et al.*, 1997).

Fig. 3.7. Management of root-knot nematodes in tomato nursery by integration of botanicals and bioagents.

Soil drenching with 5% Wellgro solution (at 200 ml/seed pan) along with *P. lilacinus* (at 2×10^4 spores/ml) gave maximum increase in seedling weight and highest reduction in root galling. Roots of tomato seedlings raised in the above treatment dipped in 5% Wellgro solution mixed with *P. lilacinus* spores (at 2×10^4 spores/ml) for 20 min. when transplanted in pots gave maximum increase in plant growth, root colonization and parasitization of egg masses of *M. incognita* by the bioagent and highest reduction in root galling and final nematode population both in soil and roots (Rao and Parvatha Reddy, 1993a).

Bare root-dip treatment of tomato seedlings in 10% castor leaf extract mixed with *P. lilacinus* spores (at 1.5×10^8 spores/ml) for 20 min. significantly increased the plant growth and reduced root galling and final nematode population. The above treatment also gave significant increase in parasitization of eggs and egg masses and propagule density of *P. lilacinus* in roots (Rao *et al.*, 1999). Similarly, root-dip treatment of tomato seedlings in 5 and 10% neem leaf extract mixed with *P. lilacinus* spores (at 6×10^4 spores/

ml) for 30 min. gave significant reduction in root galling and final nematode population. Significant increases in root colonization, parasitization of eggs and egg masses and propagule density of the bioagent in soil were also noticed in the above treatment (Parvatha Reddy *et al.*, 1998).

Application of *Pochonia chlamydosporia* (100 ml/seed pan containing 4 x 10^5 spores/ml) with neem leaves (150 g/seed pan) increased seedling weight and colonization of roots with the bioagent. Tomato seedlings raised in the above treatment transplanted in pots resulted in maximum increase in plant growth, least root galling, nematode population both in soil and roots and highest parasitization of eggs, egg masses and propagule density of *P. chlamydosporia* in roots (Parvatha Reddy *et al.*, 1999). *P. lilacinus* in combination with castor leaves reduced the nematode population up to 89% and increased plant growth and yield in tomato (Zaki and Bhatti, 1991a).

Incorporation of neem cake (20 g/pot) along with *P. chlamydosporia* (10 ml/pot containing 4 x 10^5 spores/ml) gave maximum increase in plant growth and significant reduction in root galling and nematode population both in soil and roots. The above treatment also gave highest parasitization of eggs and egg masses of *M. incognita* and maximum propagule density of the bioagent in soil and roots (Rao *et al.*, 1998b).

Integration of neem cake (40 g/plant) with *T. harzianum* (at 5 g/plant with 4 x 10^8 spores/g) was effective in increasing the plant growth and reducing root galling and final population of *M. incognita* infecting tomato. The above treatment also gave maximum reduction in number of eggs/egg mass and increased root colonization, spore density in soil and parasitization of adult females with *T. harzianum* (Rao *et al.*, 1997c).

Application of *Aspergillus niger* and *P. lilacinus* along with mustard cake gave the maximum reduction in nematode population both in root and soil with enhanced plant vigour. *A. niger* being toxic agent kills the second stage juveniles present in the rhizosphere, while *P. lilacinus* being the egg parasite invade the eggs of *M. incognita* which escaped from the toxicity of *A. niger*. As a result, there is an overall reduction in root and soil population. Further, addition of mustard cake also helps to maintain the general plant health in addition to possessing nematicidal properties (Goswami *et al.*, 1998) (Table 3.6).

Combined application of *P. penetrans* + *T. viride* + neem/castor cake each at 1/3 doses was significantly superior compared to their individual applications in terms of increased plant growth and reduced root galling, egg mass production and final populations of *M. incognita*. Similarly, total parasitization by bioagents was higher in plants treated with these bioagents in terms of number of juveniles encumbered and adult females infected with *P. penetrans* and egg masses parasitized by *T. viride*, compared to their individual applications (Rangaswamy *et al.*, 1999).

Table 3.6. Effect of *Aspergillus niger*, *Paecilomyces lilacinus* and mustard cake on biomass and multiplication of *Meloidogyne incognita* infecting tomato

Treatment	Biomass	No. of galls/plant	No. of egg masses /plant	No. of eggs/egg mass	Nematode popn./ 500 g soil
Aspergillus niger	9.8	34	18	272	470
Paecilomyces lilacinus	10.2	42	23	262	580
A. niger + P. lilacinus	12.2	26	16	264	270
A. niger + P. lilacinus + Mustard cake	12.6	22	10	255	290
Nematodes alone	9.0	98	72	320	2880
Control	14.0	---	---	---	---
CD at 5%	2.33	7.24	3.24	27.18	162.80

Nursery bed treatment with *P. fluorescens* (with 1×10^9 spores /g) and *P. chlamydosporia* (10 ml/pot containing 4×10^6 spores/ml) each at 20 g/m^2 and field application of 5 tonnes of FYM enriched with the above bioagents each at 5 kg, significantly reduced root-knot nematodes in tomato by 76% over untreated check. The yield increase was up to 21.7% with cost: benefit ratio of 1: 4.9.

Using Arbuscular Mycorrhizal Fungi and Botanicals: In nursery, integration of neem cake (at 500 g/m^2) with *Glomus mosseae* (at 250 g/m^2 containing 16 chlamydospores/g) significantly reduced *M. incognita* population in soil, root galling, egg mass production, fecundity and produced vigorous tomato seedlings with increased root colonization with *G. mosseae*. The seedlings raised in the above treatment when planted in the main field significantly reduced root galling and increased fruit yield, root colonization with *G. mosseae* and chlamydospore population in soil (Parvatha Reddy *et al.*, 1998; Rao *et al.*, 1995).

Application of calotropis leaves at 400 g/ m^2 along with *G. fasciculatum* (250 g/ m^2 containing 25-30 chlamydospores/g) in nursery beds gave significant reduction in root galling and fecundity of *M. incognita*. The above treatment also gave maximum increase in plant growth and root colonization with *G. fasciculatum* (Rao *et al.*, 1996).

Using Bio-agents and Arbuscular Mycorrhizal Fungi: Interestingly, integration of bio-agent (*P. lilacinus*) and endomycorrhizae (*G. mosseae/G. fasciculatum*) have culminated in the successful management of *M. incognita* infecting tomato. This phenomenon facilitated standardization of a strategy wherein inoculation of mycorrhizae and bio-agent in the nursery beds protected the seedlings of tomato from the attack of *M. incognita*. Further, these mycorrhizal seedlings (colonized either with *G. mosseae* or *G. fasciculatum*) can be given a root-dip treatment with spore suspension of *P. lilacinus* for 5-10 minutes for the effective management of these nematodes in the main field after transplanting (Rao *et al.*, 1993).

Integration of *G. deserticola* with *P. chlamydosporia* gave effective management of *M. incognita* in tomato, increased seedling weight, root colonization with *G. deserticola* and *P. chlamydosporia* and parasitization of eggs of *M. incognita* with *P. chlamydosporia* and reduced root galling, egg mass production and fecundity of root-knot nematodes (Rao *et al.*, 1997).

Using Bio-agents, Arbuscular Mycorrhizal Fungi and Botanicals: Strategies to integrate botanicals, bio-agents and endomycorrhizae components for the management of root-knot nematodes infecting tomato have been developed. These strategies include inoculation of *G. mosseae* in neem leaf/neem cake amended nursery beds followed by the root-dip treatment of mycorrhizal seedlings of tomato in spore suspension of *P. lilacinus* for effective management of *M. incognita* under field conditions (Rao *et al.*, 1995).

In nursery, integration of neem cake (at 500 g/m^2) with *G. fasciculatum* (200 g/m^2 containing 15 spores/g) gave maximum increase in plant growth and highest root colonization with *G. fasciculatum* and least root galling. Integration of neem cake (at 500 g/m^2) with *T. harzianum* (at 100 g/ m^2 with 4 x 10^8 spores/g) gave maximum parasitization of eggs with *T. harzianum*. In field, planting of tomato seedlings raised in nursery beds treated with neem cake + *G. fasciculatum* in pits incorporated with *T. harzianum* (at 0.5 g/ plant with 4 x 10^8 spores/g) was effective in increasing tomato fruit yield, root colonization with *G. fasciculatum* and egg parasitization with *T. harzianum* and significant reduction in root galling and multiplication rate of *M. incognita* (Parvatha Reddy *et al.*, 1998).

Using Bio-agents and Chemicals: Maheshwari *et al.* (1987) reported that application of *P. penetrans* in combination with carbofuran, aldicarb, miral, sebufos and phorate significantly improved plant growth of tomato by greatly reducing galling due to *M. javanica*.

Integration of *P. chlamydosopria* with carbofuran recorded maximum plant growth parameters and minimum gall index and nematode population. The above treatment also recorded maximum number of fruits/plant and yield/plant. Maximum parasitization of *M. incognita* with *P. chlamydosopria* was also noticed when the bioagent and the chemical were integrated (Gopinatha *et al.*, 2002).

Using Two Bio-agents: Integration of two bio-agents, *P. lilacinus* and *P. chlamydosopria* resulted in combined and complimentary effects for the successful management of *M. incognita* infecting tomato (Rao and Parvatha Reddy, 1992c).

In nursery, integration of *P. chlamydosopria* (200 g/m² containing 16 chlamydospores/g) with *P. penetrans* at 1 g mixed with 100 g of sand/m² (containing 10^5 spores/10 mg) was effective in increasing the seedling weight and root colonization with *P. chlamydosopria* and adult females of *M. incognita* adequately infected with *P. penetrans*. The tomato seedlings raised in the above treatment when transplanted in the field gave maximum reduction in root galling, fecundity, nematode population both in soil and roots and increased root colonization with *P. chlamydosporia*, infection of *M. incognita* females with *P. penetrans* and tomato fruit yield (Rao *et al.*, 1998a). Maheshwari and Mani (1988) also observed that population of *M. incognita* and *M. javanica* were suppressed effectively and yields of tomato were greater when *P. penetrans* and *P. lilacinus* were applied together.

Combined soil application of *P. lilacinus* and *Aspergillus niger* at the time of transplanting tomato is very effective in reducing root-knot nematodes.

The combination treatment with *T. harzianum* + *T. viride* each at 50 g (4 x 10^8 cfu/g) considerably increased the plant growth, yield and reduced the root galls and soil nematode population (Hassan and Sobita Devi, 2004).

Using Cultural and Chemical Methods: Combinations of deep ploughing (up to 20 cm) and nursery bed treatment with aldicarb at 0.4 g per m² and main field treatment with aldicarb/carbofuran at 1 kg a.i./ha proved effective in the control of root-knot nematodes in tomato which also registered maximum yield (102.6% higher yield over control) (Jain and Bhatti, 1985).

In tomato, application of aldicarb and carbofuran each at 1 kg a.i. per ha in combination with neem cake and urea each at 10 kg N per ha at transplanting, produced maximum yield with lowest gall index (2.5) and nematode population, 90 days after planting (Routaray and Sahoo, 1985).

Using Botanicals and Physical Methods: Integration of soil solarization of nursery beds (with LLDPE transparent film of 25 μ for 15 days) with soil incorporation of Calotropis leaves at 4 kg/1.44 m² resulted in increased plant height (29.9 cm compared to 20 cm in control), fresh shoot weight (176.49g compared to 100g in control), transplantable seedlings (546/ m² compared to 297 in control) with minimum root-knot index (1.83 compared to 3.85 in control) (Patel *et al.*, 2006). Soil solarization with clear LLDPE film (25μ) for 15 days in hot summer in combination with poultry manure at 2 t/ha proved effective in the management of nematodes and higher production of transplants in tomato nursery with cost benefit ratio of 1: 3.10.

Using Botanicals and Chemical Methods: Application of carbofuran at 1 kg a.i./ha in the nursery beds followed by neem cake at 400 kg/ha in the main field increased yield and reduced the gall index (Singh and Gill, 1998) (Table 3.7).

Table 3.7. Integrated management of root-knot nematodes in tomato with carbofuran and neem cake.

Treatment/Dose	% Reduction in root galling	% increase in yield
Carbofuran (1 kg a.i./ha) + Neem cake at 400 kg/ha	77.00	61.50
Carbofuran (1 kg a.i./ha) + Urea (23.8 kg/ha) + Neem cake at 200 kg/ha	67.90	49.20
Phenamiphos (1 kg a.i./ha) + Neem cake at 400 kg/ha	74.90	39.70
Phenamiphos (1 kg a.i./ha) + Urea (23.8 kg/ha) + Neem cake at 200 kg/ha	67.90	31.43

Integration of chopped castor leaves (40 to 60 g/kg soil) 1 or 2 weeks before transplanting with application of aldicarb or carbofuran each at 2 kg a.i./ha at transplanting, significantly reduced number of galls due to *M. javanica* and enhanced the growth of tomato (Ravi Dutt and Bhatti, 1986).

Integration of soil application of castor leaves with inorganic fertilizer (75 kg N/ha) enhanced plant growth of tomato and reduced *M. javanica* infestation (95% and 78% reduction in root galls and egg mass production, respectively) (Zaki and Bhatti, 1989).

3.1.2.2. The Reniform Nematode, *Rotylenchulus reniformis*

R. reniformis was responsible for 42.25 to 49.02% loss in fruit yield of tomato (Subramanyam *et al.*, 1990) (Fig. 3.8).

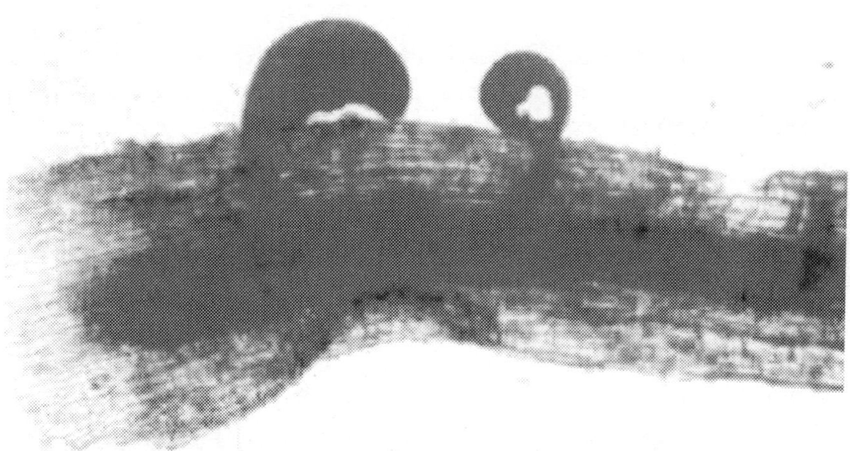

Fig. 3.8. Tomato root infected with *Rotylenchulus reniformis*.

(i) Management Methods

(a) Cultural Methods: Soil amendment with fresh glyricidia leaves, sawdust and chicken manure controlled the root-knot and reniform nematode population on tomato (Castillo, 1985). The yield increases and the nematode suppressive effect of fresh chicken manure surpassed even phenamiphos (Table 3.8).

Table 3.8. Effect of soil amendment with organic materials on *M. incognita* and *R. reniformis* population and tomato fruit yield.

Treatment	No. of nematodes/300 ml soil & 1 g roots			Yield (t/ha)
	No. of weeks after treatment			
	2	6	11	
Phenamiphos at 10 kg a.i./ha	96	366	465	31.1
Rice straw at 10 t/ha	215	500	1921	29.1
Glyricidia leaves at 10 t/ha	201	533	1885	27.3
Saw dust at 10 t/ha	342	859	958	25.3
Chicken manure at 10 t/ha	17	169	1078	32.6
Control	552	816	2515	22.2

(b) Chemical Methods: Application of DBCP (50 l/ha), metham sodium (250 l/ha) or methomyl/aldicarb/thionazin (4-16 kg/ha) was effective in reducing the reniform nematode population and in increasing tomato yields (Sivakumar *et al.*, 1979).

Reddy and Seshadri (1972) reported that thionazin at 4 kg a.i. per ha proved effective against *R. reniformis* infecting tomato.

(c) Host Resistance: Balasubramanian and Ramakrishnan (1983) observed that Kalyanpur Selection-I, Kalyanpur Selection-III, CA-121 exhibited resistant reaction.

(d) Integrated Methods: Application of neem cake in the nursery at 100 g/m^2 followed by carbofuran at 1 kg a.i./ha in the main field significantly reduced the soil population of *R. reniformis* and enhanced the fruit yield of tomato by 67% (Anitha and Subramanian, 1998).

Integration of a bio-agent, *P. lilacinus* with carbofuran at 1 kg a.i./ha was found effective in the management of reniform nematode, *R. reniformis* infecting tomato (Parvatha Reddy and Khan, 1988).

Nursery bed treatment with *P. fluorescens* (with 1 x 10^9 spores/g) and *P. chlamydosporia* (with 1 x 10^6 spores/g) each at 20 g/ m^2 and field application of 5 tonnes of enriched FYM with the above bioagents each at 5 kg significantly reduced reniform nematodes in tomato by 72% over control. The yield increase was up to 21.7% with cost: benefit ratio of 1:4.9.

3.1.2.3. The Stubby Root Nematode, Paratrichodorus minor

(i) Distribution: From India, it has been reported from Delhi and Uttar Pradesh (Prasad and Dasgupta, 1964; Siddiqi, 1962).

(ii) Hosts: P. minor has been reported from the rhizosphere of apple, citrus, sugarcane, potato, vegetables, cereals, rose and fruits from India.

(iii) Life Cycle: Males are very rare and reproduction is evidently by parthenogenesis. Life cycle on tomato is completed in 21-22 days at 22°C and in 16-17 days at 30°C with a ten fold or more increase in 60 days (Rhoade and Jenkins, 1957).

(iv) Management Methods

(a) Physical Methods: Soil solarization using clear plastic mulch or photo selective (black) polyethylene film reduce populations of *P. minor* on tomato as effectively as fumigation with a mixture of methyl bromide and chloropicrin in warm climates.

(b) Cultural Methods: *Asparagus officinalis* would not support populations of *P. minor* for more than 40 to 50 days. Tomato, normally a good host of this nematode, supported only a low population when asparagus was growing in the same pots. A glycoside (asparagusic acid) is the active principle involved which is toxic to *P. minor* and several other nematode species.

3.1.2.4. Root-knot Nematode, *Meloidogyne incognita* and Wilt, *Fusarium oxysporum* f. sp. *lycopersici* Disease Complex

Jenkins and Coursen (1957) induced wilting in *Fusarium* wilt-resistant tomato variety 'Chesapeake' only when root-knot nematodes were present along with fungal inoculum. Furthermore, when *M. hapla* was combined with the fungus, only 60% of the plants wilted, whereas *M. incognita acrita* promoted wilt in 100% of the plants.

Bhagawati and Goswami (2000) reported the interaction of *M. incognita* and *F. oxysporum* f. sp. *lycopersici* on tomato (*Lycopersicon esculentum*). It was found that when either or both pathogens were inoculated simultaneously or nematode was inoculated 10 days prior to inoculation of the fungus, the symptoms were visible by 20 days as against no such symptoms in plants where fungus was inoculated 10 days prior to nematodes. The intensity of wilt after 40-60 days of inoculation was significantly higher when simultaneously inoculated or prior inoculation of nematodes compared to prior inoculation of fungus. The number of galls and egg masses and nematode population in soil was significantly reduced when simultaneously inoculated or prior inoculation of nematodes.

(i) Management Methods

(a) Biological Methods: Application of *T. harzianum* at 50 kg/ha (2 x 10^8 cfu/g) was effective in controlling the disease complex and improving the plant growth parameters and fruit weight (Haseeb *et al.*, 2006).

(b) Integrated Methods

Bioagents and Botanicals: Tomato roots that received *P. lilacinus* along with *T. harzianum* and neem cake were free from root-knot nematodes (*M. incognita*) and did not wilt due to *F. o. f. sp. lycopersici* till harvest. The roots were also free from *Fusarium* infection. The above treatment also reduced the % of wilt (10% compared to 90% in control), root-knot index (1.8 compared to 4.4 in control) and increased root colonization with bioagents (74% compared to 0% in control), parasitization of egg masses (56% compared to 0% in control) and eggs/egg mass (44% compared to 0% in control) (Nagesh *et al.*, 2006).

Bioagents, Cultural Methods and Host Resistance: Deep ploughing and exposing soil to hot sun in summer, removal and burning of crop debris, soil application of *T. viride* and *P. lilacinus*, use of wilt resistant varieties like Utkal Pallavi, Utkal Deepti, Utkal Kumari, Utkal Urbasi, etc. help in controlling the disease complex.

Arbuscular Mycorrhizal Fungi and Botanicals: Integration of mustard cake with *Glomus etunicatum* was effective in reducing the damage caused by *M. incognita* and *F. oxysporum* f. sp. *lycopersici* on tomato (Bhagawati *et al.*, 2000).

3.1.3. Brinjal

Brinjal (*Solanum melongena*) is a widely grown vegetable crop in Asian countries. It is being cultivated in 0.530 million hectares producing 8.703 million tonnes of brinjal with an average yield of 16.4 tonnes per hectare. In India, it is adapted to a wide range of climatic conditions from north to south and east to west. In hilly regions, it is grown only in summer. Brinjal is used in a variety of culinary preparations. Pickles and industrially processed foods are also produced. It is rich in vitamin A and B and good for diabetic patients. The major brinjal producing states include West Bengal, Orissa, Bihar, Gujarat, Andhra Pradesh, Maharashtra, Karnataka and Madhya Pradesh.

Root-knot and reniform nematodes are the major nematode problems in successful cultivation of brinjal crop.

3.1.3.1. Root-knot Nematodes, Meloidogyne spp.

M. incognita was responsible for 27.30 to 48.55% loss in fruit yield of brinjal (Bhatti and Jain, 1977; Parvatha Reddy and Singh, 1981; Darekar and Mahse, 1988) (Fig. 3.9).

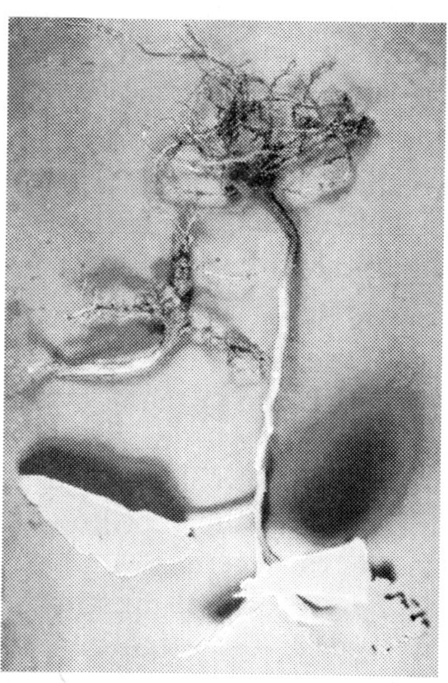

Fig. 3.9. Root-knot nematode on brinjal. Left – Healthy; Right – Infected.

(i) Interaction with Other Pathogens: M. incognita and *O.t.* var. *parasiticum* together acted synergistically by reducing the germination of brinjal to 28 per cent (Nath *et al.*, 1976). Combined effect of *M. incognita* and mycoplasma-like organisms (little leaf) on growth of brinjal was more than their individual effects (Dhawan and Sethi, 1977).

(ii) Management Methods

(a) **Physical Methods:** Burning of paddy husk or saw dust on infested nursery proved to be reasonably effective.

(b) **Cultural Methods:** Several cultural methods like selection of nematode-free nursery sites, destruction of infested roots after crop harvest, crop rotation with non-hosts (maize, wheat, sorghum) or antagonistic crops (marigold, mustard, sesame), flooding, fallowing, deep summer ploughing and harrowing either alone or in combination proved to be reasonably effective and economical to check multiplication of root-knot nematodes on brinjal.

Decrease in *M. incognita* population on brinjal occurs when the field was left fallow or following crop rotation with marigold, spinach and bottle gourd (Khan *et al.*, 1975).

Summer Ploughing: Ploughing not only leads to disturbance and instability in nematode community but also causes their mortality by exposing them to solar heat and desiccation. Generally, 2 to 3 summer ploughings each at 10 days interval during April-May (40-46°C) have been recorded to reduce 96.5% *M. javanica* population while fallowing itself during the same period registered 44.5% reduction. Additional use of plastic sheets for covering soil either in nursery beds or in field further enhanced nematode reduction. Such an approach also helps in reducing the intensity of weeds, fungi and bacteria in the soil.

The maximum reduction in *M. javanica* population and increase in fruit yield was recorded in brinjal by soil ploughing + covering with polythene sheet, followed by soil ploughing + exposure to sun (Anon, 1989a).

Crop Rotation: Rotation of brinjal with sweet potato (cv. Sree Bhadra) reduced the root-knot nematode population by 47% and increased the fruit yield by 22%. Since the sweet potato cv. Sree Bhadra is a high yielding one, the farmer will get a good crop within 3 months. Thus it becomes an additional income generating practice (Sheela *et al.*, 2002). Crop rotation with sorghum, wheat and chilli reduced the root-knot nematode population. The population of *M. incognita* larvae decrease in combination of brinjal-wheat, brinjal-chilli-wheat and red gourd-brinjal-mustard. The root-knot larvae were completely eliminated in marigold-marigold-fallow and chilli-cauliflower-cauliflower. Garlic following brinjal also brought about considerable reduction in larvae of root-knot nematode. Intercropping with marigold, onion and garlic is also recommended. The population levels of *M. incognita* on highly infested land dropped to zero after 18 months of continuous cultivation of *Panicum maximum*. Growing of brinjal after *P. maximum* was completely free from root-knot nematodes (Netscher, 1983).

Intercropping: Intercropping brinjal with marigold, garlic or resistant tomato (SL-12) improved plant growth and reduced number of galls and final soil population of nematode (Jain *et al.*, 1990). Marigold (cv. Calcutta Yellow)-brinjal cropping sequence has been recommended to the farming community for management of *M. javanica* affecting brinjal (Singh, 1991).

Trap Cropping: Growing of trap crop like knol-khol with brinjal but removed and destroyed before new generation of nematode could emerge helps in reducing the root-knot nematode population (Ayyar, 1926).

Influence of Fertilizers: Row application of fly ash at 0.6 kg/sq. m. enhanced the fruit yield by 27.7% and inhibited the reproduction of *M. incognita* (Khan and Ghadipur, 2004).

Botanicals: Nursery bed treatment with neem cake at 200 g/sq.m. gave maximum reduction in nematode population giving an yield of 20.75 q/ha (Sheela and Nisha, 2004).

Incorporation of neem leaves at 500 g/sq.m. or 250 ml of 5% neem leaf extract/ 5% neem cake extract/sq.m. in nursery beds of brinjal gave effective control of root-knot nematodes.

Spot application of mustard cake at 25g/spot and furrow application of saw dust at 210g/furrow of 3m length were found most effective in reducing the nematode infestation and increasing the yield of brinjal. Similarly, Hazarika (1990) found that spot application of castor/mustard/neem cake at 15g/spot and furrow application of the above cakes at 100g/3m furrow reduced the infestation of *M. incognita* and increased the yield of brinjal. The yield was maximum in spot application of neem cake along with maximum reduction of *M. incognita* population in soil.

(c) Chemical Methods

Nursery Bed Treatment: Treatment of brinjal nursery beds with carbofuran at 3g a.i./m^2 led to 48% increase in yield and at 6g a.i./m^2 gave 60.8% higher yield.

Main Field Treatment: Aldicarb, carbofuran, ethoprophos and phenamiphos each at 1 to 2 kg a.i. per ha were found effective in reducing the root-knot nematode population and in increasing fruit yields of brinjal (119 to 145 per cent) (Singh *et al.*, 1978).

Alam *et al.* (1973) reported that oxamyl was effective for the control of root-knot nematodes infecting brinjal. Dichlofenthion was found effective against *M. incognita* in brinjal (Alam *et al.*, 1973).

(d) Biological Methods

Antagonistic Bacteria: Application of *Pasteuria penetrans* in nursery soil of brinjal resulted in 31 and 36% suppression in root galling when seedlings were transplanted to sterilized and nematode-infested soils, respectively. The results were also indicative of easy dissemination of *P. penetrans* to the main field through infected seedlings (Walia *et al.*, 1992).

Nursery bed treatment with *Bacillus macerans* at 25 g/sq.m. + soil drench (2% solution) ten days after sowing gave maximum reduction in nematode population and maximum yield of 38.25 q/ha (Sheela and Nisha, 2004).

Avermectins effectively reduced root-knot nematode infection in brinjal under field conditions (Parvatha Reddy and Nagesh, 2002).

Antagonistic Fungi: Application of 'Royal 350' (*Arthrobotrys irregularis* cultured on oat seed medium) at 140 g/m^2 a month before transplantation of brinjal resulted in good protection against root-knot nematodes. Among the egg parasites, efficacy of *Paecilomyces lilacinus* (commercially formulated as

'Biocon' in Philippines) has been found to be comparatively higher in suppressing the population of *Meloidogyne* spp. and *R. reniformis* on brinjal. Soil application of *P. lilacinus* at 20 g/sq. m. to brinjal nursery reduced infection by *M. incognita* by 50% and improved growth of seedlings by 30%.

Application of *P. chlamydosporia* at 50 g/m^2 in the nursery beds significantly increased the plant growth parameters (42 and 50 % increase in height and weight of seedlings, respectively over control) and root colonization by the bioagent on transplants. When the above seedlings were transplanted in field, there was reduction in root galling index (5.2 compared to 7.8 in control), number of egg masses in 5 g root (20 compared to 49 in control), number of juveniles in 5 g root (62 compared to 162 in control) and number of juveniles in 100 cc of soil (67 compared to 152 in control) and increased plant growth and yield in eggplant at harvest (Naik, 2004).

(e) Host Resistance: Brinjal cvs. Vijaya and Banaras Giant were reported to be moderately resistant to *M. incognita.*

(f) Integrated Methods

Using Bio-agents and Botanicals: Integration of *P. chlamydosporia* (at 100 ml/seed pan containing 1.2 x 10^4 spores/ml) in castor cake (at 40 g/seed pan) amended soil was effective in increasing the seedling weight and colonization of roots with the bioagent. Brinjal seedlings raised in the above treatment transplanted in pots gave maximum increase in plant growth, root colonization and parasitization of eggs of *M. incognita* by *P. chlamydosporia.* The above treatment also gave least root galling and final nematode population both in soil and roots (Rao and Parvatha Reddy, 1993b).

Borkakaty (1993) observed that inoculation of *P. lilacinus* at 4 g/kg of soil in combination with mustard oil cake at 0.5 and 1.0 t/ha increased plant growth with corresponding decrease in number of galls, egg masses and eggs/egg mass of *M. incognita* on brinjal.

Application of 10% neem cake extract (at 20 ml/pot) mixed with spores of *P. lilacinus* (at 1 x 10^6 spores/ml) was effective in increasing plant growth and reducing root galling and final nematode population both in soil and roots. The above treatment also gave maximum parasitization of egg masses of *M. incognita* and spore density of *P. lilacinus* in soil (Rao and Parvatha Reddy, 1994).

Significant reduction in root galling and final nematode population of *M. incognita* were observed in brinjal seedlings which were given bare root-dip treatment in 10% neem leaf suspension mixed with spores of *P. lilacinus* (at 4 x 10^5 spores/ml) for 30 min. Significant increases were also observed in root colonization, parasitization of eggs and spore density of *P. lilacinus* in soil (Rao *et al.*, 1997b).

Root-dip treatment of brinjal seedlings in neem cake extract based formulation of *P. lilacinus* (at 5 x 10^6 spores/ml) for 20 min. and planted in pots gave significant increase in plant growth, root colonization , propagule density in soil and parasitization of eggs of *M. incognita* by *P. lilacinus* and drastic reduction in root galling, fecundity and final nematode population in soil and roots (Rao *et al.*, 1998a).

Application of castor cake extract based formulation of *T. harzianum* (at 500 ml/m² containing 9.9 x 10^3 spores/ml) to nursery beds of brinjal was effective in producing vigorous seedlings (with maximum seedling weight) with least root galling. The above treatment also increased root colonization and parasitization of *M. incognita* females by *T. harzianum* (Rao *et al.*, 1998c).

Using Arbuscular Mycorrhizal Fungi and Botanicals: Brinjal seedlings raised in nursery treated with *G. fasciculatum* (at 250 g/m² containing 16 chlamydospores/g) planted in pots amended with castor cake (applied at 10 g/kg soil 15 days before transplanting) resulted in maximum plant growth, root colonization with *G. fasciculatum* and chlamydospore density in soil and least root galling, fecundity and final nematode population in soil and roots (Rao *et al.*, 1998).

Integration of *G. fasciculatum* and neem cake at 0.5 t/ha was found to be effective in increasing plant growth parameters and yield (30 t/ha compared to 17 t/ha in control) and in reducing the root-knot index (2.6 compared to 4.8 in control) and final nematode population (173.3/250 ml soil compared to 510.8/250 ml soil in control) (Borah and Phukan, 2004).

Using Bio-agents and Arbuscular Mycorrhizal Fungi: Brinjal seedlings raised in seed pans treated with *G. mosseae* (at 100 g/seed pan containing 28-32 chlamydospores/g) dipped in *P. lilacinus* spore suspension (containing 4 x 10^5 spores/ml) for 5 min. and transplanted in pots gave maximum increase in plant growth, root colonization, propagule density in soil of both *G. mosseae* and *P. lilacinus* and parasitization of eggs of *M. incognita*. The above treatment also gave least root galling, fecundity and final nematode population in soil and roots (Rao *et al.*, 1998).

Using Bio-agents, Arbuscular Mycorrhizal Fungi and Botanicals: Strategies to integrate botanicals, bio-agents and endomycorrhizae components for the management of root-knot nematodes infecting brinjal have been developed. These strategies include inoculation of *G. fasciculatum* in the castor cake amended nursery beds followed by the root-dip treatment of mycorrhizal seedlings of brinjal in spore suspension of *P. lilacinus* for the management of *M. incognita* (Rao *et al.*, 1993b).

In nursery, soil application of neem cake at 400 g/m² along with *G. mosseae* (at 500 g/ m² containing 26-32 chlamydospores/g) and *P. lilacinus* (at

2 liters/ m^2 containing 6 x 10^5 spores/ml) resulted in production of healthy and vigorous brinjal seedlings colonized with *G. mosseae* and *P. lilacinus* and with least root galling. In field, transplanting of brinjal seedlings raised in the above treatment gave maximum reduction in root galling, egg mass production, fecundity and final nematode population in soil and roots. The above treatment also gave highest fruit yield, root colonization with *G. mosseae* and *P. lilacinus* and egg parasitization with *P. lilacinus* (Rao and Parvatha Reddy, 2001).

Using Two Bioagents: Combined soil application of *P. lilacinus* and *Aspergillus niger* at the time of transplanting brinjal is very effective in reducing root-knot nematodes.

Integration of highly toxic fungus, *Aspergillus niger* (kills most of the infective second stage juveniles) and an egg parasite, *Cladosporium oxysporum* (invades and kills the eggs in egg sac) both at half the doses significantly reduced *M. incognita* population and exhibited better plant growth than when either of the fungal bioagents in brinjal (Goswami and Singh, 2001).

In egg-plant, the nursery seedling stand index was good with seed treatment with *P. fluorescens* (50 g/kg) + soil application (10 g/m^2 seed bed) followed by soil application of neem based *T. harzianum* + *P. fluorescens* and neem based *P. fluorescens*, where the crop stand index was 4.6 and 4.0, respectively.

Using Bio-agents and Chemicals: Split application of *T. harzianum* at 50 kg/ha (10^8 cfu/g) (before transplanting and 45 days after transplanting) + carbofuran at 16.5 kg/ha was effective in increasing the brinjal fruit yield (Haseeb *et al.*, 2004).

Using Cultural and Chemical Methods: The planting of marigold combined with application of carbofuran at 1 kg a.i./ha controls *M. javanica* infestation on brinjal (Singh, 1991).

Table 3.9. Integrated management of root-knot nematodes by summer ploughing/ fallowing and nursery bed treatment in brinjal

Treatment	Reduction in nematode population (%)		Yield in kg/sq.m.	
	Hisar	Vellayani	Hisar	Vellayani
Ploughing + Fallowing for 15 days	87.00	66.20	28.98	---
Ploughing + Covering with polythene sheet for 15 days	94.90	77.00	18.84	78.00
No ploughing + No covering	77.60	---	---	---

Nursery treatment with carbofuran at 0.3g a.i./m² along with main field treatment with ploughing + exposing the field and ploughing + covering with polythene sheets for 15 days improved the plant growth and yield of brinjal by 20-32% (Sheela *et al.*, 2002). (Table 3.9).

Using Botanicals and Chemical Methods: Application of aldicarb at 1.0 kg a.i./ha in the nursery beds along with neem cake at 400 kg/ha increased yield and reduced root galling (Singh and Gill, 1998) (Table 3.10).

Table 3.10. Integrated management of root-knot nematodes in brinjal with neem cake and nematicides

Treatment	% Reduction in gall index	% Increase in yield
Aldicarb (1.0 kg a.i./ha) + Neem cake (400 kg/ha)	92.90	58.70
Carbofuran (1.0 kg a.i./ha) + Neem cake (400 kg/ha)	68.10	68.10

Using Cultural and Physical Methods: Integrating summer ploughing with soil solarization with polythene mulching effectively reduced *M. javanica* population in brinjal (Jain and Gupta, 1991). Soil solarization of nursery beds (using 100 gauge LDPE clear film for 15 days) and application of neem cake at 200 g/sq.m. proved most effective in reducing root-knot nematode infestation and increasing yield in brinjal. Solarization of nursery beds for 15 days in summer and application of poultry manure at 200 kg/ha gave maximum yield.

3.1.3.2. *The Reniform Nematode, Rotylenchulus reniformis*

(i) Management Methods

(a) Chemical Methods: Bare root dip treatment with aldicarb/ carbofuran/ turbofos at 500 or 1,000 ppm for 15 to 30 min. in brinjal economized the reniform nematode management by reducing the quantity of nematicide required (Prasad and Krishnappa, 1981).

(b) Integrated Methods: Integration of a bio-agent, *P. lilacinus* with carbofuran at 1 kg a.i./ha was found effective in the management of reniform nematode, *R. reniformis* infecting brinjal (Parvatha Reddy and Khan, 1989).

3.1.3.3. Root-knot Nematode, *Meloidogyne incognita* and Bacterial Wilt, *Ralstonia solanacearum* Disease Complex

Eggplant is prone to many soil borne diseases among which the bacterial wilt (*R. solanacearum*) in combination with root-knot nematode (*M. incognita*) takes heavy toll every year all over the world (Naik, 2004).

The root-knot nematode, *M. incognita*, present along with *R. solanacearum*, greatly increases the incidence of bacterial wilt of brinjal (Fig. 3.10). The root-knot nematode is responsible for breaking bacterial wilt resistance in "Pusa Purple Cluster" cultivar of brinjal (Parvatha Reddy *et al.*, 1979). *M. incognita* interacts with *R. solanacearum* in increasing the speed of development and severity of wilt disease of eggplant. Maximum disease occurred when the bacterium and nematode were inoculated simultaneously. Inoculation with *R. solanacearum* alone resulted in less severe disease. It was concluded that root-knot nematodes modify the plant tissues facilitating bacterial colonization (Nayar *et al.*, 1988).

Fig. 3.10. Root-knot nematode and bacterial wilt disease complex in brinjal.

(i) Management Methods

(a) Integrated Methods: Combined application of *T. harzianum* and *P. chlamydosporia* gave effective management of the disease complex and gave least root gall index (RGI), least wilt index (WI) and maximum fruit yield (1.8 RGI, 22.18% WI and 2.024 kg/3 sq.m. yield) followed by *P. fluorescens* + *P. chlamydosporia* (1.81 RGI, 25.13 WI and 1.782 kg/3 sq.m. yield) and *T. harzianum* + *P. fluorescens* (1.93 RGI, 27.28 WI and 1.680 kg/3 sq.m. yield) (Naik, 2004).

3.1.4. Chilli

Chilli (*Capsicum annuum*) is valued for its diverse commercial uses. India is a major producer, exporter and consumer of chilli. Indian chillies reach over 90 countries in the world. Bangladesh, Bahrain, Canada, Sri Lanka, Saudi Arabia, USA and UAE are the leading importing countries. In India, chilli is grown in almost all states. It is being cultivated in 0.834 million hectares producing 0.847 million tonnes of chillies with an average yield of 1.0 tonne per hectare. Andhra Pradesh has been the largest chilli-growing state followed by Karnataka and Maharashtra. Productivity of chilli is highest in Andhra Pradesh, followed by Arunachal Pradesh and Punjab. The productivity of chilli in Andhra Pradesh and Punjab is higher since the crop is raised under irrigated conditions compared with Maharashtra and Karnataka. Chilli is rich in vitamins A and C.

The root-knot nematode is the major limiting factor in successful production of chilli crop.

3.1.4.1. Root-knot Nematodes, Meloidogyne spp.

Root-knot nematodes induce much smaller galls on the roots of chillies. *M. incognita* was responsible for 24.54 to 28.00% loss in fruit yield of chillies (Singh *et al.*, 2003).

(i) Symptoms: Root galls on chilli are frequently small.

(ii) Management Methods

(a) Cultural Methods: Decrease in *M. incognita* population on chillies occurs when the field was left fallow or following crop rotation with marigold, spinach and bottle gourd (Khan *et al.*, 1975). Growing of chilli along with marigold gave highest degree of reduction in root-knot nematode population on chilli roots, followed by onion, garlic and asparagus (Trivedi and Tiagi, 1984).

Row application of fly ash at 0.6 kg/sq.m. enhanced the fruit yield by 21.3% and inhibited the reproduction of *M. incognita* (Khan and Ghadipur, 2004).

Botanicals: Soil application of decaffeinated tea leaves effectively reduced root galling on chilli plant (Trivedi and Tiagi, 1984).

(b) Chemical Methods: Aldicarb and carbofuran at 3 per cent were effective as seed treatment against root-knot nematodes infecting chilli (Kandasamy and Sivakumar, 1981).

Dichlofenthion was found effective against *M. incognita* in chillies (Alam *et al.*, 1973, Saxena *et al.*, 1974).

Aldicarb, carbofuran, ethoprophos and phenamiphos each at 1 to 2 kg a.i. per ha were found effective in reducing the root-knot nematode population and in increasing fruit yields of chillies (Handa and Mathur, 1981).

(c) Biological Methods: Pandey and Trivedi (1992) reported a significant reduction in number of galls, eggs/egg mass, hatching and final soil population of *M. incognita* infecting chillies in the presence of *P. lilacinus*.

Avermectins effectively reduced root-knot nematode infection in chilli under field conditions (Parvatha Reddy and Nagesh, 2002).

(d) Host Resistance: Chilli cvs. Pusa Jwala, Wonder Hot, Teja, Utkal Abha, Utkal Rashmi were found resistant to root-knot nematodes (Patnaik *et al.*, 2004).

3.1.4.2. Root-knot Nematode, *Meloidogyne javanica* and *Fusarium oxysporum* Disease Complex

(i) Management Methods

(a) Integrated Methods: *Pseudomonas aeruginosa* and *P. lilacinus* when used together significantly reduced infection of the disease complex on chilli (Perveen *et al.*, 1998).

3.1.5. Capsicum

Capsicum (*Capsicum annuum*), also known as bell pepper or sweet pepper, is a popular vegetable in India. It is relatively a new entrant into our country. It is mainly cultivated in Himachal Pradesh, Uttar Pradesh, parts of Gujarat, Maharashtra, Karnataka, Ranchi region of Bihar and hilly regions of Tamil Nadu. It grows well in summer season in hills and cooler season in the plains.

The root-knot nematode is an important limiting factor in successful cultivation of capsicum crop.

3.1.5.1. Root-knot Nematodes, *Meloidogyne spp.*

(i) Symptoms: Root galls caused by root-knot nematodes on sweet pepper are frequently small.

(ii) Management Methods

(a) Physical Methods: Burning of paddy husk or saw dust on infested nursery proved to be reasonably effective.

(b) Cultural Methods: Several cultural methods like selection of nematode-free nursery sites, destruction of infested roots after crop harvest, crop rotation with non-hosts (maize, wheat, sorghum) or antagonistic crops (marigold, mustard, sesame) or intercropping with marigold, onion and garlic, flooding, fallowing, deep summer ploughing and harrowing either alone or in combination proved to be reasonably effective and economical to check multiplication of root-knot nematodes on capsicum. Ploughing not only leads to disturbance and instability in nematode community but also causes their mortality by exposing them to solar heat and desiccation. Generally, 2 to 3 summer ploughings each at 10 days interval during April-May (40-46°C) have been recorded to reduce 96% *M. javanica* population while fallowing itself during the same period registered 45.5% reduction. Additional use of plastic sheets for covering soil either in nursery beds or in field further enhance nematode reduction. Such an approach also helps in reducing the intensity of weeds, fungi and bacteria in the soil.

(c) Chemical Methods: Bare-root dip treatment of capsicum seedlings in 0.1% carbosulfan/monocrotophos for 6 hr eliminates root-knot infection.

(d) Biological Methods

Antagonistic Bacteria: Field application of a nematicidal *B. thuringiensis* strain to capsicum in Puerto Rico reduced galling in roots due to *M. incognita* and increased yield significantly (Zuckermann *et al.*, 1993).

Antagonistic Fungi: Application of 'Royal 350' (*Arthrobotrys irregularis* cultured on oat seed medium) at 140 g/m^2 a month before transplantation of capsicum resulted in good protection against root-knot nematodes. Among the egg parasites, efficacy of *P. lilacinus* (commercially formulated as 'Biocon' in Philippines) has been found to be comparatively higher in suppressing the population of *Meloidogyne* spp. on capsicum.

Bio-agent *P. chlamydosporia* and *Pseudomonas fluorescens* were evaluated under field conditions for their efficacy against root-knot nematode *M. incognita* infecting capsicum. Treatment of the nursery bed with the formulation of *P. chlamydosporia* at 50g/m^2 was significantly effective in reducing the galling index, number of nematodes in roots and soil, increasing the percent parasitization of eggs by bioagent and also yields. Seed treatment with *P. fluorescens* alone and the nursery bed treatment with *P. chlamydosporia* alone were effective. The main purpose of these studies was to produce capsicum seedlings that are colonized by *P. chlamydosporia* before transplanting so that they could carry the bio-agent to the field. During this process the field soil would be enriched with the propagules of eco-friendly component of management (here it is a bio-agent) in two or three seasons (Naik, 2004).

(e) Host Resistance: Capsicum cvs. Mississipi-68, Santanka, Anaheim Chile and Italian Pickling are reported to be resistant to root-knot nematodes (Hare, 1951). Black Indica, Naharia and Pant C1 were reported to be resistant to *M. incognita*, while All Big as moderately resistant. California Wonder and Naharia were resistant to *M. javanica*, while Early California Wonder as moderately resistant. Naharia was reported to be resistant to *M. arenaria*.

(f) Integrated Methods: Integration of *P. fluorescens* and *P. chlamydosporia* in the nursery bed has proved significantly effective in reducing the root-galling index (*M. incognita*), number of nematodes in the roots and soil and increasing the yield of the capsicum crop under field conditions (Naik, 2004).

In capsicum, the seedling stand was good where the combination of neem based *P. fluorescens* and *T. harzianum* was used (4.4), followed by seed treatment with *P. fluorescens* + soil application of *T. harzianum* (Naik, 2004).

3.1.5.2. Root-knot Nematode, *Meloidogyne incognita* and Bacterial Wilt, *Ralstonia solanacearum* Disease Complex

Capsicum is prone to many soil borne diseases among which the bacterial wilt (*R. solanacearum*) in combination with root-knot nematode (*M. incognita*) takes heavy toll every year all over the world (Naik, 2004).

(i) Management Methods

(a) **Integrated Methods:** Combined application of neem based formulations of *P. fluorescens* and *P. chlamydosporia/ T. harzianum* at 40 g/m² in nursery beds and transplanting these seedlings in the main field resulted in significant reduction in disease index and root-knot index in capsicum to the tune of 70% and increased the crop yield by 37% (Rao *et al.*, 2002) (Table 3.11). Combinations of the bioagents did not affect the colonization of the individual bioagents on the roots and hence the transplants carried the bioagents to the main field. This has resulted in the effective management of the pathogens involved in the disease complex.

Table 3.11. Effect of integration of neem-based bioagents on the growth of transplants and management of disease complex and yield of capsicum

Treatment	Seedling weight (g)	Root-knot index (1-10)	Disease index (1-9)	Yield in kg/ 4 sq.m.
Seed treat. with *P. fluorescens*	421	5.6	6.4	4.3
Seed treat. with neem-based *P. fluorescens*	428	5.2	6.7	4.7
Nursery treat. with *P. fluorescens*	435	4.6	5.4	4.8
Nursery treat. with *P. chlamydosporia*	364	4.4	7.3	3.2
Nursery treat. with *T. harzianum*	374	4.8	7.0	3.4
Nursery treat. with *P. fluorescens* + *P. chlamydosporia*	463	4.1	5.2	4.0
Nursery treat. with *P. fluorescens* + *T. harzianum*	493	3.8	3.5	5.1
Control	340	8.7	8.2	2.6
CD at 5%	27.20	0.49	0.38	0.25

Naik (2004) reported that combination of *T. harzianum* along with *P. fluorescens* increased the yield (3.320 kg/3m^2 plot) followed by combination of *T. harzianum* and *P. chlamydosporia* (2.960 kg/3m^2). The least gall index was present where the combination of *T. harzianum* and *P. fluorescens* was used (1.5) followed by *T. harzianum* + *P. chlamydosporia* (1.7) and *P. fluorescens* + *P. chlamydosporia* (1.7). But among combination treatment of bio-agents, all the treatments were on par (*P. fluorescens* + *T. harzianum*, *P. fluorescens* + *P. chlamydosporia*, *T. harzianum* + *P. chlamydosporia*). Results related to the percent bacterial wilt disease (*R. solanacearum*) incidence also showed similar trend in which *T. harzianum* + *P. fluorescens* performed very well and the percent incidence was 16.90 followed by *T. harzianum* + *P. chlamydosporia* (17.20).

3.2. LEGUMINOUS VEGETABLE CROPS

3.2.1. French Bean

French bean (*Phaseolus vulgaris*), also known as common, haricot, kidney, string, salad bean, runner bean or snap bean, is an important leguminous vegetable. It is consumed as tender pods, shelled green beans and dry beans. It is a nutritious vegetable which contains proteins (1.7g), calcium (50 mg), phosphorus (28 mg), iron (1.7 mg), carotene (132 mg), Thiamine (0.08 mg), Riboflavin (0.06 mg) and vitamin C (24 mg/100 g of edible pods). It is largely grown in hilly areas of Himachal Pradesh, Jammu and Kashmir and north-eastern states during summer, winter and autumn crop in parts of Uttar Pradesh, Maharashtra, Karnataka and Andhra Pradesh.

Root-knot and reniform nematodes are the major limiting factors in successful production of French bean crop.

3.2.1.1. Root-knot Nematode, Meloidogyne spp.

M. incognita and *M. javanica* appear to be the most common root-knot species of French beans and have been reported causing damage in tropics and subtropics. Lal and Ansari (1960) reported *M. arenaria* for the first time on beans from Bihar, India. *M. incognita* was responsible for 19.38 to 43.48% loss in pod yield of French bean (Das, 1994; Patel *et al.*, 2004; Parvatha Reddy and Singh, 1981), while *M. javanica* caused 30-40% loss in yield (Sharma *et al.*, 2002).

(i) Symptoms: Gall size on roots of French bean is variable and may be nearly undetectable. The only visible symptom on the roots is the presence of large egg-masses.

(ii) Interaction with Other Pathogens: M. incognita in association with soil-borne fungi such as *R. solani* and *F. solani* was responsible for root-

rot/wilt disease complex in French bean. In the presence of nematode, the root rot due to *R. solani* and wilt due to *F. solani* increased considerably (Parvatha Reddy *et al.*, 1979; Singh *et al.*, 1981).

(iii) Management Methods

(a) Physical Methods: Root-knot nematodes can be effectively controlled by a 4-8 weeks solarization, assuming that the land will remain uncropped during summer.

(b) Cultural Methods: Intercropping of French bean with finger millet, chilli and groundnut helps in the management of *M. incognita* Race 2 (Ramappa, 1988). Growing of chilli or groundnut or finger millet in rotation with French bean significantly reduced the root-knot nematode population (Ramappa, 1988).

The root-knot nematodes would be unable to initially invade bean roots if the crop is sown at the end of winter or early in the spring, when soil temperatures are below 15°C. Escape from early root penetration would give the plant a head start. Yield would increase because the large root system can withstand the damage caused by delayed nematode invasion. Moreover, the beans would be harvested by the end of spring or early in summer, thus limiting the number of generations produced (often to only one) and overall population densities.

Botanicals: Spot application of neem cake at 16.6 g/spot 15 days before sowing was found effective in reducing root galls (64.4 compared to 122.9 in control), number of egg masses (16.0 compared to 44.4 in control) and final nematode population (253 compared to 437 in control) and in increasing plant growth and yield (3.375 t/ha compared to 0.715 t/ha in control) (Ahmed and Choudhury, 2004).

(c) Chemical Methods: Seed treatment with phenamiphos, aldicarb and carbofuran all at 1 per cent concentration were effective in controlling the root-knot nematode (*M. incognita*) infecting French bean and in increasing their pod yields (Parvatha Reddy, 1984). Seed treatment with oxamyl at 3-10% prevented development of *M. incognita*.

Aldicarb, carbofuran, ethoprophos and phenamiphos each at 1 to 2 kg a.i. per ha were found effective in reducing the root-knot nematode population and in increasing pod yields of French bean (100 to 112 per cent) (Parvatha Reddy, 1985a; Rao and Singh, 1978).

(d) Host Resistance: French bean cvs. Banat, Blue Lake Stringless, Bountiful Flat, Brown Beauty, Cambridge Countess, Gallaroy, Kenya-3, Pinto W5-114, Seafarer and Sutton's Masterpiece were reported to be resistant to *M. incognita* (Singh *et al.* (1981). Wyatt *et al.* (1983) released the first snap bean cv. Nemasnap resistant to *M. incognita*. In Kenya, the cvs. Kahuti, Red Haricot, Rono, Saginaw and Kiburn were resistant to *M. incognita* and *M. javanica* (Ngundo, 1977). Contender has been reported to be moderately resistant to *M. javanica*. French bean cvs. Kibbu, Manoa Wonder, Red Haricot, Rono and Saginaw were reported to be resistant to both *M. incognita* and *M. javanica*.

3.2.1.2. The Reniform Nematode, Rotylenchulus reniformis

The reniform nematode also damages French bean.

(i) Interaction with Other Pathogens: Vadhera *et al.*(1995) demonstrated that the incidence of root rot was maximum in simultaneous inoculation of *R. reniformis* and *F. solani* on beans. The combination treatment resulted in maximum reduction in plant growth parameters.

(ii) Management Methods

(a) Cultural Methods: Neem and karanj oil cakes at 2 tonnes per ha were most effective in reducing *R. reniformis* infecting French bean.

(b) Chemical Methods: Satisfactory nematode control has been obtained with six foliar sprays of oxamyl at 0.56 kg a.i./ha combined with a soil drench of 2.24 kg a.i./ha of the same chemical. Furrow application of 2.5 kg a.i./ha of carbofuran was also found effective.

3.2.1.3. Root-knot Nematode, *Meloidogyne incognita* and Root Rot, *Macrophomina phaseolina* Disease Complex

(i) Management Methods

(a) Cultural Methods: The beneficial effects of neem cake were observed not only against *M. incognita* and the fungus *M. phaseolina* individually, but also when both the pathogens formed a disease complex on French bean.

3.2.2. Cowpea

Cowpea (*Vigna unguiculata*), also known as southern pea and black-eye pea, is one of the most important vegetables. It is cultivated for its long, green or purplish pods to be cooked as vegetable or for dry seeds used as pulse. Its foliage is used as fodder or green manure. In India, cowpea is grown almost throughout the country.

Root-knot, reniform and pigeon pea cyst nematodes are the major limiting factors in successful production of cowpea crop.

3.2.2.1. The Root-knot Nematode, *Meloidogyne incognita*

Parvatha Reddy and Singh (1981) reported that *M. incognita* was responsible for 28.60% loss in pod yield of cowpea.

Fig. 3.11. Heavy galling of cowpea roots infected with *Meloidogyne incognita* (Courtesy: F.E. Caveness).

(i) Symptoms: Symptoms of damage induced by root-knot nematode include patches of stunted and yellowed plants. Severe damage can lead to reduced numbers of leaves and buds. At high densities severe root galling occurs (Fig. 3.11). visual symptoms of damage first occurred at 1,000 and 10,000 juveniles/500g of soil.

(ii) Interaction with Other Pathogens: The presence of heavy infestations of *M. javanica* on a cowpea cv. tolerant to wilt incited by *Fusarium oxysporum* f. sp. *tracheiphilum* caused increased wilting when compared to the susceptible cv.

(iii) Management Methods

(a) Cultural Methods: Summer ploughing was found effective for the management of root-knot nematodes and increased the cowpea yield over non-ploughed field.

The root-knot nematodes associated with cowpea can be effectively managed by the application of neem or eupatorium leaves at 15 t/ha two weeks before sowing. Neem cake incorporation in the previous crop, caused reduction in density of all nematodes in the soil on the following cowpea crop (Jain and Hasan, 1986). Cocoa pod husks incorporated at 6 t/ha caused 28% reductions in galling and 6.7% increases in yield (Egunjobi, 1985).

Rotation of cowpea with graminaceous crops or *Crotalaria spectabilis* is recommended. Populations of root-knot decreased greatly when compared to fallowed plots when *C. spectabilis* was grown as a weed free cover crop. Mulching with cowpea foliage was also highly effective in suppressing root-knot nematode populations.

M. javanica populations were lower when cowpea and maize were grown under mixed rather than under sole crop cropping systems. One crop of rice was sufficient to effectively reduce root-knot nematode infestations in succeeding susceptible cowpea crop. The reduction was greater than with rotations with non-host crops.

(b) Chemical Methods: Seed treatment with phenamiphos, aldicarb and carbofuran all at 1 per cent concentration were effective in controlling the root-knot nematode (*M. incognita*) infecting cowpea and in increasing their pod yields (Parvatha Reddy, 1984). Carbosulfon at 1,000 ppm used as seed treatment reduced juvenile penetration in cowpea plants. Posse and monocrotophos at 1,000 ppm reduced root-knot index (2.2 and 2.8, respectively) as compared to untreated check (3.5) when cowpea seeds were dipped in these chemicals for 12 hours.

Aldicarb, carbofuran, ethoprophos and phenamiphos each at 1 to 2 kg a.i. per ha were found effective in reducing the root-knot nematode population and in increasing pod yields of cowpea (40 to 50 per cent) (Parvatha Reddy, 1985a; Rao and Singh, 1978). An increase in seed yield to the tune of 61.8% was reported with application of phenamiphos at 2.5 kg a.i./ha (Jain and Hasan, 1984).

(c) Biological Methods: Cowpea seed treatment with *Paecilomyces lilacinus* spores was reported to be effective in reducing the root-knot infestation level (Midha, 1985). Application of certain organic materials, viz. subabool leaves, neem cake or saw dust prior to the application of *P. lilacinus* not only enhanced the growth and multiplication of the fungus resulting into high percentage of egg mass infection but also effectively reduced the soil incidence in cowpea (Hasan, 1988). Application of *P. lilacinus* at 15 kg/ha along with certain organic materials effectively reduced the root-knot nematode incidence in cowpea (Hasan and Jain, 1992).

(d) Integrated Methods: Hasan and Jain (1992) reported that soil application of *P. lilacinus* cultured on sorghum seeds together with certain organic matter effectively reduced the incidence of *M. incognita* and increased the crop yield of cowpea.

Summer ploughing along with seed treatment with carbosulfon 3% w/w or seed soaking in monocrotophos at 0.1% for 6 hr. gave effective control of root-knot nematodes and increased the cowpea yield.

3.2.2.2. *The Reniform Nematode, Rotylenchulus reniformis*

The emergence of cowpea seedlings was delayed by 7 days and seedling stand was reduced to the tune of 6 to 11% due to *R. reniformis* at one nematode/g of soil (Nanjappa *et al.*, 1978). *R. reniformis* was responsible for 13.20 to 32.00% loss in pod yield of cowpea (Palanisamy and Sivakumar, 1981; Hasan and Jain, 1998).

(i) Histopathology: R. reniformis feeds on cortical tissue of cowpea.

(ii) Races: The nematode has two races based on their ability to parasitize cowpea, castor or cotton with race A reproducing on all three hosts and race B only on cowpea (Dasgupta and Seshadri, 1971).

(iii) Management Methods

(a) Physical Methods: Under field conditions, soil solarization significantly reduced soil population of the reniform nematode up to 15 cm deep and increased the yield of cowpea.

(b) Cultural Methods: Soil application of neem seed kernel at 2.5 t/ha was most effective in reducing the number of females/plant (46.2 compared to 80.7 in check), number of egg masses/plant (26.0 compared to 45.5 in check), number of eggs and larvae/egg mass (46.2 compared to 80.7 in check) and final soil population/200 cc soil (143.5 compared to 831.0 in check) of the reniform nematode (Ram and Baheti, 2004).

Summer ploughing was found effective for the management of reniform nematodes and increased the cowpea yield over non-ploughed field. Solarization was considered as an effective method for reducing the nematode densities to a depth of 15 cm.

The reniform nematodes associated with cowpea can be effectively managed by the application of neem or eupatorium leaves at 15 t/ha two weeks before sowing.

The reniform nematode densities were suppressed when cowpea was grown intercropped with maize.

(c) Host Resistance: Cowpea cvs. V-16 and Pusa Phalguni are reported to be resistant to *R. reniformis* (Thaker and Patel, 1984).

(d) Integrated Methods: Summer ploughing along with seed treatment with carbosulfan 3% w/w or seed soaking in monocrotophos at 0.1% for 6 hr. gave effective control of reniform nematodes and increased the cowpea yield.

3.2.2.3. *The Pigeon pea Cyst Nematode, Heterodera cajani*

The cyst nematode has been found associated with cowpea in a number of regions of India.

(i) Symptoms: The cyst nematode retarded emergence of leaves and reduced the number of flowering buds, flowers, growing pods and yield.

(ii) Management Methods

(a) Cultural Methods: Cowpea rotated with paddy rice may be less affected by the cyst nematode, because of the negative effect of flooding on nematode densities.

(b) Host Resistance: Cowpea cv. CO6 was found resistant to the pigeon pea cyst nematode. The cowpea cv. Barsati Mutant has been reported to be tolerant to the cyst nematode (Sharma and Sethi, 1976).

3.2.3. PEA

Pea (*Pisum sativum* var. *hortense*) is an important vegetable crop widely grown throughout the world. As a cool season crop, it is extensively grown in temperate zone; but restricted to cooler altitudes in the tropics and winter season in the sub-tropics. A rich source of proteins (25%), amino acids and sugars (12%), green peas are an all time favourite vegetables. It is rich in vitamins A,B,C and minerals. It is being cultivated in 0.276 million hectares producing 1.972 million tonnes of peas with an average yield of 7.1 tonnes per hectare. Garden pea is grown in Uttar Pradesh, Himachal Pradesh, West Bengal, Jammu and Kashmir, Punjab and Haryana.

Root-knot, reniform and lance nematodes are the important limiting factors in successful cultivation of pea crop.

3.2.3.1. Root-knot Nematodes, Meloidogyne spp.

M. incognita was responsible for 20.00 to 50.61% loss in pod yield of peas (Parvatha Reddy, 1985b; Upadhayay and Dwivedi, 1987; Sharma, 1985).

(i) Symptoms: The infested fields show patches in which peas are stunted, chlorotic and have few flowers which produce small and often empty pods. The root systems are reduced in size, exhibit large galls and poor nodulation. Senescence also tends to occur earlier.

(ii) Management Methods

(a) Cultural Methods: Intercropping with marigold or mustard with pea reduced the damage of root-knot nematodes.

Peas escape nematode attack, in the subtropics when sowing is postponed to mid-autumn, when temperatures drop.

(b) Chemical Methods

Seed Treatment: Bhagawati and Phukan (1990) found that carbofuran at 3% w/w as seed treatment was effective in reducing galls and egg masses in roots of pea and increased yields. Seed treatment with phenamiphos, aldicarb and carbofuran all at 1 per cent concentration were effective in controlling the root-knot nematode (*M. incognita*) infecting peas and in increasing their pod yields (Parvatha Reddy, 1984).

Main Field Treatment: Aldicarb, carbofuran, ethoprophos and phenamiphos each at 1 to 2 kg a.i. per ha were found effective in reducing the root-knot nematode population and in increasing pod yields of peas (58 to 85 per cent) (Parvatha Reddy, 1985a; Rao and Singh, 1978).

Drill application of carbofuran at 1.5 kg a.i./ha below the seed level reduces the incidence of *M. incognita* in pea.

(c) Host Resistance: Pea cv. Wando is reported to be highly resistant to *M. incognita* and moderately resistant to *M. javanica*.

3.2.3.2. The Reniform Nematode, Rotylenchulus reniformis

Dalal and Vats (1998) reported that *R. reniformis* was responsible for 15.80% loss in pod yield of peas.

(i) Interaction with Other Pathogens: Vats and Dalal (1997) reported maximum and early wilting in pea due to prior inoculation of *R. reniformis* followed by *F. o. f. sp. pisi.*

(ii) Management Methods

(a) Cultural Methods: Intercropping with marigold or mustard with pea reduced the damage of reniform nematodes.

3.2.3.3. The Lance Nematode, Hoplolaimus uniformis

(i) Interaction with Other Pathogens: In peas, early yellowing and root rot disease is dependant upon the presence of both *Hoplolaimus uniformis* and *F.o.f.sp. pisi* race 3.

3.2.4. Cluster Bean

Cluster bean (*Cyamopsis tetragonoloba*) is also popularly called as 'guar' in India. Its tender pods being edible and are used in preparation of curry. The green pods are rich in vitamins A and C. Since it is drought tolerant, it can be grown in summer. It is also grown as a forage and green manure crop. Its seeds reported to be good cattle feed.

Root-knot nematode and *Fusarium* wilt disease complex is the major limiting factor in successful production of cluster bean crop.

3.2.4.1. Root-knot Nematode, *Meloidogyne javanica* and Wilt, *Fusarium solani* Disease Complex

(i) Management Methods

(a) Integrated Methods: *Pseudomonas aeruginosa* and *Paecilomyces lilacinus* when used together significantly reduced infection of the disease complex on cluster bean (Perveen *et al.*, 1998).

3.3. CRUCIFEROUS VEGETABLE CROPS

3.3.1. Cabbage and Cauliflower

Cabbage (*Brassica oleracea* var. *capitata*) is an important winter vegetable of cole group. It is a rich source of vitamins A,B and C and also contains minerals. It is being cultivated in 0.290 million hectares producing 6.147 million tonnes of cabbages with an average yield of 21.2 tonnes per hectare. It covers about 4% of total area under vegetables. India comes next to China in cabbage production. It is now grown almost throughout the year. West Bengal, Bihar, Orissa, Assam, Maharashtra, Gujarat and Uttar Pradesh are major cabbage-growing states.

Cauliflower (*Brassica oleracea* var. *botrytis*) is the most popular winter vegetable among cole crops. It is rich in vitamin C. It is being cultivated in 0.238 million hectares producing 4.507 million tonnes of cauliflowers with an average yield of 18.9 tonnes per hectare. West Bengal, Orissa, Bihar, Maharashtra, Haryana, Assam and Gujarat are major cauliflower-growing states. With the development of new varieties, it is now being grown in non-traditional areas like Andhra Pradesh, Tamil Nadu and Kerala.

Stunt and root-knot nematodes are the major limiting problems in successful cultivation of cabbage and cauliflower crops.

3.3.1.1. The Stunt Nematode, *Tylenchorhynchus brassicae*

Siddiqi *et al.* (1972) encountered *T. brassicae* as the most prevalent species around the roots of cabbage and cauliflower in Uttar Pradesh.

(i) Distribution and Hosts: The populations of *T. brassicae* were exceptionally high in Aligarh, Kanpur, Ghazipur and Rampur districts of U.P. In general, all crucifers are good hosts of *T. brassicae*, but cabbage, cauliflower and tomato proved to be most efficient hosts. The multiplication rate on cabbage (cv. Suttons Earliest), cauliflower (cv. Patna) and Knol-khol (cv. Purple Vienna) was 13, 18 and 29, respectively. Common weeds such as *Chenopodium album*, *Trianthema monogyna* and *Portulaca oleracea* were found to be hosts of *T. brassicae*.

(ii) Symptoms: T. brassicae caused reduction in fresh weight of plants and suppressed root growth, resulting in reduced water absorption.

(iii) Life Cycle: The optimum soil temperature (30°C) and soil moisture (25-30%) for reproduction of *T. brassicae* was same for both cabbage and cauliflower. Further, high temperature (35-40°C) and low moisture are highly unfavourable for survival of *T. brassicae* in the absence of host. However, at moderate temperature and moisture conditions, the nematode is capable of surviving even up to 240 days.

(iv) Interaction with Other Pathogens: T. brassicae alone did not affect the percentage emergence of cauliflower seedlings. *R. solani* was highly destructive by itself, but the adverse effect on the emergence of seedlings was greater when nematode infestation occurred along with fungal infection (Khan *et al.*, 1971).

(v) Histochemistry: An increase in total soluble phenols, ortho-dihydroxy phenols, free amino acids and proteins, but not in total carbohydrates, was found in the roots of cabbage and cauliflower infected with 10,000 *T. brassicae* (Alam *et al.*, 1976). Total phenols, ortho-dihydroxy phenols and free amino acids increased with increasing seedling age at inoculation (Khan *et al.*, 1980).

(vi) Management Methods

(a) Cultural Methods: There was sudden decline in population of *T. brassicae* when cabbage and cauliflower were rotated with wheat (Siddiqi *et al.*, 1973). Intercropping of cabbage and cauliflower with neem reduced the multiplication rate of *T. brassicae* and improved the plant health (Siddiqi and Saxena, 1987a,b). The mustard (rabi), radish (summer), sesame (kharif) sequence considerably reduced the population of *Tylenchorhynchus* spp. (Haque and Gaur, 1985).

Botanicals: Seed dressing with latex of *Calotropis gigantea, C. procera, Euphorbia milii, E. nerifolia* and *E. tirucalli* significantly inhibited *T. brassicae* multiplication on cabbage and cauliflower and improved plant growth (Siddiqi and Alam, 1989a).

Bare-root dip of cabbage and cauliflower seedlings in leaf extracts of neem and Persian lilac (Siddiqi and Alam, 1989b), interculture of seedlings of marigold, neem and Persian lilac reduced multiplication rate of *T. brassicae* and improved plant growth.

Amendment of soil with oil cakes of neem, groundnut, mustard and castor (equivalent to 100 kg N/ha) was effective in reducing the population of *T. brassicae* around the roots of cabbage and cauliflower (Siddiqi *et al.*, 1976). Neem cake was the best.

(b) Chemical Methods: Aldicarb, carbofuran, ethoprophos and phenamiphos each at 1 to 2 kg a.i. per ha were found effective in reducing the stunt nematode population and in increasing flower yields of cauliflower (Ahuja, 1983)

Aldicarb at 0.5 kg a.i. per ha and carbofuran at 1.5 kg a.i. per ha were effective in reducing *T. brassicae* and *T. dubius* population and in increasing yields of cabbage (Varma *et al.*, 1978).

(c) Host Resistance: *Brassica rapa* cv. Sarson (yellow mustard) was found to be least susceptible to *T. brassicae*, and could be grown on infected soil in preference to *B. nigra* (black mustard) (Khan *et al.*, 1986).

3.3.1.2. The Root-knot Nematode, Meloidogyne incognita

(i) Interaction with Other Pathogens: Maximum wilt score due to *F. oxysporum* f. sp. *conglutinans* (4.0) and wilt symptom index (80.0) was observed in cauliflower plants which received *M. incognita* 14 days before the inoculation of wilt fungus (Pathak and Keshari, 2004).

3.3.1.3. Root-knot Nematode, *Meloidogyne incognita* and Club Root, *Plasmodiophora brassicae* Disease Complex

(i) Management Methods

(a) Integrated Methods: PGPR strains (*Pseudomonas fluorescens, Bacillus subtilis*) combined with fungal biocontrol agents (*Trichoderma viride, T. harzianum*) were found to be effective in reducing the nematode-fungal disease complex in cabbage (Loganathan *et al.*, 2001). The bioformulation mixture of *P. fluorescens, T. viride* and chitin effectively reduced the disease complex in cabbage and cauliflower both under greenhouse and field conditions (Samiyappan, 2003) (Table 3.12).

Table 3.12. Efficacy of bioformulation mixtures against club root disease – root-knot nematode complex in cabbage under greenhouse conditions

Treatment	Club root index	Nematode incidence	
		Population	Root-knot index
Trichoderma viride	25.99 (30.65)	129	2.66
Pseudomonas fluorescens	28.20 (32.07)	112	2.33
T. viride + *P. fluorescens*	25.33 (30.22)	114	2.33
T. viride + Chitin	25.44 (30.29)	108	2.33
P. fluorescens + Chitin	25.66 (30.43)	111	2.00
T. viride + *P. fluorescens* + Chitin	22.22 (28.12)	108	2.00

Chitin alone	31.70 (34.26)	139	3.00
Carbendazim	19.90 (26.49)	264	4.66
Carbofuran	39.90 (39.17)	106	1.66
Cabendazim + Carbofuran	15.00 (22.79)	103	1.66
Plasmodiophora brassicae alone	48.90 (44.37)	0.033	0.133
Meloidogyne incognita alone	0.03 (0.60)	280	5.00

3.4. CUCURBITACEOUS VEGETABLE CROPS

3.4.1. Cucumber

India is considered to be the home of cucumber (*Cucumis sativus*) . Cucumber is cultivated for fresh consumption or as pickling cucumber for preservation, marinated with vinegar, salt, dill or other spices. It is an important salad crop cultivated both in north and south and lower as well as higher hills in India.

The root-knot nematode is the most important pest on cucumber that is of some economic importance.

3.4.1.1. *Root-knot Nematodes, Meloidogyne spp.*

(i) Symptoms: The root-knot nematodes cause severe galling on cucumber roots (Fig. 3.12).

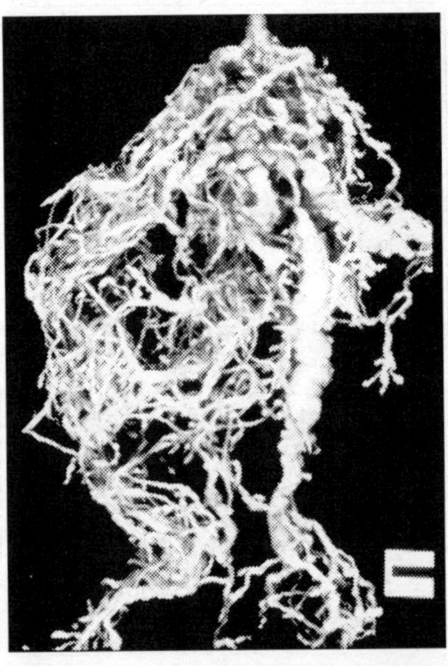

Fig. 3.12. Cucumber roots severely galled due to root-knot nematode infection.

(ii) Management Methods

(a) Physical Methods: The most important nematode pests controlled by greenhouse steaming are root-knot nematodes attacking cucumbers .

(b) Chemical Methods: Drill application of carbofuran at 1.5 kg a.i./ha below the seed level reduces the incidence of *M. incognita* in cucurbitaceous crops.

(c) Integrated Methods: Combined inoculation of AMF and *Pseudomonas fluorescens* had positive effect on root-knot nematode control on cucumber (Jakobsen, 1999).

3.4.1.2. The Dagger Nematode, *Xiphinema americanum* and Tobacco Ring Spot Virus Disease Complex

(i) Management Methods

(a) Chemical Methods: Benomyl application inhibited feeding of *X. americanum* on cucumber, which was reflected in a marked reduction of its transmission ability of tobacco ring spot virus (McGuire and Good, 1970).

3.4.2. Pumpkin

Pumpkin (*Cucurbita moschata*) occupies a prominent place among vegetables owing to its high productivity, nutritive value, good storability, long period of availability, better transport qualities and extensive cultivation in sub-tropical and tropical parts of the world. Fruits are rich in vitamin A. It is also consumed as processed and stock feed. The fruits can be used in preparing sweets, candy, fermented into beverages. Yellow or orange-fleshed pumpkins are rich in carotene. In India, it is grown mainly in Assam, West Bengal, Tamil Nadu, Karnataka, Madhya Pradesh, Uttar Pradesh, Orissa and Bihar.

The root-knot nematode is the most important pest on pumpkin.

3.4.2.1. Root-knot Nematodes, *Meloidogyne* spp.

The very broad leaved plant like pumpkin which is infected with root-knot nematodes may show day time wilting and develop much larger galls. In cucurbits, the roots react to the presence of *Meloidogyne* spp. by the formation of large, fleshy galls.

(i) Management Methods

(a) Physical Methods: Pulverizing the planting pit soil and exposing to sunlight by repeated raking and heaping 10 to 15 cm thick dry trash and burning it before application of compost helps in reducing the nematode population. Frequent cultivation may be given in May-June to expose the soil to sun.

(b) Cultural Methods: Application of 250g of fresh neem or karanj cake at the time of sowing reduces the nematode population in soil. Crop rotation with rice, oats, wheat and taramira is also effective.

(c) Chemical Methods: Seed treatment with carbosulfan (Marshall 25 ST) at 6% w/w is effective against root-knot nematodes. Application of carbofuran at 0.2g a.i./pit at sowing is also effective against root-knot and reniform nematodes.

(d) Host Resistance: Pumpkin cvs. Jaipuri and Dasna were found resistant to root-knot nematodes.

(e) Integrated Methods: Addition of 250 g fresh neem cake or karanj cake enriched with *Trichoderma harzianum / Paecilomyces lilacinus* per planting pit effectively controlled the nematodes.

3.4.2.2. Root-knot Nematode, *Meloidogyne javanica* and Root Rot/ Wilt, *Macrophomina phaseolina, Fusarium oxysporum, F. solani* Disease Complex

(i) Management Methods

(a) Integrated Methods: *Pseudomonas aeruginosa* and *Paecilomyces lilacinus* when used together significantly reduced infection of the disease complex on pumpkin (Perveen *et al.*, 1998).

3.4.3. Pointed Gourd

Pointed gourd (*Trichosanthes diocea*) or parwal is widely cultivated in Bihar, Uttar Pradesh, West Bengal, Orissa, Assam, Madhya Pradesh and Gujarat. Recently, it has been introduced in and around Hyderabad and Bangalore.

The root-knot nematode is the major limiting factor in successful production of pointed gourd.

3.4.3.1. The Root-knot Nematode, *Meloidogyne incognita*

Verma (2001) reported that *M. incognita* was responsible for 30-40% loss in yield of pointed gourd.

(i) Management Methods

(a) Cultural Methods: Selection of healthy looking roots for planting is recommended. Planting of marigold, *Tagetes patula* at 3 to 5 plants/pit will manage attack of root-knot nematodes on pointed gourd.

(b) Chemical Methods: Rooted cuttings should be given a 6 hr dip in 0.1% carbosulfan 25 EC or triazophos 40 EC before planting. In the ratoon crop, carbofuran/phorate should be applied at 0.2g a.i./plant.

(c) Integrated Methods: Integration of *Paecilomyces lilacinus* (50 g) + *Trichoderma harzianum* (100 g) + neem cake (250 g) + marigold (3 plants/pit) increased plant growth parameters and yield (6.6 kg/plant compared to 1.4 kg/plant in control), and reduced root galling (14.3/5 g roots compared to 75.0/ g roots in control), number of egg masses (40/5 g roots compared to 150/ g roots in control) and final nematode population (66.6/100 g soil compared to 86660/ 100 g soil in control) (Verma *et al.*, 2005).

3.4.4. Bottle Gourd

Bottle gourd (*Lagenaria siceraria*) is grown for immature fruits used for culinary purposes. Tender fruits are rich in vitamin B. It is also used for preparation of different types of sweets. Fruit pulp is very good source of fibre-free carbohydrates and fruit pericarp for crude fibre. The oil extracted from kernels of seed, is a fine cooking medium. In India, it is cultivated in Uttar Pradesh, Punjab, Gujarat, Assam, Meghalaya and Rajasthan.

The root-knot nematode is an important limiting factor in successful cultivation of bottle gourd.

3.4.4.1. *Root-knot Nematodes, Meloidogyne spp.*

(i) Management Methods

(a) Cultural Methods: Application of press mud at 10 t/ha (ICBR 1: 18.95) or mustard cake at 3.333 t/ha (ICBR 1: 16.10) or neem cake at 4.286 t/ha one week in advance of sowing gave effective management of root-knot with increased yield of bottle gourd (Dahiya *et al.*, 1998).

(b) Chemical methods: Seed treatment with aldicarb and carbofuran at 3 per cent was effective against root-knot nematodes infecting bottle gourd (Kandasamy and Sivakumar, 1981). Carbosulfan and benfuracarb at 3% w/w as seed treatment increased yield and reduced root galling in bottle gourd.

3.4.5. Ridge Gourd

Ridge gourd (*Luffa acutangula*) or ribbed gourd is a monoecious viny vegetable. It has some medicinal value and has all the nutrients. The fibre from well grown fruits has commercial value, as it can be used for making brushes, for cleaning. It is mainly cultivated in Andhra Pradesh, Tamil Nadu, Karnataka, Gujarat, Assam, West Bengal and Konkan region of Maharashtra.

The root-knot nematode is the major limiting factor in successful production of ridge gourd.

3.4.5.1. Root-knot Nematodes, Meloidogyne spp.

(i) Management Methods

(a) Host Resistance: Ridge gourd cvs. Panipati and Meerut Special were found resistant to root-knot nematodes.

3.4.6. Watermelon

Watermelon (*Citrullus lanatus*) is an important cucurbitaceous summer vegetable grown for ripe fruits. Though it can be grown in garden land, it is a major river-bed crop of Uttar Pradesh, Rajasthan, Gujarat, Maharashtra and Andhra Pradesh. An excellent desert fruit, it is relished by rich as well as poor. The fruit juice makes an excellent refreshing and cooling beverage after adding a pinch of salt and black pepper. The fruits contain 92% water, 0.2% protein, 0.3% minerals and 7.0% carbohydrates in a 100g edible flesh.

The root-knot nematode is the major limiting factor in successful production of watermelon.

3.4.6.1. Root-knot Nematodes, Meloidogyne spp.

Hasan and Jain (1998) reported that *M. incognita* was responsible for 18-33% loss in yield of watermelon.

(i) Management Methods

(a) Host Resistance: Watermelon cvs. Shehjanpuri was found resistant to *M. incognita*, while Dixie Queen was found resistant to *M. javanica*.

3.5. MALVACEOUS VEGETABLE CROPS

3.5.1. Okra

Okra (*Abelmoschus esculentus*) is an annual vegetable crop grown in tropical and sub-tropical regions. It is being cultivated in 0.358 million hectares producing 3.524 million tonnes of okra with an average yield of 9.8 tonnes per hectare. Bihar, West Bengal, Orissa, Andhra Pradesh, Gujarat, Maharashtra and Assam are major okra-growing states. Tender, green fruits are cooked in curry and soup. The root and stem are used for clearing cane juice in preparation of 'gur'. High iodine content of fruits helps to control goitre while leaves are used in inflammation and dysentery. It is a good source of vitamins A, B and C. The fruits also help in cases of renal colic, leucorrhoea and general weakness. The dry seed contains 13-22% good edible oil and 20-24% protein. The oil is used in soap, cosmetic industry and as vanaspati while protein is used for fortified preparations. The crushed seed is fed to cattle for more milk production and the fibre is utilized in jute, textile and paper industry.

Root-knot and reniform nematodes have been recognized as the major nematode pests of okra.

3.5.1.1. Root-knot Nematodes, Meloidogyne spp.

M. incognita was responsible for 28.08 to 90.90% loss in fruit yield of okra (Bhatti and Jain, 1977; Parvatha Reddy and Singh, 1981), while *M. javanica* caused 20.20 to 41.20% loss in yield (Jain *et al.*, 1986).

(i) Interaction with Other Pathogens: Root-knot nematodes cause collar rot disease complex in association with *Pythium aphanidermatum* and *Ozonium texanum* var. *parasiticum* in okra.

(ii) Management Methods

(a) Cultural Methods: Decrease in *M. incognita* population in okra occurs when the field was left fallow (Khan *et al.*, 1975). The maximum reduction in *M. javanica* population and increase in fruit yield was recorded in okra by soil ploughing + covering with polythene sheet, followed by soil ploughing + exposure to sun (Anon, 1989).

Crop Rotation: Decrease in root-knot nematode population in okra occurs following marigold, spinach and bottle gourd (Khan *et al.*,1975). Out of 15 different cropping sequences studied, tomato-onion-resistant tomato cv. Hisar Lalit-okra was found to be the best in terms of root-knot nematode reduction and economic returns (Kanwar, 1990). Crop rotations having okra-garlic-cluster bean and okra-coriander-resistant tomato cv. Hisar Lalit effectively managed the root-knot nematode population (Anon, 1993b).

Rotation of okra with sweet potato (cv. Sree Bhadra) reduced the root-knot nematode population by 21% and increased the fruit yield by 20.57% (Sheela *et al.*, 2002). Crop rotation with cabbage, marigold and wheat significantly reduced the root-knot nematode population. Crop rotation with cereals and marigold helps in reducing the nematode population in soil. In cropping sequence involving cabbage-marigold-wheat-kochia, the population of root-knot nematode larvae reduced to such an insignificant level that growing okra following the above sequence did not bring about increase in number of larvae (Alam *et al.*, 1977).

Intercropping: Atwal and Mangar (1969) showed that root exudates from sesame (*Sesamum orientale*) have nematicidal properties against *M. incognita*. When okra was grown in *M. incognita* infested soil it was only slightly attacked and there were fewer nematodes compared with when okra was grown in the absence of sesame.

Intercropping of okra with sesame decreased root-knot nematode penetration in okra and also delayed the life cycle of *M. incognita* (Tanda and Atwal, 1988). Similarly, intercropping of okra with marigold, margosa and Persian lilac have been recorded to reduce nematode population and root galling.

Botanicals: Singh and Sitaramaiah (1966) applied finely divided oil cakes to root-knot infested soil and noted a reduction in disease incidence in okra. Soil application of neem cake at 2 t/ha recorded lowest root-knot index (1.6), followed by fresh neem leaves at 5 t/ha (1.6) and silk worm litter at 2.5 t/ha (1.7). Significant highest yield was also obtained with neem oil cake (6.2 t/ha) followed by neem leaves (6.1 t/ha) (Sahoo *et al.*, 2004b).

Best results in respect of root-knot reduction due to *M. javanica* and increase in yield of okra were obtained by amending the soil with saw dust at 2.5 tonnes per ha 3 weeks before planting and then applying N through urea at 120 kg per ha (Singh and Sitaramaiah, 1971b).

Spot application of groundnut cake and sunflower cake each at 0.25 t/ha, saw dust and coal ash each at 5 t/ha significantly reduced the root-knot nematode population (Bhosle *et al.*, 2006).

Application of press mud at 15 t/ha, a week prior to sowing is recommended for effective management of root-knot nematodes in kharif season with cost: benefit ratio of 1: 3.1. Application of fresh neem or karanj cake at 1 t/ha at the time of final land preparation is recommended.

Furrow application of neem cake at 100g/2m furrow, spot application of neem cake at 15g/spot and spot and furrow application of mustard and castor cakes gave 58.88, 48.36, 35.86, 35.53, 34.87 and 31.91% increase in yield, respectively and reduced the nematode infestation (Talukdar, 1993).

Bala and Sukul (1987) observed that Eugenol, an active nematicidal principle from *Ocimum sanctum* showed systemic effect in reducing root-knot infection in okra. Ahuja and Mukhopadhyaya (1984) observed that application of neem leaves and seeds of Persian lilac maximally decreased the root-knot index due to *M. incognita* and significantly increased the yield of okra fruits.

(b) Chemical Methods: Seed treatment with aldicarb and carbofuran both at 6 and 12 per cent was effective against the root-knot and reniform nematodes infecting okra (Sivakumar *et al.*, 1976). Jain and Bhatti (1981) reported effective control of *M. javanica* on okra by seed treatment with phenamiphos at 2, 4 and 6 per cent.

Alam *et al.* (1973) reported that oxamyl was effective for the control of root-knot nematodes infecting okra. Aldicarb, carbofuran, ethoprophos and phenamiphos each at 1 to 2 kg a.i. per ha were found effective in reducing the root-knot nematode population and in increasing fruit yields of okra (56 to 87 per cent)

Drill application of carbofuran at 1.5 kg a.i./ha below the seed level reduces the incidence of *M. incognita* in okra.

(c) Biological Methods

Antagonistic Bacteria: Root-knot nematodes in okra can be managed by the application of *Bacillus macerans* at 1.2×10^8 cells/sq.m. before sowing.

Antagonistic Fungi: *P. lilacinus* reduced root-knot nematode egg masses by 91% after 60 days and by 96% after 90 days, while aldicarb reduced egg masses by 90 and 91% after 60 and 90 days after inoculation, respectively in okra. In general, *P. lilacinus* and aldicarb treatments greatly suppressed the numbers of nematode galls and egg masses on okra plants. However, it was evident that *P. lilacinus* was more effective in reducing nematode egg masses than aldicarb (Ibrahim and Rezk, 1988).

The soil incorporation of *P. lilacinus* (in rice hulls and bran soil mix) was effective in reducing the gall index (2.7 as against 5.0 in control), nematode population in soil (77.3% over control) and in increasing the number of okra fruits (121.7 as against 86.0 in control) and fruit weight (3.26 kg as against 2.11 kg in control) (Davide and Zorilla, 1986).

(d) Host Resistance: Okra cv. Red Wonder was reported to be resistant, while cv. Parbhani Kranthi was found moderately resistant to *M. incognita*. Clemson Spineless was reported to be moderately resistant to *M. arenaria*.

(e) Integrated Methods

Using Bio-agents and Botanicals: Integration of *Paecilomyces lilacinus* with neem cake gave effective control of *M. incognita* on okra.

Application of 5 per cent inoculum of *Arthrobotrys conoides* to the pot soil amended with FYM, effectively reduced the larval penetration of *M. incognita* and root galling was reduced by 30-40 per cent in okra (Srivastava and Swarup, 1986).

Soaking of okra seeds in 10% castor cake suspension mixed with spores of *P. lilacinus* (1.5×10^6 spores/ml) for 30 min. and sowing in soil drenched with 10% castor cake suspension at 20 litres/6 sq.m.) was effective in reducing root galling, final nematode population of *M. incognita* and increasing the fruit yield, root colonization, propagule density in soil and parasitization of eggs by *P. lilacinus* (Rao *et al.*, 1997a).

Using Botanicals and Chemicals: An integrated management of *M. incognita* infecting okra using neem or karanj oil cake at 0.5 t/ha along with carbofuran at 1 kg a.i./ha gave maximum reduction in root galling with consequent increase in okra fruit yield (Parvatha Reddy and Khan, 1991).

Application of subabool (*Leucaena leucophila)* leaves at 40 g/kg soil which are allowed to decompose for four weeks before sowing + carbofuran at 1 kg a.i./ha while sowing of okra seeds resulted in minimum galling (35.4 galls/plant) in comparison to control (64 galls/plant) and better plant growth parameters in okra (Paruthi *et al.*, 1987).

Using Bio-agents, Chemicals and Botanicals: The combined treatment with *P. lilacinus* at 4 g/kg soil + carbosulfan 25 EC at 0.2% + poultry manure at 2.5 t/ha + FYM at 2.5 t/ha gave maximum increase in plant growth parameters and yield (9.2 t/ha compared to 2.0 t/ha in control), and decrease in number of galls (83.5 compared to 183.6 in control), egg masses per root system (23.6 compared to 70.5 in control) and final nematode population in soil (200 compared to 585 in control) (Das and Sinha, 2005).

Using Cultural and Chemical Methods: Summer ploughing + seed treatment with carbofuran at 3% a.i. w/w + main field treatment with aldicarb at 1 kg a.i./ha led to 76-79% decrease in nematode population and 35.1% higher yield over untreated check. Further, Summer ploughing + mulching transparent polythene sheet + seed treatment with carbosulfon at 3% a.i. w/w led to 32.5% higher yield. Summer solarization + treated seeds + use of neem cake at 400 kg/ha was most effective treatment and gave 50% higher okra yield.

Using Botanicals and Physical Methods: Integration of soil solarization for 15 days in summer and application of neem cake at 200 kg/ha is effective in the management of root-knot nematodes and in getting higher yields.

Using Physical and Chemical Methods: Soil solarization with single layer of polyethylene mulch for 20 days during June and application of carbofuran at 0.5 kg a.i./ha gave least root galling (42/plant compared to 245/plant in control) (Sharma *et al.*, 2005).

3.5.1.2. *The Reniform Nematode, Rotylenchulus reniformis*

(i) Interaction with Other Pathogens: Okra plants succumbed to wilt at an early stage when the reniform nematode, *R. reniformis* and *R. solani* were present together (Kumar and Sivakumar, 1981).

(ii) Management Methods

(a) Cultural Methods: Application of potash in combination with phosphorus or nitrogen or potash alone checks the reniform nematode multiplication on okra to a great extent (Sivakumar and Meerazainuddin, 1974).

(b) Chemical Methods: Treatment of okra seeds with carbofuran at 1.25g a.i. w/w economized the nematode management by reducing the quantity of nematicide required (Sivakumar *et al.*, 1973).

(c) Biological Methods: Seed treatment with avermectin 100% recorded maximum fruit yield and highest reduction in *R. reniformis* population. Avermectin treated plants also recorded significantly increased total phenol content and enzymatic activities such as peroxidase and IAA oxidase (Jayakumar *et al.*, 2004).

(d) Integrated Methods: Application of aldicarb at 1 kg a.i./ha + neem cake at 0.5 t/ha followed by carbofuran at 1 kg a.i./ha + neem cake at 0.5 t/ha proved most effective in reducing the *R. reniformis* population and in increasing the growth of okra plants (Krishna Rao *et al.*, 1987).

Deep ploughing (20 cm) followed by fallowing for one month, fallowing for one month after weeding, integration of aldicarb application at 0.8 kg a.i. per ha at sowing after either of the cultural practices or deep ploughing (20 cm) together with carbofuran or aldicarb seed treatment resulted in the control of the reniform nematode (*Rotylenchulus reniformis*) and better yield of okra (Lakshmanan and Sivakumar, 1981).

3.5.1.3. Root-knot Nematode, *Meloidogyne incognita* and Wilt, *Fusarium oxysporum* f. sp. *vasinfectum* Disease Complex

(i) Management Methods

(a) Biological Methods: Among botanicals and biocontrol agents evaluated, *Trichoderma harzianum* was the most effective treatment in improving plant growth followed by neem seed powder, neem cake and *Aspergillus niger*. Highest suppression of root colonization of the fungus (*F. o. f. sp. vasinfectum*) was achieved in plants treated with neem seed powder (7.5), neem cake (12.5), *A. nizer* (15.0) and *T. harzianum* (17.5). The greatest suppression of root-knot development was achieved by *T. harzianum* followed by neem seed powder, neem cake and *A. niger* (Abuzar and Haseeb, 2006).

3.5.1.4. The Root-knot Nematode, *Meloidogyne incognita* and Root Rot, *Rhizoctonia solani* Disease Complex

(i) Management Methods

(a) Biological Methods: *Paecilomyces lilacinus* can be effectively used against *Rhizoctonia solani* – *M. incognita* complex of okra (Shahzad and Ghaffar, 1989).

3.6. BULBOUS VEGETABLE CROPS

3.6.1. Onion and Garlic

Onion (*Allium cepa*) is one of the most important commercial vegetables. The bulbs are consumed both in mature and tender stages. Even the leaves and stalks of inflorescence are used as vegetables. It is being cultivated in 0.594 million hectares producing 7.515 million tonnes of onions with an average yield of 12.7 tonnes per hectare. It is grown in western, northern as well as in southern India. Maharashtra, Gujarat, Bihar, Karnataka, Andhra Pradesh, Madhya Pradesh, Haryana and Rajasthan are major onion-growing states of India.

Garlic (*Allium sativum*) is mainly used for flavouring vegetables and meat dishes. It is carminative, stimulant and cures digestive disorders. It has been found to be very effective in curing diabetes and high blood pressure. It is being cultivated in 0.107 million hectares producing 0.436 million tonnes of garlic with an average yield of 4.1 tonnes per hectare. Madhya Pradesh, Gujarat, Orissa, Maharashtra, Uttar Pradesh and Rajasthan are major garlic-growing states. More than 50% production comes from Madhya Pradesh and Gujarat only.

Stem and bulb, root-knot and reniform nematodes have been recognized as the limiting factors in successful production of onion and garlic.

3.6.1.1. The Stem and Bulb Nematode, Ditylenchus dipsaci

The stem and bulb nematode cause great damage to the onion and garlic crops in temperate areas of many countries. The nematode is primarily observed in heavy soils. The nematode problem becomes serious with abundant precipitation or a high soil humidity. The nematode causes serious damage at inoculum level of 10 or more nematodes/500g of soil. *D. dipsaci* was responsible for 'bloat' disease of onion.

(i) *Symptoms:* The leaves of the infected onion plants become abnormally thickened and twisted; they become flabby, droop, and later turn yellow. In a more severe infestation, the sprouts die before reaching two-leaf stage. The inner scales of the bulbs are infested more often than the outer scales which frequently rupture. The bulbs eventually become soft and begin to rot.

The leaves of garlic plants turn yellow and die but no thickening or twisting is evident (Fig. 3.13). The infected garlic bulbs become loose (the cloves separate from the base) and cannot be used for seed even if they tolerate winter storage well because the plants which form often die during the vegetative period.

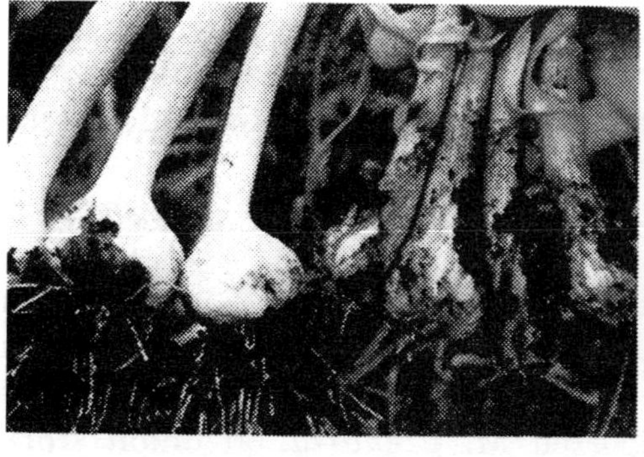

Fig. 3.13. The stem and bulb nematode (*Ditylenchus dipsaci*) on garlic.
A – Garlic crop in field infected with *D. dipsaci*. B – Left, healthy; Right, infected.
(Courtesy: Sasser, 1971).

(ii) Life Cycle: The fourth stage juveniles penetrate the stem and leaf tissue through the stomata. Egg laying begins at temperatures of 1-5°C with the optimum at 13-18°C. *D. dipsaci* completes one generation in 19-23 days at 15°C. Nematode activity stops at 36°C. The nematode prefers the cool moist climatic conditions existing in the upland tropics and wet winter seasons in the subtropics. *D. dipsaci* can parasitize plants in both heavy and light soils, although a higher incidence of infestation seems to occur on heavy soils.

(iii) Spread and Survival: The nematode can survive in the soil without a host plant for more than one year and the fourth juvenile stage can survive in anabiosis for many years. The nematode can be spread by transportation in infested bulbs, plant residue and adhering soil. Seed-borne infections also are responsible for long distance spread in onion. Other hosts and weeds are responsible for maintaining infestations between onion and garlic.

(iv) Management Methods

(a) Physical Methods: The nematodes can be exterminated by soaking the plants in water containing 0.5% formalin at 24°C, followed by a hot water treatment (with addition of 0.5% formalin at 49°C for 20 min.).

Soaking of onion sets or garlic cloves in cold water for 3 to 4 days will also kill the nematodes.

(b) Cultural Methods: The diseased plants and their remnants should be gathered and destroyed.

Bulbous plants, pea, carrot and beans should be excluded from the crop rotation system in order to control the onion race. Onion and garlic may be rotated with wheat, barley, maize, rape, white clover, red clover, lucerne or lettuce to overcome the stem and bulb nematode damage.

(c) Chemical Methods: Seed stocks should be disinfected with methyl bromide (30g of methyl bromide/m³ for 20 hr). Garlic cloves may also be disinfected by fumigation with methyl bromide (30g of methyl bromide/m³ for 8 hr) before autumn sowing.

Row application of thionazin at 4.5g/metre row also gives effective control of the stem and bulb nematodes.

3.6.1.2. Root-knot Nematodes, Meloidogyne spp.

Vadivelu and Rajendran (1986) reported *M. arenaria* on onion from Palladam district of Tamil Nadu. The infected plants showed patchiness and unthrifty growth with 'neck twisting' symptoms. The root system was very much reduced and bulbs were sleek and elongated.

(i) Symptoms: Symptoms of root-knot on onion are very discrete, the main symptom being the presence of the protruding egg masses.

(ii) Management Methods

(a) Physical Methods: Hot water treatment of onion bulbs at 43.5°C for 120 min eliminates the nematode infection.

(b) Cultural Methods: The lowest populations of plant parasitic nematodes were recorded in onion-cowpea-fallow cropping sequence (Vetrivelkalai and Subramanian, 2006).

(c) Integrated Methods: Root dip treatment of onion seedlings in *P. lilacinus*+neem cake suspension for 1 hr before transplanting proved effective.

3.6.1.3. The Reniform Nematode, Rotylenchulus reniformis

(i) Management Methods

(a) Host Resistance: Onion cv. Evergreen is reported to be resistant to the reniform nematode.

3.7. ROOT VEGETABLE CROPS

3.7.1. Radish

Radish (*Raphanus sativus*) is a root vegetable which can be grown as an intercrop with other crops. Its leaves are rich in vitamins A and C. It can be eaten raw and also cooked. It has some medicinal value. Radish is cultivated throughout India. West Bengal, Bihar, Uttar Pradesh, Punjab, Assam, Haryana, Gujarat and Himachal Pradesh are major radish-growing states.

The stunt nematode is recognized as the major limiting factor in successful production of radish crop.

3.7.1.1. Stunt Nematodes, *Tylenchorhynchus* spp.

(i) Management Methods

(a) Cultural Methods: The mustard (rabi), radish (summer), sesame (kharif) rotation sequence considerably reduced the population of *Tylenchorhynchus* spp. (Haque and Gaur, 1985).

3.7.2. Carrot

Carrot (*Daucus carota*) is another root vegetable rich in vitamin A and some minerals. It is grown all over India. Main carrot growing states are Uttar Pradesh, Assam, Karnataka, Andhra Pradesh, Punjab and Haryana. It is taken raw as well as in cooked form. It is made into pickles and sweetmeat. Carrot juice is a rich source of carotene (a precursor of vitamin A) and contains appreciable quantities of thiamine and riboflavin. It is sometimes used for colouring butter and other foods. Green carrot leaves are highly nutritive, rich in protein, minerals and vitamins and used as fodder and also for preparation of poultry feed.

The root-knot nematode is the major limiting factor in successful cultivation of carrot crop.

3.7.2.1. Root-knot Nematodes, Meloidogyne spp.

Devi (1993) reported that *M. incognita* was responsible for 56.64% loss in yield of carrots. Root-knot nematodes develop characteristic forking of the roots in carrots (Fig. 3.14).

Fig. 3.14. Forking of carrots incited by root-knot nematodes
(Courtesy: AICRP on Nematodes, Hisar).

(i) Management Methods

(a) Cultural Methods: The population of *M. incognita* larvae decreases in combination of spinach-marigold-carrot and spinach-radish-carrot (Hasan and Jain, 1998). Late sown carrots suffer less damage due to *M. incognita* in California, USA.

Botanicals: The oil cakes (neem, groundnut, mustard and castor) were effective in reducing parasitic nematode population (*Hoplolaimus, Tylenchorhynchus, Meloidogyne* and *Helicotylenchus* species) and in increasing yields of carrot.

In a field trial, saw dust, neem cake, poultry manure at 1.0 and 1.5 t/ha and water hyacinth at 1 and 2 t/ha increased yield of carrot (28-66%) by reducing the infestation of *M. incognita*. The maximum reduction of *M. incognita* in soil was recorded in neem cake (38.5%) followed by saw dust (23%), poultry manure (22%) and water hyacinth (14%) at higher dose over control (Devi, 1993).

(b) Biological Methods: Application of *P. lilacinus* (containing 1×10^6 spores at 10 g/ 2.5 sq. m.) recorded least gall index (1.6 as against 5.0 in control), reduced nematode population by 45.23%, reduced forking by 92.4% and increased the yield by 147.6% (Sivakumar, 1998).

(c) Host Resistance: Carrot cv. Arka Suraj is reported to be tolerant to the root-knot nematode (Swamy and Sadashiva, 2004). Among 103 germplasm screened, 10 genotypes namely New Kuroda, Selendro, Perfecta Elita, Zeno, Nantes Strong Top, Nantes, F-524, Danvers Half Long, Avenger and HYDC-4 were found resistant to *M. incognita* (Veere Gowda *et al.*, 2001).

(d) Integrated Methods: A soil fumigant at the beginning of two consecutive carrot crops, followed by two year onion production and at last a cover crop sudax in the course of 5 year rotation can control *M. hapla* problem (Bird, 1981).

Seed treatment with *T. harzianum* at 10 g (with 1×10^6 cfu/g)/kg and *P. fluorescens* at 10 g (with 1×10^9 cfu/g)/kg and subsequent field application of 5 tonnes of enriched FYM with *T. harzianum* (with 1×10^6 cfu/g) and *P. fluorescens* (with 1×10^9 cfu/g) per ha significantly reduced root-knot and reniform nematodes in carrot roots by 79 and 75%, respectively. The yield increase was to the tune of 29.8% with a cost: benefit ratio of 1: 13.6.

3.7.3. Turnip

Turnip (*Brassica rapa*) is grown in temperate, sub-tropical and tropical regions of India. It is extensively cultivated in Bihar, Haryana, Himachal Pradesh, Punjab and Tamil Nadu. Turnip is rich in vitamins A, B, C and Calcium and iron. It can be eaten raw.

Root-knot and reniform nematodes are recognized as the major limiting factors in successful production of turnip crop.

3.7.3.1. *Root-knot Nematodes, Meloidogyne spp.*

(i) Management Methods

(a) Cultural Methods: The oil cakes (neem, groundnut, mustard and castor) were effective in reducing parasitic nematode population (*Hoplolaimus, Tylenchorhynchus, Meloidogyne* and *Helicotylenchus* species) and in increasing yields of turnip.

3.7.3.2. *The reniform nematode, Rotylenchulus reniformis*

(i) Management Methods

(a) Host Resistance: Turnip cvs. Purple Top and White Globe are reported to be resistant to the reniform nematode.

3.8. LEAFY VEGETABLE CROPS

3.8.1. Lettuce

Lettuce (*Lactuca sativa*) is a very common cool season salad crop. Its leaves are rich in vitamin A (900 IU), C (10 mg), choline (178 mg) and minerals – calcium (50 mg) and phosphorus (28 mg).

Root-knot and reniform nematodes are recognized as the major limiting factors in successful cultivation of lettuce crop.

3.8.1.1. *Root-knot Nematodes, Meloidogyne spp.*

The root-knot nematode cause large decreases in yield of lettuce.

(i) Management Methods

(a) Physical Methods: The most important nematode pests controlled by greenhouse steaming are root-knot nematodes on lettuces.

(b) Cultural Methods: Planting of 4 and 5 seedlings of *Tagetes erecta* in a pot around a lettuce plant reduced root galling by 56 and 72% and nematode population by 73 and 94%, respectively (Pervez *et al.*, 1988).

(c) Chemical Methods: Soil fumigation with 98% methyl bromide + 2% chloropicrin (Dowfume MC-2) at 1 lb/100 cu. ft. of soil was best, both in terms of nematode control (0 galls as compared to 65 galls in control) and yield increase (82.5g/plant as compared to 46.4g/plant in control).

3.8.1.2. The Reniform Nematode, Rotylenchulus reniformis

(i) Management Methods

(a) Physical Methods: Under field conditions, soil solarization significantly reduced soil population of the reniform nematode up to 15 cm deep and increased the yield of lettuce.

3.9. MUSHROOMS

Mushrooms have achieved significant importance due to their nutritive and medicinal values. Mushrooms, the edible fleshy fungi, represent one of the few valuable sources of nutritive and palatable food. They are rich in proteins (35-48%), vitamins and minerals. They are weight reducers and useful for diabetic patients. Though, these fungi have been naturally occurring and being utilized in India since time immemorial, the commercial cultivation is only of recent times. It was only in 1964 that the first successful crop of mushroom was grown at Solan in Himachal Pradesh. Since then, research on various aspects of mushroom cultivation has been strengthened and today it is one of the important cottage industries, not only in the state of Himachal Pradesh but its commercial cultivation has also been initiated in most of the other states and union territories of India.

At present, it is estimated that 40,000 tonnes of mushrooms are produced annually. The major share (85%) is contributed by button mushroom (*Agaricus bisporus* and *A. bitorquis*). Oyster mushroom (*Pleurotus* spp.) is popular in southern states and Maharashtra and milky mushroom (*Calocybe indica*) in Tamil Nadu, Andhra Pradesh and Karnataka.

3.9.1. Mushroom Nematodes

An intensive survey of mushroom farms in India revealed that 84.4% nematode infestation, which led to extremely poor flushes in initial stages followed by total crop failures (Seth, 1984). Nematodes are most dangerous pests of mushrooms which once entered into the crop cannot be eradicated without completely destroying the crop. Their presence in the beds simply means very poor yields or total crop failures.

In India, so far eight species of *Aphelenchoides* besides *Ditylenchus myceliophagus* have been found to damage mushroom mycelium (Table 3.13).

Table 3.13. Fungal feeding nematodes associated with mushrooms from India

Nematode Order	Fungal feeding nematodes	References
Tylenchida	*Ditylenchus myceliophagus*	Thapa *et al.*, 1981
Aphelenchida	*Aphelenchoides agarici*	Seth & Sharma, 1986
	A. asterocaudatus	Bahl & Prasad, 1985
	A. composticola	Khanna & Sharma, 1988
	A. neocomposticola	Khanna & Sharma, 1988
	A. minor	Seth & Sharma, 1986
	A. myceliophagus	Seth & Sharma, 1986
	A. sacchari	Sharma *et al.*, 1981
	A. swarupi	Seth & Sharma, 1986

3.9.1.1. Aphelenchoides agarici

This species was recorded from mushroom farms in Himachal Pradesh (Seth & Sharma, 1986). It is one of the most destructive nematode species and multiplies freely on mushroom mycelium. An initial inoculum level of 10 nematodes has been shown to cause 100% mycelial depletion within 25 days at 28°C resulting over 800 times multiplication rate under laboratory conditions (Khanna & Sharma, 1988). It completes its life cycle within 8 days at 28±1°C and the ratio between female to male was 1. 5:1 (Seth, 1984).

3.9.1.2. Aphelenchoides composticola

It is one of the most widely distributed species present throughout the mushroom growing countries of the world. In India, it is known to occur in Himachal Pradesh and Punjab. This nematode also has a short life cycle of 8 days at 23°C and multiplies very fast on mushrooms (Cayrol, 1967). An initial inoculum of 10 nematodes completely depleted the mycelium within 40 days resulting in about 1000 times multiplication rate (Khanna & Sharma, 1988). *A. composticola* was responsible for 35-60% loss in yield of mushroom (Laqman Khan, 2001).

3.9.1.3. Aphelenchoides myceliophagus

It is recorded from Solan district in Himachal Pradesh. It is also one of the highly destructive nematode species infesting mushroom showing the pathogenicity status of the level of *A. composticola* (Khanna & Sharma, 1988).

3.9.1.4. Aphelenchoides neocomposticola

It was reported from Shimla district of Himachal Pradesh. The nematode, though highly pathogenic in nature, was found to be less damaging than *A. agarici*, *A. composticola* and *A. myceliophagus* (Khanna & Sharma, 1988).

3.9.1.5. Aphelenchoides sacchari

It was reported by Sharma *et al.* (1981) and found to be highly pathogenic to white button mushroom, *Agaricus bisporus*. An average of 50-100% yield loss in sporophores in white button mushroom has been reported due to this nematode species (Sharma *et al.*, 1984). It is quite prevalent on mushrooms in India (Thapa *et al.*, 1981). Singh *et al.* (2003) reported that *A. sacchari* was responsible for 40.6 to 100% loss in yield of mushroom.

3.9.1.6. Ditylenchus myceliophagus

This nematode is widely distributed in almost all mushroom growing belts of the world. In India, it is found in Himachal Pradesh, Jammu and Kashmir and Punjab (Thapa *et al.*, 1981). It is one of the highly destructive species. At an inoculum level of 10-1000 individuals could cause 15-70% mycelial depletion within 60 days (Sharma *et al.*,1985). However, it is considered to be less pathogenic than *A. composticola* and *A. sacchari* (Sharma *et al.*, 1981).

3.9.2. Nature of Damage

The mycophagus nematodes thrust their hollow stylet into the hyphal cell. The mycelium bleeds and its sap is sucked upon by to and pro movement of the stylet. After feeding upon one cell the nematode shifts to another cell with the help of the moisture film present in the compost.

3.9.2.1. Symptoms

Initially whiteness of spawn slowly starts turning brown followed by sparse and patchy appearance of the mycelium. This mycelium is stingy in nature. The compost surface sinks. Alternately, medium and low quantity of sporophores (fruiting bodies) is obtained in succession of flushes. There are extremely poor yields in the later flushes finally leading to total failure of sporophore production thus reducing the total yields as well as duration of the crop.

3.9.2.2. Losses

Economic losses due to decline in yields are generally very high. Extent of losses mainly depends upon the time of nematode infestation, nematode species involved and population of the nematodes. Losses caused by *A. sacchari* are only so far worked out and have been given in Table 3.14 (Sharma *et al.*,1984).

Table 3.14. Yield losses in mushroom due to *Aphelenchoides sacchari*

Time of inoculation	Initial inoculum	Yield loss (%)
Spawning	1×10^1	98.2
	1×10^2	100.0
	1×10^3	100.0
	1×10^4	100.0
Casing	1×10^1	90.6
	1×10^2	92.1
	1×10^3	87.3
	1×10^4	100.0

3.9.3. Management Methods

3.9.3.1. Physical Methods

Use of heat has been the most successful method of nematode control in mushroom cultivation during composting and casing process as well as at the end of cropping. The air as well as bed temperature should be maintained at 60°C for at least 2 hours during peak heating of compost. Thorough pasteurization of casing soil with steam for at least 4 hours gave good control of nematodes. Treatment of used trays and handling tools in boiling water for 1-2 minutes or in disinfectants like formalin is recommended.

An effective method of casing pasteurization, based on the use of solar heat, was developed wherein the thinly spread casing is exposed to the sun under a polythene cover (150 gauge) for 5-6 hours (Grewal and Grewal, 1988).

Wood and Goodey (1957) found that r-rays in between 48,000 and 96,000 r inactivated *Ditylenchus myceliophagus* in mushroom compost.

3.9.3.2. Cultural Methods

Incorporation of neem/pongamia/coconut cake at 2 kg/100 kg of compost during first or second turning is recommended. Dry leaf powder of neem can be mixed with compost at 2-5% concentration. This has been found useful not only to check nematode multiplication in mushroom growing trays but also helps in proper mycelial growth and improved sporophore yields (Khanna and Sharma, 1994). Addition of oil cakes like neem, karanj, coconut, castor and groundnut have been reported to control nematodes and increase button production (Rao *et al.*, 1992). Further, incorporation of fresh leaves of Pedilanthus, eucalyptus and pongamia in the compost during first or second turning was observed to reduce the population of mushroom nematodes and to increase the yield of button mushrooms. The incorporation of dried leaves of neem, castor and *Cannabis sativa* resulted in the suppression of *A. composticola* below economic injury levels and increased mushroom yields by 6.5 to 19% (Grewal, 1989).

3.9.3.3. Chemical Methods

Thionazin, applied at 80 ppm concentration, is quite effective in controlling the myceliophagous forms without producing any adverse effect on mycelial growth and also does not leave any residue in sporophores.

3.9.3.4. Biological Methods

Nematode trapping fungus, *Arthrobotrys robusta* commercially available as 'Royal 300' (strain Antipolis) in France has been recommended for the management of mushroom nematodes. *Arthrobotrys oligospora* and *A. superba* have also been found to reduce nematode populations and increase the yields. In India, *Arthrobotrys irregularis* and *Candelatrella musiformis* isolated from spent compost have shown promising results in checking nematode population (Khanna and Sharma, 1990). *A. irregularis* was also found to be highly effective against *Aphelenchoides composticola*. Further studies revealed that *A. irregularis* did not compete with *Agaricus bisporus* mushroom and checked the nematode population significantly and increased the yield.

Grewal and Sohi (1988) mass produced *Arthrobotrys conoides* on wheat grain and, after incorporating it into compost at spawning, reduced a mixed nematode population of *A. composticola* and *Rhabditis* sp. by 42.5% in 25 days.

The bacterium, *Bacillus thuringiensis* is found to suppress the population of *Ditylenchus myceliophagus* (Cayrol, 1974).

3.9.3.5. Host Resistance

Agaricus edulis has been reported to be resistant to *D. myceliophagus*. Thapa *et al.* (1987) reported that *Pleurotus sajor caju* and *Stropharia rugoso annulata* were found resistant to both *A. sacchari* and *D. myceliophagus*. K 30 strain of *A. bitorquis* showed resistance to *A. sacchari* only. Resistance in *Pleurotus sajor caju* against *A. sacchari, A. agarici* and *A. composticola* has been reported by several workers.

3.9.3.6. Integrated Methods

- Cropping should be done in purposely built mushroom houses with proper ventilation.

- Cleanliness should be maintained inside and surroundings of mushroom farm so that flies, mites and other contaminants are avoided by spraying 2% formalin solution.

- Composting platform must be cemented to prevent the direct contact of compost with soil.

- No person or worker should be allowed to enter into the farm without proper disinfestation of their hands and feet.

- Wooden trays and handling tools should be disinfected before use by dipping them in boiling water or in 5% formalin for 2 minutes before use.

- Compost should be steamed at 60°C for minimum of 2 hours before spawning.

- Casing material should be well sterilized before use. Formalin at 5% concentration can be used for this purpose.

- Irrigation water should be nematode-free.

- Proper cook-out of mushroom houses with steam at 70°C for 5-6 hours or at 80°C for 30-60 minutes is essential after the crop is over to avoid carrying of infestation to subsequent crops.

- The dry neem leaf powder may be mixed with compost at 2-5% concentration which not only check nematode multiplication in mushroom growing trays but also helps in proper mycelial growth and improved sporophore yield.

Table 3.15. Biological control of plant parasitic nematodes in vegetable crops.

Vegetable crop	Nematode	Effective bioagent/s	Reference
Tomato	*M. incognita*	*P. lilacinus*	Ekanayake & Jayasundara, 1994; Cannayane & Sivakumar, 2001; Sharma *et al.*, 1988
			Goutam *et al.*, 1995
			Sharma & Trivedi, 1989
		T. viride	Rekha Arya & Saxena, 1998
			Xiujuan *et al.*, 2000

	T. harzianum & *T. viride*	Nagesh *et al.*, 2001
	T. virens	Susan *et al.*, 2002b
	P. chlamydosporia	Gopinatha *et al.*, 2002 Rao *et al.*, 1998b
	G. fasciculatum	Mishra & Shukla, 1997 Hemavathi, 2002 Kantharaju *et al.*, 2002 Bagyaraj *et al.*, 1979 Suresh & Bagyaraj, 1984 Nagesh *et al.*, 1999
	G. mosseae	Sundarababu & Suguna, 1998
	P. fluorescens	Verma *et al.*, 1999 Santhi & Sivakumar, 1995
	P. penetrans	Rameshchand & Gill, 2002 Somasekhar & Gill, 1991 Walia & Dalal, 1994 Parvatha Reddy *et al.*, 1997
	P. stutzeri *B. thuringiensis*	Khan & Tarannum, 1999
	A. chroococcum	Chahal & Chahal, 2003
	S. avermitilis	Parvatha Reddy & Nagesh, 2002
M. javanica	*P. lilacinus*	Zaki, 1998 Khan & Esfahmi, 1990 Zaki & Bhatti, 1991a
	G. fasciculatum	Sundarababu *et al.*, 1993
	P. penetrans	Giannakiou *et al.*, 1999 Maheshwari & Mani, 1988 Rao *et al.*, 1997 Mankau, 1975
R. reniformis	*P. lilacinus*	Alan & Walters, 1994 Parvatha Reddy & Khan, 1988
	G. fasciculatum	Sitaramaiah & Sikora, 1982
	P. fluorescens	Niknam & Dhawan, 2001b
	B. subtilis	Niknam & Dhawan, 2001a

Brinjal	*M. incognita*	*P. lilacinus*	Pandey & Dwivedi, 2001
			Saikia & Das, 2001
			Trivedi, 1992
			Rao *et al.*, 1998d
		P. chlamydosporia	Naik, 2004
		G. fasciculatum	Sharma & Trivedi, 1989
			Jothi & Sundarababu, 2000
		G. mosseae	Borah & Phukan, 2000
			Jothi & Sundarababu, 2000
			Rao & Parvatha Reddy, 2001
		P. penetrans	Karuna *et al.*, 2001
			Singh *et al.*, 1998
		A. chroococcum	Chahal & Chahal, 1988
		S. avermitilis	Parvatha Reddy & Nagesh, 2002
	M. javanica	*P. lilacinus*	Zaki & Maqbool, 1991
		A. chroococcum	Bansal & Verma, 2002
		P. lilacinus	Parvatha Reddy & Khan, 1989
	R. reniformis	*P. lilacinus*	Pandey & Trivedi, 1992
		S. avermitilis	Parvatha Reddy & Nagesh, 2002
Chilli	*M. incognita*	*G. mosseae*	Sundarababu & Suguna, 1998
Bell pepper	*M. incognita*	*T. virens*	Susan *et al.*, 2002a
		P. chlamydosporia	Naik, 2004
Potato	*G. rostochiensis*	*T. harzianum*	Saifullah, 1996
	M. incognita	*P. fluorescens*	Mani *et al.*, 1998
	M. incognita & *G. pallida*	*P. lilacinus*	Jatala *et al.*, 1980
	G. rostochiensis & *Pratylenchus* sp.	*P. lilacinus*	Davide & Zorilla, 1995
	G. rostochiensis & *G. pallida*	*Cylindrocorpon destructus, F. solani, F. oxysporum*	Morgon Jones & Rodriguez Kabana, 1985
Okra	*M. incognita*	*P. lilacinus*	Davide and Zorilla, 1986.
		T. harzianum & *T. viride*	Lal *et al.*, 2002

		G. fasciculatum	Sharma & Trivedi, 1994
		G. mosseae	Sundarababu & Suguna, 1998
		P. fluorescens	Sobita Devi & Dutta, 2002
	M. javanica	*P. lilacinus*	Walia *et al.*, 1991
Carrot	*M. hapla*	*P. lilacinus*	Sivakumar, 1998
	R. reniformis	*P. fluorescens*	Rao & Shylaja, 2004
Pea	*M. incognita*	*B. thuringiensis*	Chahal & Chahal, 2003
Sugar beet	*H. schactii*	*P. fluorescens*	Oostendorp & Sikora, 1989

Chapter 4

ORNAMENTAL CROPS

India is known for growing of traditional flowers such as jasmine, marigold, chrysanthemum, tuberose, crossandra and aster. Commercial cultivation of cut flowers – rose, orchids, gladiolus, carnation, gerbera, anthurium and lilies – has also become popular. The important flower-growing states are Tamil Nadu, Karnataka, Andhra Pradesh, Haryana, Maharashtra, West Bengal, Gujarat and Delhi.

With the liberalization of Indian economy, the domestic floricultural trade, which hitherto remained an unorganized sector, became an organised sector with the active participation of corporate giants. The establishment of state-of-art hi-tech floricultural units in early 90's for intensive cultivation of cut flowers contributes about Rs. 200 crores of valuable foreign exchange today. Various government agencies are actively involved in the promotion of this sector and are highly committed to resolve the teething problems.

The world demand for floricultural products is approximately US $ 40 billions and cut flowers contribute nearly 60% (US$ 25 billions) to the global trade. India's share in the world flower market is about 0.7% which is contributed mainly from open cultivation.

Floriculture is estimated to cover an area of 116,000 ha with production of 655,000 MT of loose flowers. The area under cut flowers has increased significantly in recent years and the production has reached 1952 million numbers. Another 500 ha of climatic control greenhouse are available for growing quality flowers for export. Many corporate houses (more than 100) have set up 100% export oriented units with an investment of more than Rs. 1000 crores. With this, India is emerging as one of the powerful forces in world flower trade. India is bestowed with natural advantages like favourable weather conditions, soil, cheap labour which are likely to make a real mark in international trade.

Floriculture is an intensive type of agriculture wherein the income per unit area is much higher than any other agricultural product if it is done in a scientific way. The area used for production of ornamental plants is low, but the cash returns are high because of the intensive cropping. The flowers grown in large areas include jasmine, rose, marigold, chrysanthemum, crossandra, champak, tuberose, gladiolus and aster. New crops like lilies, tulips, anthurium, carnation, gerbera, etc., are also finding potential place in the foreign market. With the rapid development in urban planning and hotel industry, there has been an increasing demand for cut flowers and ornamentals. The cut flowers followed by live plants, cut foliage, seeds and bulbs have got an immense potential for export to Western countries where the cost of production is comparatively high. India exports roses, gladioli, carnations, chrysanthemums, lilies and jasmines to Gulf, Germany, France, England, Netherlands and Singapore. Our share in the export market has registered a sharp increase from Rs. 14.45 crores in 1991-92 to Rs. 210.99 crores in 2004-05.

Nematodes and their impact on Ornamental plants

Monoculture, the common feature of the ornamental crops, provides congenial environment for one or the other nematode, the population of which if unchecked may touch an alarmingly high level and may even wipe out the whole crop. Apart from this, even a low nematode population may render the planting stock like tuberose and gladioli bulbs, unmarketable. The infested planting material may also act as a carrier of some nematode genera/species of quarantine significance to newer areas. In USA, losses in ornamentals due to nematodes were estimated to the tune of $ 60 million annually (Hague, 1972). Throughout the world in ornamental crops, plant nematodes are responsible for 11.1% losses (Sasser and Freckman, 1987). However, in India, there is no precise information about the losses in ornamental crops alone.

Ornamentals like other crops, harbour a multitude of nematode fauna of which some nematodes based on their intrinsic potentials have caused tremendous economic losses. Nematodes like *Ditylenchus dipsaci* has been reported to be responsible for completely wiping out the narcissus industry of Britain (Hague, 1972). *Meloidogyne* spp. on gladioli, *Aphelenchoides ritzemabosi*, *Belonolaimus longicaudatus* and *Trichodorus* sp. on chrysanthemum in USA, *Xiphinema americanum* on roses in USA and *Xiphinema diversicaudatum* in Western Europe have been listed as the major pests resulting in considerable growth reduction (Hague, 1972).

A perusal of literature indicates that little information is available from India on these noxious pests. Hence, keeping in view the importance of these pests, a review at this stage would help to identify the progress made as also to pin-point our major deficiencies and gaps from the view point of practical approach to management of these pests.

4.1. ROSE

Rose (*Rosa* spp.) is the most ancient and popular flower grown the world over. In India, it is commercially grown for cut flowers and rose oil. Rose flowers without stems and loose flower petals are used in traditional markets for making garlands, for offering in temples, while the florist shops sell cut roses with stems mainly for bouquets and floral arrangements. In recent times, about 60 units have been established under joint ventures around Bangalore, Pune, Nasik, Mumbai, Hyderabad, Gurgaon (Haryana), Chandigarh and Saharanpur (Uttar Pradesh) for growing roses in greenhouses for export of cut flowers to Japan, The Netherlands, Germany and other European countries. Besides, the Damask rose (*Rosa damascena*) and Eduard rose (*R. bourboniana*) are cultivated for rose *attar* and other products – *gulkand, gulabjal,* and *pankkurj*. It is cultivated in about 6,000 ha area in the states of West Bengal, Uttar Pradesh, Gujarat, Haryana, Jammu and Kashmir, Tamil Nadu, Karnataka, Rajasthan, Madhya Pradesh, Andhra Pradesh and Punjab.

The lesion and root-knot nematodes are the major problems in rose cultivation. The unproductive flowers in roses were attributed to the attack of the ectoparasites – *Hemicycliophora labiata* and *Xiphinema basiri* (Muthukrishnan *et al.* (1975).

4.1.1. Lesion Nematodes, *Pratylenchus* spp.

The lesion nematodes are the major problems in rose cultivation. *P. zeae* was associated with poor growth of rose plants.

4.1.1.1. Economic Importance and Losses

Sundarababu and Vadivelu (1988a) observed the pathogenic potential of *P. zeae* on Edward rose. Prasad and Dasgupta (1964) found *P. pratensis* in rose plants exhibiting stunted growth.

4.1.1.2. Symptoms

The affected plants showed chlorotic symptoms, poor and stunted growth with necrotic lesions on rose roots. *P. zeae* was responsible for 69.6, 36.4, 59.6 and 33.3% reduction in shoot length, shoot weight, root length and root weight, respectively. Affected roots show patchy discolouration in cortical region, which coalesce and lead to root rotting.

4.1.1.3. Management Methods

(i) Physical Methods: Hot water treatment of rose seedlings at 50°C for 5 min. or 48°C for 10 min. eliminates *P. zeae* infestation.

(ii) Cultural Methods: Interplanting of roses with African marigold helps in reducing the population of *P. penetrans*.

(iii) Chemical Methods: Application of aldicarb at 4 oz/1000 sq. ft. was effective in reducing *P. penetrans* population and in increasing flower production by 16% (Johnson and McClanahan, 1974).

(iv) Host Resistance: Ohkawa and Saigusa (1981) found that *Rosa indica* (cv. Major) and *R. multiflora* (cv. 60-5) were resistant to *P. penetrans* and *P. vulnus*, respectively.

4.1.2. Root-knot nematodes, *Meloidogyne* spp.

M. incognita was recorded on rose (*Rosa indica*) from Aligarh (Alam *et al.*, 1973) and *M. javanica* from Delhi (Prasad and Dasgupta, 1964).

4.1.2.1. Management Methods

(i) Physical Methods: Hot water treatment of rose roots at 45.5°C for 60 min. eliminates root-knot nematode infestation.

(ii) Chemical Methods: Dipping of bare roots of rose in 0.1% solution of phenamiphos for 30 min. gave considerable reduction in root-knot infection (Dale, 1973).

(iii) Host Resistance: Ohkawa and Saigusa (1981) reported that rose cv. Mavetti to be resistant against *M. hapla*.

(iv) Integrated Methods: Pre-treatment of soil with dazomet (at 25 g/m^2) followed by soil amendment with neem cake (at 1 kg/ m^2) along with *Pochonia chlamydosporia* (at 2 x 10^{10} spores/ m^2) recorded reduced root galling and increased flower yield (Nagesh and Janakiram, 2004).

4.2. CARNATION AND GERBERA

Carnation (*Dianthus caryophyllus*) flower is valued for its excellent keeping quality, wide array of colours and forms, and ability to rehydrate after continuous transportation. Moderate climatic control measures that are economical can deliver quality carnations at the internationally competitive prices year round. In India, Carnation and gerbera are commercially grown mostly under cover around big cites like Solan, Shimla, Mandi, Kullu, Chandigarh, Ludhiana, Guragaon, Kalimpong, Pune, Bangalore and Delhi. Carnation has tremendous potential for export market.

Gerbera (*Gerbera jamesonii*), commonly known as Transvaal daisy, Barberton daisy or African daisy is an important flower grown throughout the world for beds, borders, pots and rock gardens. The flowers are of various colours and suit very well for floral arrangements. The cut blooms also have long vase life. In India, it is distributed in the temperate Himalayas from

Kashmir to Nepal at altitudes of 1300 to 3200 meters above sea level. Gerbera belongs to the family Asteraceae and is native to South African and Asiatic regions. To meet the quality standards of export market, it has to be grown under naturally ventilated low cost polyhouses.

The root-knot nematode is recognized as the major limiting factor in successful production of carnation and gerbera crops.

4.2.1. The Root-knot Nematode, *Meloidogyne incognita*

The root-knot nematode, *M incognita* is one of the serious limiting factors in commercial cultivation of carnation and gerbera under polyhouse conditions. Nagesh and Parvatha Reddy (1994) reported that *M incognita* was responsible for 27 and 31% loss in flower yield of carnation and gerbera, respectively.

4.2.1.1. Economic Importance and Losses

Most of the highly fetching exotic cultivars of Carnation and Gerbera from Europe have shown 40 to 60% mortality in polyhouse beds due to root-knot nematode infection in and around Bangalore (Nagesh and Parvatha Reddy, 1996a).

4.2.1.2. Symptoms

The root-knot infected Carnation and Gerbera plants exhibit stunted growth, leaf yellowing and premature dropping and root galling (Fig. 4.1).

Fig. 4.1. Root galling on carnation due to root-knot nematodes.

4.2.1.3. Interaction with Other Pathogens

M. incognita interacts with soil-borne fungi such as *Fusarium oxysporum* f. sp. *dianthi* and *Rhizoctonia solani* in causing wilt/root rot complex in Carnation and Gerbera.

4.2.1.4. Management Methods

(i) Cultural Methods: Application of neem cake at 1 kg per m^2 gave effective control of *M. incognita*.

(ii) Chemical Methods: Application of carbofuran at 2 kg a.i. per ha gave effective control of *M. incognita*.

(iii) Biological Methods: An experiment carried out for the biological control of root-knot nematode *M. incognita* in carnation (*Dianthus caryophyllus*) showed that, though the growth parameters were found to be at par in plants treated with *P. lilacinus* and *P. chlamydosporia*, significantly higher reduction of root galling was achieved with *P. chlamydosporia*. Thus, it was concluded that under polyhouse conditions, *P. lilacinus* and *P. chlamydosporia* were better bio-agents for the control of root-knot nematode compared to *T. harzianum* (Shylaja, 2004).

(iv) Host Resistance: Carnation cvs. Desco, Castelaro, Kappa, Rara, Logupink, Target and Antalia were highly resistant to *M. incognita*.

Screening of the four commercially available varieties of carnation for their resistance to root-knot nematode, *M. incognita,* revealed that *cv.* Ivonne was the most susceptible while, Randez was moderately tolerant. The other two varieties, Dark Randez and Sunshine were observed to be moderately susceptible (Shylaja, 2004).

(v) Integrated Methods

Bioagents and Botanicals: Integration of neem cake at 0.5 kg per plant with *T. viride* at 100 g per m^2 was effective in the management of root-knot nematodes and increased flower yields.

Integration of *P. lilacinus/T. harzianum* at 0.5 l per m^2 (aqueous spore suspension containing 2 x 10^4 spores per ml) with neem cake at 0.5 kg per m^2 or fenamiphos at 2 g a.i. per m^2 increased plant growth parameters and flower yield of carnation and gerbera. The above treatments also increased root-knot egg parasitization by the parasitic fungi (Nagesh and Parvatha Reddy, 1996a) (Fig. 4.2).

Three antagonistic fungi viz., *T. harzianum, V. lecanii* and *P. lilacinus* at 2 x 10^4 spores/ml in combination with neem cake reduced *M. incognita* population in both soil and roots of carnation.

Fig. 4.2. Management of root-knot nematodes on carnation by integration of bioagents and botanicals. Left – Untreated; Right – Treated.

Application of *P. lilacinus* at 0.5 g/kg of soil along with neem cake at 1.0 t/ha efficiently suppressed the nematode population and checked its build up, enhancing the plant growth parameters resulting in better production with increased flower stalk length and flower diameter in carnation. The plants also came to flower early (Johnson, 2000).

Arbuscular Mycorrhizal Fungi and Bioagents: Combined inoculation of AMF and *Pseudomonas fluorescens* had positive effect on root-knot nematode control on carnation (Anusuya and Vadivelu, 1999).

Bioagents, Botanicals and Chemicals: Pre-plant treatment of beds with dazomet (at 40 g/m^2) followed by the application of neem cake (at 1 kg/m^2 15 days later) along with antagonistic fungi, *P. chlamydosporia/ P. lilacinus* (at 2 x 10^{12} spores/ m^2) significantly reduced root-knot nematode population (*M. incognita*), mortality of plants and suppressed the nematode infection for 2 years. The antagonistic fungi established better in the beds treated with dazomet. The above treatment also reduced root galling, nematode multiplication rate and plant mortality, and increased spike/stem length, flower yield and root colonization with the bioagents (Nagesh and Parvatha Reddy, 2005).

4.2.2. Root-knot nematode, *Meloidogyne incognita* and Wilt, *Fusarium oxysporum* f. sp. *dianthi* Disease Complex

In an experiment carried out to study the role of root-knot nematode, *M. incognita* in predisposing carnation to *Fusarium* wilt, it was observed that the appearance of the wilt symptoms were accelerated when *M. incognita* was inoculated two weeks prior to *F. oxysporum* f. sp. *dianthi*. The rate of wilting was observed to be 2.1 and 4.6, respectively during 12th and 25th week of observation, while the root galling index (RGI) values were recorded to be

4.15 and 4.75, respectively. It was observed that maximum mortality was recorded when *M. incognita* was inoculated 2 weeks prior to *F. oxysporum* f. sp. *dianthi* which was 33.4% and 79.2%, respectively, at 12th and 25th week of observation. The plant growth was also reduced significantly due to prior inoculation of *M. incognita* (Shylaja, 2004).

Ferraz and Lear (1976) studied the interaction of four plant parasitic nematodes (*Circonemoides curvatum, Pratylenchus dianthus, Rotylenchus robustus* and *Meliodogyne hapla*) and *F. oxysporum* f. sp. *dianthi* in carnations. They reported that plants wilted earlier and more severe symptoms were present with the nematode-fungus combination than with the fungus alone. All nematodes except *R. robustus* showed synergistic interaction with *Fusarium*.

4.2.2.1. Management Methods

(i) Integrated Methods: The studies carried out to evaluate combination of bio-agents for the biological control of wilt (*F. o.* f. sp. *dianthi*) and root-knot nematode (*M. incognita*) disease complex in carnation, revealed that a combination of *P. chlamydosporia* + *P. lilacinus* gave the best control of wilt and root knot disease complex (Shylaja, 2004).

4.3. CHRYSANTHEMUM

Garden chrysanthemum (*Chrysanthemum* spp.), also known as 'Queen of the East', is a highly versatile and accommodating ornamental with a wide range of type, size and colour in flowers. In India, large flowered varieties are grown for cut flower, making garland, wreaths and 'veni', religious offerings and for bedding and potting purposes. Its commercial cultivation is being done in Maharashtra, Rajasthan, Madhya Pradesh and Bihar. Small flowered varieties are commercially grown for loose flowers in Tamil Nadu, Karnataka and Maharashtra. In India, chrysanthemum is grown in about 4,000 ha.

The lesion and the bud and leaf nematodes are major problems in chrysanthemum cultivation.

4.3.1. The Lesion Nematode, *Pratylenchus coffeae, P. chrysanthus*

Edward *et al.* (1969) reported *P. chrysanthus* associated with root rot of chrysanthemum.

4.3.1.1. Symptoms

Rashid and Khan (1975) reported that *P. coffeae* is responsible for heavy root damage which subsequently led to poor growth of chrysanthemum.

Stunting of plants with premature yellowing and drying of leaves are common above ground symptoms. Flower size is also reduced.

4.3.1.2. Histopathology

P. coffeae infects and remains confined to the cortical cells. Parenchymatous cells are completely destroyed due to nematode feeding.

4.3.1.3. Interaction with Other Pathogens

The presence of *P. coffeae* has been found to aggravate the disease severity of *Pythium aphanidermatum* and *Rhizoctonia solani* (Hassan, 1988). Edward *et al.* (1969) reported *P. chrysanthus* associated with root-rot of chrysanthemum.

4.3.1.4. Management Methods

(i) Integrated Methods: Application of neem cake enriched with *T. harzianum* at 1 t/ha is effective in reducing the nematode population.

4.3.2. Bud and Leaf Nematode, *Aphelenchoides ritzemabosi*

A. ritzemabosi is a major pest and causes considerable damage to the foliage of chrysanthemum, zinnia, salvia, aster and dahlia (Gill and Sharma, 1976).

4.3.2.1. Symptoms

In early infection, *A. ritzemabosi* feeds ectoparasitically on buds and growing points causing retardation of growth and distorted, deformed leaves with roughened feeding areas. Later it invades leaves and feeds on and destroys cells of parenchymatous tissue of the mesophyll resulting in leaf spots or blotches, easily seen first on the underside; gradually these areas turn brown and then black as infestation spreads. The veins hinder nematode movement within leaves and thus the characteristic symptoms take the form of interveinal discolouration (Fig. 4.3). These are pale green at first, gradually becoming yellow and then dark brown or black. The infested leaves shrivel, hang down as the leaf dies. The nematodes escape through stomata and migrate in surface water films to infect terminal flower buds which produce deformed and under-sized blossom.

4.3.2.2. Life Cycle

In chrysanthemum leaves, a female lays about 25-30 eggs in a compact group. Eggs hatch in 3-4 days and the larvae take 9-10 days to reach maturity. The life cycle takes 10-13 days (Wallace, 1960).

Fig. 4.3. Chrysanthemum leaves infected with *Aphelenchoides ritzemabosi* (Courtesy: Sasser, 1971).

4.3.2.3. Management Methods

(i) Physical Methods: Hot water treatment of stools at 46°C for 5 min. eliminates nematode infestation.

Sufficient chrysanthemum stools were immersed in hot water at 48°C to bring the temperature down to 46.6°C, no further heating being necessary, and stools were removed after 5.5 min. when the temperature was down to 46.1°C.

(ii) Cultural Methods: Thorough cleaning up and burning of all infected plant material is essential in all control programmes. Plants propagated from top cuttings selected from plants growing under dry conditions offer an opportunity to secure relatively clean stock for foundation plantings. Avoiding wetting and overlapping of leaves of adjacent plants, banding the base of stems with petroleum jelly or tree-banding grease and destroying all plant remains and fallen leaves are some of the practices recommended for reducing the nematode inoculum. For outdoor plantings, the ground can be kept free of nematodes by a weed free fallow of 2-3 months during winter.

(iii) Chemical Methods: Dipping of chrysanthemum stools in 0.03% parathion for 20 min. was most effective for eliminating *A. ritzemabosi*. Foliar application of methyl parathion, chloropyriphos and quinalphos each at 0.05% were effective in the control of *A. ritzemabosi* infecting zinnia and chrysanthemum (Gill, 1981; Gill and Walia, 1980). Drenching with 0.02% thionazin at 6.8 litres/m^2 is also effective.

(iv) Host Resistance: Chrysanthemum cvs. Amy Shoesmith, Orange Beauty and Orange Peach Blossom are comparatively resistant.

4.3.3. Root-knot Nematodes, *Meloidogyne* spp.

M. arenaria and *M. javanica* have been reported on chrysanthemum (Sen and Dasgupta, 1977; Chandwani and Reddy, 1967).

4.3.3.1. Management Methods

(i) Biological Methods: Talc and pesta granules of *P. lilacinus* applied at 10 and 15 kg/ha to root-knot infested field reduced root galling (RKI=1.4 to 2.7), nematode multiplication rate (Rf=1.4 to 2.12) and enhanced bioagent propagules in rhizosphere (700-1070), bioagent infectivity (32-52%) and chrysanthemum flower yield by 23 to 28% (Nagesh *et al.*, 2003).

4.3.4. The Sting Nematode, *Belonolaimus longicaudatus*

4.3.4.1. Interaction with Other Pathogens

Chrysanthemum roots inoculated with *B. longicaudatus* and *Pythium aphanidermatum* develop symptoms of root rot earlier and more extensively than those inoculated with the fungus alone.

4.4. CHINA ASTER

China aster (*Callistephus chinensis*) is one of the important annual flower crops. It ranks next to chrysanthemum and marigold among traditional flowers. In recent years it has gained more popularity due to its multifarious uses including as cut flower. The flowers fetches very good price and returns to the farmer when their production coincides with their demand during particular season/festival. It is more popular among small and marginal farmers due to its easy cultivation. It is grown for cut flowers, herbaceous borders, bedding, bouquet and pots and cultivated in most parts of the country. It is commercially grown in Karnataka, Tamil Nadu, Andhra Pradesh, Maharashtra (Pune) and West Bengal. In Bangalore and Pune alone it is being grown in an area of 500 and 400 ha, respectively.

The root-knot nematodes are recognized as the major limiting factor in the successful production of china aster crop.

4.4.1. Root-knot Nematodes, *Meloidogyne* spp.

The root-knot nematodes are a major limiting factor both in nursery and main field (Krishnappa *et al.*, 1980; Khan and Parvatha Reddy, 1992).

4.4.1.1. Management Methods

(i) Host Resistance: China aster cv. Shashank was found resistant, while Poornima was moderately resistant to root-knot nematodes. There were significantly higher total phenols and total proteins and higher activity of polyphenol oxidase in infected roots of resistant variety, compared to that in susceptible variety (Kamini) during early stages of infection. As the infection advanced the activity of phenyl alanine ammonia lyase was higher in the roots of resistant variety (Nagesh *et al.*, 1995).

4.5. GLADIOLUS

Gladiolus (*Gladiolus* spp.) is very much liked for its majestic spikes containing attractive, elegant and delicate florets. These florets open in sequence over a longer duration and hence have a good keeping quality of cut spikes. It is cultivated for cut flowers, garden and interior decoration, making bouquets, bedding, rockeries and pots. West Bengal, Maharashtra, Uttar Pradesh, Punjab, Haryana and Andhra Pradesh are the major gladiolus-growing states.

Root-knot nematodes are recognized as the major limiting factors in successful production of gladiolus crop.

4.5.1. Root-knot Nematodes, *Meloidogyne* spp.

The root-knot nematodes are major limiting factors in the production of gladiolus.

4.5.1.1. Economic Importance and Losses

Parvatha Reddy *et al.* (1979) reported *M. incognita* causing heavy damage to gladiolus.

4.5.1.2. Symptoms

Severe galling on roots results in yellowing of leaves which subsequently leads to stunted growth. The nematode invades roots, daughter corms and cormels which develop after flowering.

4.5.1.3. Survival

The nematodes would survive in corm tissue in the soil as a source of inoculum for next season (Parvatha Reddy *et al.*, 1979).

4.5.1.4. Interaction with Other Pathogens

Incidence of gladiolus scab caused by *Pseudomonas marginata* was found to increase the population of *M. javanica* (Elgoorani *et al.*, 1976).

4.5.1.5. Management Methods

(i) Physical Methods: Hot water treatment of corms at 58°C for 30 min. eliminates root-knot nematode infection.

(ii) Cultural Methods: Nematode-free planting material should be used for planting. Crop rotation or intercropping with marigold helps in reducing the nematode population.

(iii) Chemical Methods: Dipping of gladiolus corms in thionazin or fensulfothion solution (0.5 g a.i. per litre) gave reduced root-knot nematode infestation (Overman, 1970). Application of Vorlex (35 gal/acre) by broadcast method also resulted in better flower yield (Overman, 1967).

(iv) Integrated Methods: Soil application of neem cake enriched with *T. harzianum* at 1 t/ha is effective.

Application of 5 tonnes of FYM enriched with *P. fluorescens* (with 1×10^9 cfu/g) per ha significantly reduced *M. incognita* by 61-76%, *R. reniformis* by 65-70% in the roots of gladiolus and increased the flower yield by 19-22%.

4.5.2. Root-knot nematode, *Meloidogyne incognita* and Wilt, *Fusarium oxysporum* f. sp. *gladioli* Disease Complex

4.5.2.1. Management Methods

(i) Biological Methods: Application of *P. fluorescens* effectively controlled the disease complex leading to significant improvement in plant growth and flowering (Mustafa and Khan, 2004).

(ii) Integrated Methods: Gladiolus plants treated with *P. lilacinus* + *T. harzianum* + neem cake and *P. lilacinus* + *T. viride* + neem cake combinations not only controlled *M. incognita* infection but also *Fusarium* wilt (*F. oxysporum* f sp. *gladioli*) till the harvest of flower spikes. The corms and cormels obtained from plants treated with these combinations were free from *Fusarium* infection. Fungal colonization of galled roots was maximum in *P. lilacinus* + *T. viride* combination (94%) followed by *P. lilacinus* + *T. harzianum* combination (69%). Similarly, the parasitization of eggs was maximum

in *P. lilacinus* + *T. viride* combination (44%) followed by *P. lilacinus* + *T. harzianum* combination (38%), while egg mass parasitization was same in both the combinations (58%) (Nagesh *et al.*, 1998).

4.6. TUBEROSE

Tuberose (*Polianthes tuberosa*) is one of the most important bulbous ornamentals grown both for cut and loose flowers, and also for the extraction of its highly valued natural flower oil. It is also grown for garden decoration in pots, beds and borders. It is commercially grown in Karnataka, West Bengal, Maharashtra, Punjab, Andhra Pradesh and Tamil Nadu. At present, the total area under tuberose cultivation in the country is estimated at about 3,000 ha.

Root-knot and bud and leaf nematodes are recognized as the major limiting factors in successful production of tuberose crop.

4.6.1. Root-knot Nematodes, *Meloidogyne* spp.

Jayaraman *et al.*(1975) found that root-knot nematodes as the major factors in tuberose decline.

M. incognita, M. javanica and *M. arenaria* have been reported as the major limiting factors in successful tuberose cultivation in Tamil Nadu (Sundarababu and Vadivelu, 1988b), while *M. incognita* and *M. javanica* were potential pests in Karnataka (Khan and Parvatha Reddy, 1992) (Fig. 4.4).

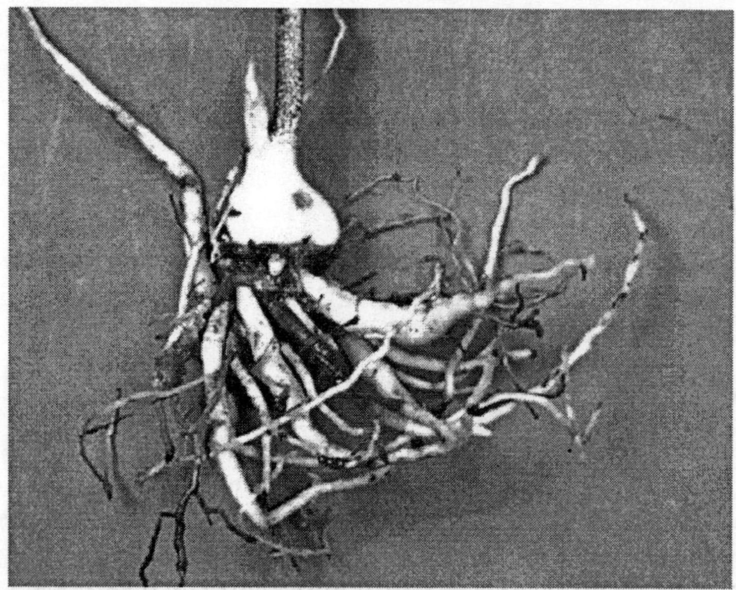

Fig. 4.4. Root-knot nematode on tuberose.

4.6.1.1. Economic Importance and Losses

M. incognita was responsible for 13.25, 9.87, 14.30, 13.78 and 28.58% reduction in plant weight, number of flowers, spike length, spike weight and number of bulblets, respectively (Khan and Parvatha Reddy, 1994).

4.6.1.2. Symptoms

Affected plants exhibit stunting, yellowing and drying up of leaves and rotting of bulbs (Jayaraman *et al.*, 1975). Further, the emergence of side shoots from the bulbs was also affected and the numbers were conspicuously less. In severely infected plants, the emergence of spike is suppressed and 65% reduction in top growth occurred (Sundarababu and Vadivelu, 1988b).

4.6.1.3. Management Methods

(i) Physical Methods: Hot water treatment of bulbs at 49°C for 60 min. eliminates root-knot nematode infection.

(ii) Cultural Methods: Soil application of neem cake at 1 t/ha is effective.

(iii) Chemical Methods: For the control of *M. incognita* on tuberose, aldicarb, fensulfothion and phorate each at 1.5 to 2.5 kg a.i. per ha were effective and increased flower yield. Application of carbofuran at 2g a.i./plant also checked the nematode effectively.

(iv) Host Resistance: Tuberose cv. Shringar (Single type) was found resistant, while Suvasini (Double type) was tolerant to *M. incognita* (Nagesh *et al.*, 1995).

Among the six varieties of tuberose (*Polyanthes tuberosa*) screened for their resistance to root-knot nematode *M. incognita, cv.* Mexican Single was tolerant under screenhouse experiments in sterilized soil, while under field conditions it was found to be moderately tolerant. Both, under screenhouse and field conditions, *cv.* Vybhav and Suvasini were found to be moderately susceptible, while *cv.* Shringar and Pearl Double were found to be moderately tolerant under screenhouse conditions, but moderately susceptible under field conditions. In case of *cv.* Prajwal, it was observed to be moderately susceptible under screenhouse conditions, but susceptible under field conditions (Shylaja, 2004).

(v) Biological Methods: The experiment carried out for the biological control of root-knot nematode (*M. incognita*) in tuberose showed that, *P. chlamydosporia* and *T. harzianum* treatments gave significantly higher control of nematodes than *P. lilacinus* (Shylaja, 2004).

(vi) Integrated Methods

Arbuscular Mycorrhizal Fungi and Botanicals: Integration of arbuscular mycorrhizal fungi (AMF) such as *Glomus mosseae* or *G. fasciculatum* with neem cake or/and aldicarb gave effective management of root-knot nematodes infecting tuberose (Khan and Parvatha Reddy, 1994).

Bioagents and Botanicals: Integration of *P. lilacinus* with neem cake gave effective control of *M. incognita* on tuberose. Treatment of tuberose bulbs with neem cake extract mixed with *P. lilacinus* spores significantly reduced *M. incognita* infection and multiplication besides stimulating the plant growth. Soil application of neem cake enriched with *T. harzianum* at 1 t/ha is also effective.

Split application (at planting and 45 DAP) of neem cake at 1 kg/1.5 m^2 in combination with *P. lilacinus* (at 25 g/1.5 m^2 containing 18 x 10^8 spores/g) significantly reduced root galling, soil and root population of *M. incognita* and its multiplication rate. The above treatment also significantly increased number of spikes, spike length and weight, number of flowers/spike, parasitization of egg masses and root colonization by *P. lilacinus* (Nagesh *et al.*, 1998).

Bulb dip treatment with 5% neem leaf extract mixed with spores of *P. lilacinus* (2 x 10^4 spores/ml) for 30 min. combined with soil drench with 5% neem leaf extract (at 50 ml/plant) 30 DAP gave significant increase in plant growth, flower yield and parasitization of eggs and egg masses and least root galling and reproduction factor (Nagesh *et al.*, 1997).

Two Bioagents: Combined application of *P. chlamydosporia* + *T. harzianum* gave significantly higher control of *M. incognita* on tuberose (Shylaja *et al.*, 2004). Soil application of *T. harzianum* (10^9 cfu/g) + *P. lilacinus* (10^6 cfu/g) gave effective control of root-knot nematodes.

4.6.2. Bud and Leaf Nematode, *Aphelenchoides besseyi*

4.6.2.1. Management Methods

(i) Physical Methods: Cent percent mortality of *A. besseyi* was obtained by immersing tuberose bulbs in hot water bath maintained at 50°C for 30 minutes (Khan *et al.*, 2004).

(ii) Host Resistance: Tuberose cvs. Suhasini, Shringar, Prajwal and Hyderabad Double were tolerant to *A. besseyi* (Khan *et al.*, 2004).

(iii) Integrated Methods: Pre-soaking of tuberose bulbs in water overnight followed by hot water treatment at 50°C for 20 minutes + spraying of the crop twice with monocrotophos 36 EC at 0.15% (at sprouting and 30 days after first spray) in the first year and 3 sprays in second and third year crop at 15-

20 days interval was found most effective in reducing % nematode infestation as well as disease indices (Khan *et al.*, 2004).

4.6.3. Root-knot nematode, *Meloidogyne incognita* and Wilt, *Fusarium oxysporum* f. sp. *dianthi* Disease Complex

In an experiment undertaken to assess the impact of *Fusarium* wilt-root knot disease complex on the flower yield in tuberose, it was observed that the disease complex drastically reduced the yield of flowers in *P. tuberosa*, thereby bringing down the production of flowers in this commercial crop. It was found that the damage was maximum in presence of both pathogens *viz; M. incognita* and *F. oxysporum* f. sp. *dianthi*, when compared to the presence of either one of these pathogens (Shylaja, 2004).

4.6.3.1. Management Methods

(i) Integrated Methods: In course of the experiments carried out to evaluate combination of bio-agents for the control of wilt (*F. o.* f. sp. *dianthi*) and root-knot nematode (*M. incognita*) disease complex in tuberose, the best result was obtained in plants treated with *P. chlamydosporia + T. harzianum* (Shylaja, 2004).

4.7. CROSSANDRA

Crossandra (*Crossandra undulaefolia*) is mainly cultivated commercially for its loose flowers in Tamil Nadu, Karnataka and Andhra Pradesh. It is also grown in Sri Lanka, Tropical Africa and Madagascar. It is a perennial herb and belongs to the family Acanthaceae.

Pratylenchus delattrei, Meloidogyne incognita and *Longidorus africanus* are the major nematode pests of crossandra.

4.7.1. The Root-knot Nematode, *Meloidogyne incognita*

In India, *M. incognita* was first reported on crossandra by Rajendran *et al.* (1976).

4.7.1.1. Economic Importance and Losses

The nematode is responsible for 25.62% and 21.64% loss in number of flowers and weight of flowers, respectively (Khan and Parvatha Reddy, 1994).

4.7.1.2 Occurrence and Distribution

Root-knot nematodes are found in all crossandra growing areas like Tamil Nadu, Karnataka and Andhra Pradesh.

4.7.1.3. Symptoms

Infected plants are stunted with dried peripheral branches bearing smaller chlorotic leaves almost turning to white in later stages (Rajendran *et al.*, 1976). Roots exhibit severe galling. Inflorescences are small and sometimes fail to produce flowers.

4.7.1.4. Interaction with Other Pathogens

M. incognita interact synergistically with wilt fungi *Fusarium oxysporum* and *F. solani* and results in premature death of crossandra plants (Khan and Parvatha Reddy, 1992). *F. solani* and *F. oxysporum* are invariably associated with the lesions caused by *P. delattrei* in crossandra (Khan and Parvatha Reddy, 1992; Srinivasan and Muthurksihanan, 1975). Srinivasan (1974) found that wilt symptoms appeared earlier when nematodes were added two weeks prior to *F. solani*.

4.7.1.5. Management Methods

(i) Cultural Methods: Deep ploughing and fallowing in summer for about a month helps in reducing the nematode population. Planting crossandra after graminaceous crops should be avoided. Incorporation of FYM in soil and intercropping of crossandra with marigold or pangola grass would reduce the root-knot nematode infection.

Intercropping of crossandra with enemy plant like marigold significantly reduced the root-knot and lesion nematode population both in soil and roots (Khan and Parvatha Reddy, 1994).

(ii) Chemical methods: Application of carbofuran at 2 to 3 kg a.i. per ha would be beneficial.

(iii) Biological Methods: The parasitic fungus, *Paecilomyces marguandi* grown on paddy seeds at 2 g per kg soil gave effective control of root-knot nematode infecting crossandra (Khan and Parvatha Reddy, 1994).

Among four arbuscular mycorrhizal fungi evaluated, *Glomus fasciculatum* and *G. mosseae* colonized crossandra roots better with comparatively lower nematode infection and multiplication, thus promoting better root health and plant growth compared to *G. intradices* and *Aucalospora laevis*.

(iv) Integrated Methods

Bioagents and Botanicals: Incorporation *V. lecanii* with neem cake facilitated the effective management of *M. incognita* on crossandra. Application of neem cake enriched with *T. harzianum* at 2 kg/m^2 (2 t/ha) in nursery beds gave effective control.

Combinations of *V. lecanii* and *P. lilacinus* (2 x 10⁴ spores/ml each) with 5% neem leaf extract resulted in significantly higher plant growth parameters and crossandra flower yield (Nagesh and Parvatha Reddy, 1995). Root galling, nematode multiplication factor were least and the parasitization of eggs and egg masses was highest in *V. lecanii* + 5% neem leaf extract.

Arbuscular Mycorrhizal Fungi and Botanicals: *G. fasciculatum* and *G. mosseae* in combination with neem cake gave better control of nematodes over carbofuran treatment. The root colonization of AM fungi increased significantly in the presence of neem cake which in turn improved their efficacy in reducing *M. incognita* population in crossandra roots (Nagesh *et al.*, 1998). Integration of neem, karanj and castor cakes with AMF, *G. mosseae* significantly enhanced plant growth parameters and flower yield of crossandra, root colonization and sporulation of AMF. The above treatments also reduced root-knot nematode multiplication and root-galling (Nagesh and Parvatha Reddy, 1997).

Bioagents, Arbuscular Mycorrhizal Fungi and Botanicals: Integration of a bioagent (*V. lecanii*), endomycorrhiza (*G. mosseae*) with botanicals improved the growth of crossandra and reduced the population of *M. incognita*. *G. mosseae* reduced the requirement of phosphatic fertilizer and favoured the antagonistic potential of *V. lecanii* against *M. incognita*.

Two Bioagents: Combined application of *P. lilacinus* and *Pasteuria penetrans* enhanced plant growth and flower yield of crossandra besides reducing root galling due to *M. incognita* (Nagesh *et al.*, 1995).

Bioagents and Botanicals/Chemicals: Integrated management of root-knot nematodes infecting crossandra was achieved by rational combination of a biocontrol agent (*T. harzianum* or *P. chlamydosporia*) with a nematicide (aldicarb) or oil cake (neem cake) (Khan and Parvatha Reddy, 1994).

4.7.2. The Lesion Nematode, *Pratylenchus delattrei*

The lesion nematode was first recorded by Srinivasan (1974) on crossandra from Tamil Nadu.

4.7.2.1. Economic Importance and Losses

Heavily infected plants do not produce tertiary spikes and thereby the flower yield is reduced by nine-fold (Srinivasan and Muthukrishnan, 1975).

4.7.2.2. Occurrence and Distribution

P. delattrei has been reported on crossandra from Tamil Nadu and Rajasthan.

4.7.2.3. Symptoms

P. delattrei was frequently associated with crossandra crop exhibiting stunting, chlorosis of leaves and wilting. The leaves showed mottled appearance which turned to brown and becoming pinkish eventually. Later, the leaf tips dried and entire leaves turned yellow and shed. The plants were devoid of side shoots and remained completely defoliated. The roots exhibited brown to black lesions. Heavily infested plants did not produce tertiary spikes and thereby flower yield is reduced (Srinivasan and Muthukrishnan, 1975).

4.7.2.4. Host Range

Besides crossandra, *P. delattrei* has been recorded on sorghum, maize, red gram, okra, sunnhemp and wheat.

4.7.2.5. Life Cycle

The total life cycle of this nematode from egg to adult stage ranges between 46 to 55 days. Eggs are laid inside the roots and all stages from larvae to adult are infective.

4.7.2.6. Histopathology

Nematode moves intracellularly and remains in cortical region. Its infection results in dissolution of cell walls which leads to cavity formation in cortical region. Cell wall gets thickened and cell contents become more dense and granular.

4.7.2.7. Interaction with Other Pathogens

Fusarium solani and *F. oxysporum* are invariably associated with the lesions caused by *P. delattrei* (Khan and Parvatha Reddy, 1992). Srinivasan and Muthukrishnan (1975) also opined that crossandra decline in Tamil Nadu occurs due to *P. delattrei* and *F. solani* complex. Srinivasan (1974) found that wilt symptoms appeared earlier when nematodes were added two weeks prior to *F. solani*.

4.7.2.8. Survival

The nematode can survive up to eight months in host-free moist soil, whereas in desiccated soil it could survive up to three months.

4.7.2.9. Management Methods

(i) Cultural Methods: Intercropping of crossandra with enemy plant like marigold significantly reduced the lesion nematode population both in soil and roots (Khan and Parvatha Reddy, 1994). Application of farm yard

manure has been found to reduce the lesion nematode population in crossandra.

(ii) Chemical control: Nursery bed treatment with phorate, aldicarb, fensulfothion and carbofuran each at 5g per m^2 gave good and healthy crossandra seedlings free from *P. delattrei* (Vadivelu and Muthukrishnan, 1979).

Root dip treatment for 8 hr in 0.05 to 0.1% solution of aldicarb, carbofuran, fensulfothion, phenamiphos or phorate protected crossandra seedling from *P. delattrei* for 30 days after transplanting (Vadivelu and Muthukrishnan, 1979).

Srinivasan and Muthukrishnan (1976) found that carbofuran at 2.5 kg a.i. per ha was effective under field conditions. Application of phorate at 1 g per plant a week after planting checked the lesion nematodes effectively and increased the flower yield (Vadivelu and Muthukrishnan, 1979).

(iii) Biological Methods: Application of *T. viride* at 10 g/kg soil was found effective to maximize the flower yield and suppressing the lesion nematode population both in soil and roots (Jothi *et al.*, 2004).

4.7.3. The Needle Nematode, *Longidorus africanus*

In India, this nematode was reported for the first time on crossandra by Vadivelu *et al.* (1976) from Tamil Nadu.

4.7.3.1. Symptoms

In crossandra, *L. africanus* causes stunting of plants, branching, swelling and darkening of root tips (Muthukrishnan *et al.*, 1977).

4.7.3.2. Host range

Sorghum, snap bean, lima bean, sugar beets are reported as best hosts, while wheat, cotton, okra, cucumber, egg plant and tomato as good hosts (Kolodge *et al.*, 1987).

4.7.3.3. Histopathology

L. africanus initiates hyperplasia in parenchymatous cells of peripheral region which subsequently progress towards the centre, forking of roots and clubbing of root terminal. Due to this, meristem is pushed laterally which ultimately leads to formation of darkly stained cells (Muthukrishnan *et al.*, 1977).

4.7.3.4. Management Methods

(i) Cultural Methods: Pea, carrot, squash, spearmint and onion are reported to be poor hosts; while cauliflower, cabbage and radish as non-hosts.

Either of these crops can be fitted as intercrops or in crop rotation scheme. Rotation of crossandra with spearmint which is a poor host of *L. africanus* would reduce the needle nematode population (Kolodge *et al.*, 1987).

Table 4.1. Biological control of plant parasitic nematodes in ornamental crops

Crop	Nematode	Effective bioagent/s	Reference
Tuberose	*M. incognita*	*P. chlamydosporia, P. lilacinus, T. harzianum*	Shylaja, 2004
Crossandra	*M. incognita*	*P. lilacinus, V. lecanii*	Nagesh & Parvatha Reddy, 1995
	M. incognita	*P. margaundi*	Khan & Parvatha Reddy, 1994
	M. incognita	*G. mosseae*	Nagesh *et al.*, 1996b
Carnation	*M. incognita*	*P. lilacinus*	Shylaja, 2004

Future Lines of Investigation on Nematode Pests of Ornamental Crops

More emphasis needs to be placed on the following aspects in future:

- There is considerable lacuna in our knowledge about nematodes associated with ornamental crops. Intensive and systematic surveys of ornamental crops should be conducted with the dual objectives of determining the incidence, prevalence and severity of such diseases and the geographical distribution of the nematode involved. This would go a long way in developing an advisory diagnostic service for the farmers.

- Adequate emphasis should be given to studies on the biology and host-parasite relationship of major nematode pests which may lead to formulation of control methods on sound basis.

- Studies have to be carried out on biotypes and on intraspecific variation in nematodes already known to be of economic importance in ornamental production in India.

- Techniques for precise determination of damage thresholds of populations and assessment of crop losses have to be standardized.

- Work on disease complexes involving nematodes be intensified with adequate collaboration between nematologists and plant pathologists.

- Data has to be generated on crop rotation as a method of nematode management. The effect of deep summer ploughing, field sanitation and other cultural practices, hot water treatment of planting material, are the areas which needs more attention in the coming years.

- Regulatory and quarantine measures against introduction of exotic nematode pests needs to be strengthened.

- The use of neem products for the management is highly economical since it can be used for seed dressing, nursery bed treatment, seedling bare-root dip treatment or foliar application. Neem products form ideal components in the integrated nematode management. Their effectiveness has also been proved against disease complexes involving nematodes and fungi. Research on the use of neem products for the management of nematodes on ornamental plants needs intensification.

- There is a great need to develop varieties which are resistant or tolerant to nematodes. Varietal screening and subsequent breeding programmes should be intensified. Fundamental investigations in relation to biochemical and physiological basis of resistance may also be taken up.

- Attempts should be directed towards biological control. *Paecilomyces lilacinus, Pochonia chlamydosporia, Pasteuria penetrans*, VAM fungi have been identified all over the world as potential biocontrol agents against plant nematodes. The possibility of using these biocontrol agents against nematodes infecting ornamental crops should be explored.

- Development of integrated nematode management models for ornamental crops should be taken up. Research on the use of tolerant/ resistant cultivars, parasitic fungi and bacteria, organic amendments/ plant products and vesicular arbuscular mycorrhizae should be evaluated and if acceptable to the farmers should be popularized through demonstration plots.

- Creation of increased awareness among farmers regarding nematode problems on ornamentals crops through extension services, literature, audio-visuals in local languages, demonstrations in farmer's fields and training programmes; strengthening of various extension programmes in nematology under lab-to-land and training and visit systems have to be explored.

Chapter 5

MEDICINAL CROPS

In medicinal plant sector, the WHO has estimated that about 80% of the population in developing countries rely on traditional medicines mostly plant drugs for primary health-care needs. Even modern medicines contain about 25% drugs derived from plants. A large number of people are earning their livelihood from different activities of medicinal plants such as collection, cultivation, marketing, processing, etc. The area under cultivation of the above crops is low, but the cash returns are high because of intensive cropping.

In recent times, due to increasing realization of health hazards and toxicity caused by synthetic drugs (side effects), there has been a renewal of interest in the use of plant based drugs throughout the world. A campaign is going on worldwide to utilize more and more plant derived chemicals in the human health care/personal care system. As a result, the demand for plant-based raw materials has increased enormously in both national and international markets. Herbal plants are finding diverse uses as raw materials, not only for medicines but also as biopesticides.

Medicinal crops form one of the important groups due to their demand in various pharmaceutical industries and also to earn foreign exchange by way of export. The potential for earning foreign exchange by India from the exports of medicinal and aromatic plants is estimated over US$ 3,000 million per annum. The export of these plants and their products has a tremendous potential in advanced countries like Europe, USA and Japan. The international market for medicinal plants related trade is estimated at US$ 60 billion/year having a growth rate of 7% per annum. Destination of foreign tourists is now aimed at India for health treatment. Kerala and Gujarat have become their favourite destinations since Ayurvedic treatment is very popular in these two states.

Unless we start cultivating the important medicinal plants, it will not be possible to meet the increased demands. However, cultivation of medicinal plants has to be taken up in polyculture models rather than in monoculture models. There is a huge task to be carried out at every level to make the Indian system of medicine a viable, sustainable and modern. For this task, an urgency is felt which can only be established through regular and systematic scientific research on collection, conservation, evaluation and development of good agricultural practices to compete in the world market.

Plant parasitic nematodes are one of the important limiting factors in the production of medicinal crops. Monoculture, the characteristic feature of these crops, provides congenial environment for one or the other nematode, the population of which if unchecked may touch an alarmingly high level and may even wipe out the whole crop. Besides, new agro-technologies like cropping system approach, high fertilizer regimes and irrigation schedule also provides favourable conditions for preponderance of one or the other nematode species.

5.1. ASHWAGANDHA

Ashwagandha (*Withania somnifera*) is a plant of immense medicinal importance growing all over north-western and central India. Roots of this plant are the major source of alkaloids and steroidal lactones (withanoids), which are intensively used in various pharmaceutical industries. It has adaptogenic, immuno-modulator, aphrodisiac, anti-stress and mildly sedative properties. Its roots yield valuable drugs which are employed in rheumatic pain, inflammation of joints, nervous disorders, cough, cold, epilepsy, and cardio and nerve tonic. At present, 4,000 – 5,000 ha area is under its cultivation mainly in Madhya Pradesh and neighbouring districts (Kota and Charu) of Rajasthan.

Root-knot nematodes are recognized as the major limiting factors in successful production of ashwagandha crop.

5.1.1. The Root-knot Nematode, *Meloidogyne incognita*

The root-knot nematode, *M. incognita* is an important limiting factor in the successful cultivation of ashwagandha.

5.1.1.1. Symptoms

The infected plants were chlorotic, stunted, less branched with fewer and smaller leaves. Roots of such plants were severely galled.

5.1.1.2. Management Methods

(*i*) *Integrated Methods:* Integration of neem cake with *T. harzianum,* vermicompost with *T. harzianum,* and cow urine with *T. harzianum* considerably reduced the root-knot nematode development and enhanced plant growth and yield. Maximum root-knot suppression was noticed in vermicompost with *T. harzianum* followed by mentha distillate with *Glomus aggregatum*. Highest increase in plant yield was recorded when the soil was amended with mentha/curry leaf distillates along with *T. harzianum/ G. aggregatum* (Pandey and Kalra, 2005b) (Table 5.1).

Table 5.1. Effect of organic materials and bioagents on plant growth of Ashwagandha infected with *Meloidogyne incognita*

Treatment	Dry weight of plant (g)	Root-knot index (RKI)	Total nema popn (soil + roots)
Untreated – Uninoculated	8.5	---	---
Untreated – Inoculated	5.2 (-38.8)	4.00	6460
Carbofuran	8.0 (-5.9)	1.66	3012
Neem comp + Davana	23.8 (+62.4)	1.66	2820
Neem comp + Curry leaf	7.4 (-12.9)	1.33	2261
Neem comp + Vermicompost	12.0 (+41.2)	1.66	2649
Neem comp + *T. harzianum*	12.8 (50.6)	1.66	2674
Neem comp + *G. aggregatum*	14.3 (+68.2)	3.00	2964
Neem comp + *Mentha* distillate	14.2 (67.1)	1.66	2409
Davana + Curry leaf	10.8(+27.1)	1.33	2030
Davana + Vermicompost	14.1 (65.9)	3.00	2594
Davana + *T. harzianum*	9.6 (+12.9)	1.66	2110
Davana + *G. aggregatum*	10.3 (+21.2)	1.33	2050
Davana + *Mentha* distillate	13.2 (+55.3)	1.33	2150
Curry leaf + Vermicompost	14.0 (+64.7)	1.66	2440
Curry leaf + *T. harzianum*	15.0 (+76.5)	3.33	3252
Curry leaf + *G. aggregatum*	13.4 (+57.6)	3.00	2635
Curry leaf + *Mentha* distillate	15.3 (+80.0)	1.66	2404
Vermicompost + *T. harzianum*	14.4 (+69.4)	0.66	1400
Vermicompost + *G. aggregatum*	13.6 (+60.0)	1.99	2480
Vermicompost + *Mentha* distillate	13.9 (+63.5)	1.33	2000
G. aggregatum + *Mentha* distillate	15.6 (+83.5)	1.00	1784
CD (P = 0.05)	0.71	0.01	465.70

5.2. SARPAGANDHA

Sarpagandha (*Rauvolfia serpentina*) is a perennial native Indian herb. Its roots are used for controlling high blood pressure and certain forms of insanity in Ayurvedic system of medicine since ancient times. It has received worldwide recognition after isolation of its bioactive reserpine alkaloid in allopathic medicines. In addition, its sedative property is utilized by Ayurvedic physicians in treatment of insomania, epilepsy and asthama. It is commercially cultivated in Madhya Pradesh, Uttar Pradesh, West Bengal, Assam and Orissa. The root is its economic part, containing 55 alkaloids of which reserpine, deserpidine, ajmalicine, serpentine and yohimbine being pharmacologically active and important. Of these, ajmalicine, reserpine and serpentine are highly potent in medicine. The commercial root crop contains 1.4-3.0% of total alkaloids.

The root-knot nematode is recognized as the major limiting factor in successful cultivation of sarpagandha crop.

5.2.1. Root-knot Nematodes, *Meloidogyne* spp.

5.2.1.1. Management Methods

(*i*) *Integrated Methods:* Soil application of neem cake enriched with *T. harzianum* at 1 t/ha is recommended.

5.3. COLEUS

Coleus (*Coleus forskohlii*) belongs to the Tulsi family Lamiaceae and is indigenous to India. It is used in Ayurvedic medicines and as a condiment. It shot into prominence in modern medicine with the isolation of the diterpinoid, forskolin, from its tubers. The therapeutical properties of forskolin in the treatment of glaucoma, congestive cardiomyopathy, asthma and certain cancers combined with its use in cosmetics has enhanced importance of coleus in modern medicine. Coleus proved to be the exclusive source of forskolin and therefore, the sudden spurt in the demand for its tubers led to exploitative collection from natural stand. The drug is claimed to improve appetite, facilitate digestion, increase vitality and is useful in anemia and inflammation. Recently it is also being used in the treatment of alcoholic addiction. Forskolin has a multiple biological activities like positive iontropic, antihypertensive, branchospasomolytic, antithrombotic, platelet aggregation inhibiting and adenylate cyclase stimulation. The species is now considered endangered and a prime candidate for cultivation, more especially captive cultivation.

Root-knot nematodes are recognized as the major limiting factors in successful production of coleus crop.

5.3.1. The Root-knot Nematode, *Meloidogyne incognita*

The root-knot nematode infestation was reported on coleus from Kerala and Orissa. The dry weight of the tubers was reduced by 20% due to root-knot nematodes. The percentage of starch on fresh weight basis showed drastic reduction (16%) in the infested tubers. Senthamarai *et al.* (2006c) reported that *M. incognita* was responsible for 70.2% loss in tuber yield of coleus.

5.3.1.1. Symptoms

The galls on coleus roots are very big and pronounced. The root-knot nematode damage often leads to crop failure. The infested tubers swell in size with irregular surface and cracking of the skin. When the infestation is severe, rotting sets in even before harvest. Infested tubers rot after harvest and rarely reach market.

5.3.1.2. Interaction with Other Pathogens

Simultaneous inoculation of *M. incognita* and *Macrophomina phaseolina* as well as nematode followed by fungus 15 days later, caused 100% root rot disease and significant reduction in plant growth parameters in Coleus (Senthamarai *et al.*, 2006a).

5.3.1.3. Management Methods

(i) Cultural Methods: Sree Bhadra, a high yielding variety of yam released by Central Tuber Crops Research Institute, Trivandrum, is identified as a resistant trap crop for the root-knot nematode. The nematodes were able to penetrate the root but giant cell formation was not induced and as a result the nematodes die inside the roots. Planting this variety in root-knot nematode infested field helped in clearing the field free of nematodes and giving good tuber yield. Subsequently growing of susceptible crop like coleus escaped nematode damage (Mohandas, 2001).

(ii) Chemical Methods: Treatment with DBCP and fensulfothion reduced damage to 4.0 and 5.7%, respectively as against 21% in control. The above treatments also increased the production of healthy tubers to the extent of 53 and 47%, respectively (Pillai, 1976).

(iii) Biological Methods: Soil application of *Trichoderma harzianum* and *Pseudomonas fluorescens* each at 2.5 kg/ha recorded increased plant growth and reduced *M. incognita* population (Senthamarai *et al.*, 2006b).

(iv) Integrated Methods: Integration of soil solarization in the nursery for 15 days with 150 gauge LDPE film and application of *Paecilomyces lilacinus* + neem cake or *P. lilacinus* + *Bacillus macerans* in the main field are the best treatments in increasing plant height (64.3 and 60.3 cm compared to 40.0 cm in control), number of leaves (593.3 and 583.3 compared to 310.0 in control), weight of tubers/plant (560.0 and 546.6 g compared to 350.0 g in control), yield (11.5 and 11.3 kg/plot compared to 6.9 kg/plot in control) and in reducing root galls (0.3 and 1.0 compared to 50.6 in control), nematode population in soil (25.0 and 30.0/100 ml soil compared to 196.6/100 ml soil in control) and roots (1.0 and 1.6/5 g roots compared to 79.0/5 g roots in control) (Nisha and Sheela, 2006).

5.4. KACHOLAM

Kacholam (*Kaempferia galanga*), also known as Acangi/Chandramuli, is a glabrous annual medicinal and aromatic herb with very fragrant underground rhizomatous plant. It belongs to the family Zingiberaceae. The plant is generally abundant in China, Sub-tropical Himalayas, Nepal and Kumaon.

Roots are bitter, thermogenic, acrid, carminative, aromatic, depurative, diuretic, expectorant, digestive, antihelmintic, febrifuge and stimulant. Rhizomes are used in dyspepsia, leprosy, skin diseases, rheumatism, asthma, cough, bronchitis, wounds, ulcers, malarial fever, inflammatory tumour and nasal obstruction. The rhizomes also contain resimic acid. Leaves are used for pharyngodynia, ophthalmia, swellings, fever and rheumatism. Oil showed tranquillizing activity of short duration.

The root-knot nematode is recognized as the major limiting factor in successful production of kacholam crop.

5.4.1. The Root-knot Nematode, *Meloidogyne incognita*

The root-knot nematode was reported to cause severe damage to the medicinal plant Kacholam (Sheela *et al.*, 1996). *M. incognita* was responsible for 64% loss in yield of kacholam (Sheela and Rajani, 1998).

5.4.1.1. Management Methods

(i) Physical Methods: Hot water treatment of rhizomes at 55°C for 5 min was effective in reducing the nematode population (more than 74%) and giving 81% increase in yield (Nisha and Sheela, 2004).

(ii) Cultural Methods: Rhizome treatment with neem leaf extract (4%) + garlic extract (1%) and application of neem cake at 200 g/m² gave maximum reduction of nematode population in soil (69%), roots (21.5 galls/plant compared to 83.25 galls/plant in control) and root galling (RKI=1.25 compared to 2.25 in control) (Nisha *et al.*, 2001).

Mulching of kacholam with green leaves of neem and chromolaena at 5 kg/m² 15 days before planting rhizomes, increased the rhizome yield by 160.7% and 143.3% over control, respectively. The above treatments also reduced root galling (1 as compared to 5 in control), the nematode population both in soil (81.8% and 80.4%, respectively) and roots (89.60% and 86.14%, respectively) (Sreeja *et al.*, 2001) (Table 5.2).

Table 5.2. Effect of mulching of green leaves on root-knot nematodes and yield of kacholam

Treatment	Root-knot index	Nema popn. (100g soil)	% redn. over control	Nema popn. (5g root)	% redn. over control	Rhizome yield (kg/4 m²)	% increase over control
Neem leaf	1	45.44 (6.74)*	81.80	7.00 (2.76)	89.60	5.6	160.7
Glyricidia	2	52.48 (7.24)	79.05	14.67 (2.84)	78.21	3.9	79.6
Mangium	5	178.26 (13.35)	28.50	33.67 (5.87)	49.99	2.6	21.6
Clerodendron	2	95.35 (9.92)	60.60	22.70 (4.82)	6.33	3.4	56.4
Calotropis	2	85.86 (9.27)	65.60	16.00 (4.03)	76.24	3.6	69.8
Chromolaena	1	48.96 (6.99)	80.40	9.33 (3.18)	86.14	5.3	143.3
Control	5	249.37 (15.79)	---	67.33 (8.26)	---	2.2	---
CD (P=0.05)	---	(1.02)	---	(1.45)	---	1.28	---

*Figures in parentheses are values after square root transformation.

(iii) Biological Methods: Rhizome treatment with *Pseudomonas fluorescens*, *Trichoderma viride* and AMF each at 3% w/w significantly reduced the nematode population (more than 74%) and improved plant growth and yield (Nisha and Sheela, 2004). Application of *Glomus fasciculatum* at 200 spores/plant gave maximum reduction of nematode population in soil (43.55%), roots (28.5 galls/plant compared to 83.25 galls/plant in control) and root galling (RKI=1.75 compared to 2.25 in control) (Rajani *et al.*, 1998).

5.5. CHINESE POTATO

Chinese Potato (*Solenostemon rotundifolius*), also known as coleus, is one of the minor tubers used for edible purpose. Its tubers are used as vegetable.

It is grown in Kerala, Tamil Nadu and Karnataka. It is also cultivated on a small scale in north-eastern states.

Root-knot nematodes are recognized as the major limiting factors in successful production of Chinese potato crop.

5.5.1. The Root-knot Nematode, *Meloidogyne incognita*

5.5.1.1. Management Methods

(*i*) *Physical Methods:* Hot water treatment of Chinese Potato tubers at 53°C for 10 minutes eliminated *Meloidogyne* spp. infection.

(*ii*) *Biological Methods:* Dipping of plants in 3% solution of *Bacillus macerans* + soil drenching with 2% solution of *B. macerans* 7 days after planting recorded maximum reduction in nematode population and root-knot count (14 galls/g of roots) and gave maximum increase in yield (Sheela *et al.*, 2004).

5.6. HENBANE

Henbane (*Hyoscyamus niger, H. muticus, H. albus*) is a popular ancient herb. Henbanes are one of the chief sources of tropane alkaloid viz. hyoscine, hyoscyamine and atropine. The hyoscine and its derivatives have been used in pharmaceutical preparations, since they are having anticholingeric, antispasmodic and mydriatic properties. It is used in the treatment of asthama and whooping cough. It also provides relief in gripping pain in intestinal disorders. In India, its small scale cultivation is done in Malwa region in Madhya Pradesh.

The root-knot nematode is recognized as the major limiting factor in successful cultivation of henbane crop.

5.6.1. The Root-knot Nematode, *Meloidogyne incognita*

Henbane (*H. niger* and *H. muticus*) were severely infested with *M. incognita* and *M. javanica*. Even 3-4 larvae/g of soil cause significant damage to the crop (Pandey, 1990).

5.6.1.1. Symptoms

Up to 60-70% plants were chlorotic and stunted showing a patchy appearance with fewer smaller leaves and flowers (Haseeb and Pandey, 1989a) (Fig. 5.1). Varying sizes of galls were found in the root system.

Fig. 5.1. Root-knot nematode on henbane. Left – Healthy; Right – Infected
(Courtesy: Haseeb, 1992).

5.6.1.2. Management Methods

(i) Cultural Methods: It was observed that essential oils of palmarosa, citronella, basil and menthol mint were quite effective in reducing *M. incognita* population and improving plant growth. The oil of palmarosa at 2ml/pot was most effective (Pandey *et al.*, 2003) (Table 5.3).

Table 5.3. Effect of essential oils of aromatic crops on growth and root galling of henbane

Treatment	Plant dry weight (g)	No. of leaves/plant	Root-knot index
Control	4.4	39.0	4.00
Menthol mint - 0.5 ml	5.3	50.6	2.33
Menthol mint - 1.0 ml	8.4	63.6	1.66
Menthol mint - 2.0 ml	10.4	72.6	1.33
Basil - 0.5 ml	9.6	86.6	2.00
Basil - 1.0 ml	11.4	123.3	1.33

Basil - 2.0 ml	9.0	116.0	1.33
Citronella – 0.5 ml	10.0	112.0	2.33
Citronella – 1.0 ml	11.3	94.3	1.66
Citronella – 2.0 ml	9.8	88.0	1.33
Palmarosa – 0.5 ml	9.6	79.7	2.00
Palmarosa – 1.0 ml	12.5	83.6	1.66
Palmarosa – 2.0 ml	12.9	106.3	1.00
CD (P=0.05)	0.555	12.645	0.030

(ii) Biological Methods: Negative impact of *Glomus aggregatum, G. fasciculatum* and *G. mosseae* on reproduction of *M. incognita* infecting *H. niger* has been observed (Pandey *et al.*, 1999). Significant reduction in *M. incognita* population was observed when henbane plants were inoculated with *Pseudomonas fluorescens* or *G. aggregatum* resulting in higher herbage yield and root colonization with the bioagents.

5.7. BRAHMI

Brahmi (*Bacopa mannieri*) is an important medicinal plant and chief source of bacoside A and B which are extensively used in formulation of different drugs useful for improving intellect, asthma and epilepsy.

Root-knot nematodes are recognized as the major limiting factors in successful production of brahmi crop.

5.7.1. The Root-knot Nematode, *Meloidogyne incognita*

5.7.1.1. Symptoms

Stunting in plant growth, chlorosis and gall formation in the root system give a simple indication of presence of root-knot nematodes. A negative correlation exists between increasing population levels of *M. incognita* and plant growth and yield characteristics.

5.7.1.2. Management Methods

(i) Integrated methods: Neem compound, distillation waste of menthol mint, curry leaf, davana and vermicompost and bioinoculants such as *Glomus aggregatum* and *Trichoderma harzianum* alone or in combination were root-knot nematode suppressive and enhanced the growth and yield of brahmi (Pandey *et al.*, 2002) (Table 5.4).

Table 5.4. Effect of organic materials and bioagents on plant growth, yield and root galling in brahmi infected with *Meloidogyne incognita*

Treatment	Total dry weight (g/pot)	Root-knot index
Control	47.1	3.66
Carbofuran	68.2	2.33
Davana	96.0	0.66
Neem compound	102.6	1.00
Neem cake	100.8	2.33
Linseed cake	64.0	2.66
Sesamum cake	68.3	2.00
Glomus aggregatum	69.5	3.33
G. fasciculatum	69.0	2.66
Trichoderma harzianum	77.3	2.66
CD (P=0.05)	3.271	0.041

5.8. *AMMI MAJUS*

Ammi majus (Family: Apiaceae) is an important medicinal plant commonly used in the treatment of vitilago leucoderma. It is also used in formulation of suntan lotion.

5.8.1. The Root-knot Nematode, *Meloidogyne incognita*

The root-knot nematode is a major constraint in the cultivation of *Ammi majus* (Pandey, 2002).

5.8.1.1. Management Methods

(i) Cultural Methods: Treatment with *Costus speciosus* dried shoot powder at 10 g/kg soil showed the least nematode population (Rf=0.4) and root-knot index (RKI=0.6) followed by *Spilanthes acmella* (Rf=0.6, RKI=0.6), *C. speciosus* rhizome powder (Rf=0.5, RKI=1.0) and neem cake (Rf=0.5, RKI=1.0) (Pandey, 2002) (Table 5.5).

Table 5.5. Effect of dried powder of botanicals on growth and root galling in *Ammi majus*

Treatment	Total dry weight (g)	Total nema popn.	Reproduction factor	Root-knot index
Control	13.4	21620	4.3	4.0
Carbofuran	16.9	7000	1.4	2.6
Citronella leaves	21.1	17010	3.4	3.6
Lemon grass leaves	22.2	3890	0.8	2.0
Spilanthes acmella shoot	23.1	3000	0.6	0.6
Costus speciosus shoot	22.7	1816	0.4	0.6
C. speciosus rhizome	21.1	2716	0.5	1.0
Neem cake	22.7	2580	0.5	1.0
CD (P=0.05)	---	323.113	0.221	0.89

5.9. KHASI KATERI

Khasi kateri (*Solanum viarum*) is a steroid-bearing perennial, tall bush distributed all over Assam, Sikkim and Manipur between 1,600-2,000m elevations. Its berry pulp is rich in solasodine alkaloid, which is a starting chemical for production of steroids. It is used in production of contraceptive pills, corticosteroids and sex hormones. It is cultivated in 3,000-5,000 ha, mainly in Maharashtra and Andhra Pradesh. It is also cultivated in West Bengal, Assam, Madhya Pradesh and Gujarat on a small scale.

The root-knot nematode is recognized as the major limiting factor in successful production of khasi kateri crop.

5.9.1. The Root-knot Nematode, *Meloidogyne incognita*

M. incognita infected *S. viarum* plants exhibited severe leaf yellowing and wilting symptoms (Shetty and Reddy, 1984). It was responsible for 33.6 and 11.5 per cent reduction in plant height and root weight.

5.9.1.1. Management Methods

(i) Chemical Methods: Bare root-dip treatment of *S. viarum* seedlings with phenamiphos and aldicarb sulfone both at 1000 ppm for 2 hr was effective against *M. incognita* infection (Shetty and Reddy, 1985).

(ii) Host Resistance: Among four species of *Solanum* tested for their reaction to *M. incognita, S. torvum* and *S. seaforthianum* gave resistant

reaction which was reflected in the reduction in number of galls, egg masses and fecundity of females (Shetty and Reddy, 1985b).

5.10. DIOSCOREA

Various species of Dioscorea (*Dioscorea floribunda*) are the source of diosgenin, a steroidal sapogenin, and is the most commonly used precursor for synthesis of certain drugs which include cortisones, sex hormones and oral contraceptives.

Root-knot and dry rot nematodes are recognized as the major limiting factors in successful production of dioscorea crop.

5.10.1. Root-knot Nematodes, *Meloidogyne incognita, M. javanica*

Dioscorea floribunda has been reported to be severely parasitized by *M. incognita* and *M. javanica*.

5.10.1.1. Symptoms

Severe infestation of *D. alata* and *D. spinosa* with *M. incognita* in Kerala was reported (Raveendran and Nadakal, 1975). The economic threshold and economic injury level of *M. incognita* were found to be 250 and 1250 nematodes per plant of *D. rotundata* cv. Igave, respectively. Tubers from pots inoculated with more than 1250 nematodes per plant were so heavily galled that the market value was reduced by 40%. The quality of tubers, an important consideration in growing of this crop, is also affected (Atu *et al.*, 1983).

5.10.1.2. Interaction with Other Pathogens

A disease complex involving *Fusarium oxysporum* f. sp. *dioscorea* and *M. incognita/P. coffeae* has been reported on *D. rotundata*. A positive interaction between *M. incognita* and *F. oxysporum* f. sp. *dioscorea* occurs in *D. rotundata* cv. Habanero.

5.10.2. The Dry Rot Nematode, *Scutellonema bradys*

S. bradys is responsible for 'dry rot' of yam tubers in Nigeria and Puerto Rico. Hutton (1978) considered *Pratylenchus coffeae* to be the most serious pest of yam in Jamaica, causing 'dry rot' or burning. This condition is characterized by cracking in the skin underlined by a brown, corky rot in the storage tissues. Rot progresses deeper into the yam tissues following harvest and prior to planting or consumption and is more pronounced towards the stem and yam tubers.

5.11. SAFED MUSLI

Safed musli (*Chlorophytum borivilianum*), known as minor forest produce, is an important medicinal plant belonging to family Liliaceae. This plant is widely distributed throughout India especially in Uttar Pradesh, Bihar, Rajasthan, West Bengal, Madhya Pradesh, Gujarat and Maharashtra. It is a major source of carbohydrates (42%), protein (8.9%), root fibres (3-4%) and saponins (2-17%). Ancient books suggest that various parts of this plant were used in the cure of different human diseases without any side effects.

The root-knot nematode is a major limiting factor in successful production of safed musli.

5.11.1. The Root-knot Nematode, *Meloidogyne incognita*

Safed musli has been found to be greatly affected with root-knot infection caused by *M. incognita*. A severe loss due to the disease was found under field conditions.

5.11.1.1. Symptoms

The infested field showed stunting, drying and falling of leaves and severe galling on the root system.

Chapter 6

AROMATIC CROPS

In recent years, there has been an increased interest in the cultivation of aromatic plants to meet the requirements of cosmetic, flavouring and perfumery industries. The aromatic plants provide raw material for the production of flavours, herbal cosmetics, perfumery etc. aromatic plants and their products, particularly the essential oils, are now becoming one of the more important export items from many developing countries of Asia. Chemically essential oils are terpenes, which act as carriers of the aromatic substances. Most essential oils also contain camphors, and the more odiferous compounds present in them consist of oxygen derivatives of terpenes, alcohols, esters, aldehydes and ketones. India has enjoyed a pre-eminent position in the manufacture of superior perfumes and aromatics by using essential oils. The important aromatic plants are lemon grass, vetiver, patchouli, palmarosa, citronella mints, geranium, lavender, basil, jasmine etc.

In the world market, India stands in second position. Most of the essential oils produced though marketed within the country, but a sizeable amount has also been exported. Among the essential oils exported from India are Japanese mint oil, ambrette seed essence, sandal wood oil, citronella oil, lemon grass oil, palmarosa oil, ginger oil, clove oil, eucalyptus oil, vetiver oil, pepper oil, etc. The annual exports of the derivatives of aromatic plants are to the tune of Rs. 600-700 million.

With the increase in cultivated area under aromatic crops in recent years and introduction of new species and varieties of aromatics, problems due to plant parasitic nematodes are becoming prominent.

6.1. MINTS

Mints (*Mentha* spp.) are a group of aromatic herbs belonging to the family Lamiaceae, which are considered to be the most important cash crops in Indo-Gangetic plains. Cultivation of mints has been done in a large scale in many tropical and sub-tropical countries of the world including India, China, Brazil, Japan and USA. Indian farmer grow it as a bonus crop on an area of more than 1,50,000 ha in central and northern parts of Indo-Gangetic plains. The crop fit well in the traditional food based cropping system with other crops like paddy, wheat, potato, mustard, maize, okra, carrot, onion, spinach, pigeon pea, cowpea, etc. Similarly it can be grown as an intercrop with sugarcane and some legumes. In India, the highest area under commercial cultivation of mint has been in Uttar Pradesh (Terai region), but during last few years it has spread through out the country, especially in north and north-western states including Madhya Pradesh.

Among different oil yielding species of mint, Japanese mint (*M. arvensis*) occupies the highest area under mint cultivation followed by peppermint (*Mentha spicata*), spear mint (*M. piperata*), Bergamot mint (*M. citrata*) and Scotch spearmint (*M. cardiaca*).

Japanese mint essential oil is used as a source of menthol, menthyl acetate, menthone and terpenes which finds wide use in medicine, perfumes, food and cosmetics. The oil of peppermint, which contains less menthol but has a sweeter aroma and taste, is used in toothpaste, chewing gums, candies, high grade liquors, medicines and other pharmaceutical preparations. Spearmint oil having carvone and limonene, is mainly used for flavouring whereas, Bergamot mint oil is mainly used in perfumery and flavouring due to higher content of linayl acetate and linalool.

Among the recognized species of plant parasitic nematodes, root-knot and root lesion nematodes have been considered most important.

6.1.1. Root-knot Nematodes, *Meloidogyne incognita, M. javanica, M. hapla*

Root-knot disease of Menthol/Japanese mint, Spearmint, Scotch Spearmint, Peppermint and Bergamot mint caused by *M. incognita* and *M. javanica* was observed for the first time in Lucknow and Terai region of Uttar Pradesh which reduce the herbage yield and oil content (Haseeb and Pandey, 1989b). *M. incognita* dominates over *M. javanica* in mixed infection. The quality of mint oil was also adversely affected due to nematode infection. The root-knot nematode, *M. incognita* caused 40.2% loss in herbage yield and 46.6% loss in oil yield in menthol mint (Pandey, 2001).

6.1.1.1. Symptoms

Initial symptoms of the disease include occasional yellowing of leaves which in a month time spread to a large portion of the foliage. Growth ceases soon after yellowing. Leaves turn yellow and thin, scorching easily and eventually turning brown. Initially symptoms appear in patches as the reduced plant growth with smaller leaf size and temporary wilting under slight stress of water during hot sun. As the crop grow, especially after first harvest, symptoms become more severe and appears as stunted growth with yellowing of leaves, while veins remain green.

The below-ground symptoms are the large numbers of small sized galls with large egg masses on the roots. Under severe infection, initiation of lateral roots and rootlets on suckers is checked. As a result, uptake of nutrients is inhibited, which in turn produces deficiency symptoms on the aerial portion of the plants (Haseeb and Pandey, 1989b).

6.1.1.2. Interaction with Other Pathogens

The severity of root-knot disease of Japanese mint was increased under field conditions in the presence of *Rhizoctonia solani*. The *M. incognita-R. solani* disease complex caused severe reduction in growth and oil yield at lower inoculum densities, and at higher inoculum levels, young plants could not survive.

6.1.1.3. Management Methods

(i) Physical Methods: Hot water treatment of *M. arvensis* suckers at 48°C for 30 min gave complete eradication of *M. incognita* without any adverse effect to suckers and roots.

(ii) Cultural Methods

Nematode-free Planting Material: The nematodes which damage mints are generally endoparasitic (*M. incognita* and *P. thornei*) in nature and they can easily spread through suckers (propagating materials) from infested planting sites to uninfested planting sites. Therefore, it is essential and utmost important to use nematode-free planting material to check the spread of nematodes (Pandey *et al.*, 2003).

Crop Rotation: Cultivation of some non-host crops in rotation with Japanese mint, water logging, rice cultivation during rainy season caused reduction in the population of root-knot nematodes. Crop rotation sequences of Menthol mint with maize, potato, mustard, pigeon pea, paddy and wheat were effective in reducing the root-knot nematode population and in increasing the yield (Table 6.1).

Table 6.1. Effective crop rotation sequences for the management of root-knot nematode in Menthol mint.

Crop rotation sequence	Root-knot index	Net benefit (in rupees)
Maize – potato – menthol mint	Mild	83,000
Paddy – potato – menthol mint	Mild	80,000
Maize – mustard – menthol mint	Mild	76,000

Pigeon pea – menthol mint	Mild	72,000
Paddy – menthol mint	Mild	75,000
Paddy – wheat – menthol mint	Mild	68,000

Organic Amendments: The severity of the nematode was reduced to a considerable extent in clayey soils. The high organic matter content in the soil suppressed the nematodes and enhanced the growth and oil yield. Soil application of neem/castor/mahua oil cakes at 1.0 g N/kg of soil was found effective against *M. incognita* and resulted in increase of oil yield and also plant growth parameters (Pandey *et al.*, 1992). Incorporation of neem cake 15 days before transplanting of *M. spicata* has been found to be the best approach for managing the population of *M. hapla* (Khan and Khanna, 1992).

The application of hydro-distillation wastes of *M. arvensis, M. piperita, Cymbopogon martini* (palmarosa) and *C. winterianus* (citronella) in the soil (1 month before transplantation) was found effective in reducing the nematode population under field conditions.

Application of different oil seed-cakes improved the growth of Japanese mint under field conditions coupled with increased oil yield and reduced root-knot and lesion nematode population. However, best results were achieved when the soil was treated with neem cake. The use of vermicompost and different distillation wastes were found to enhance the growth and yield of different mint species and reduced nematode population significantly.

Maximum reduction in *M. incognita* population in Menthol mint was recorded in neem cake treated soil followed by mustard cake and *Trichoderma harzianum*. Higher herbage yield (21 t/ha) was observed in the soil treated with oil cakes in comparison to control (14.66 t/ha). The crop produced had significantly higher oil content (2.7%) and yield (151 kg/ha) in neem cake treatment followed by mustard cake (2.7% oil and 143.6 kg/ha yield), *T. harzianum* (2.5% oil and 116.4 kg/ha yield) and *Glomus aggregatum* (2.4% oil and 112 kg/ha yield). The untreated crop gave 1.8% oil and 70.4 kg/ha yield (Pandey *et al.*, 1998) (Table 6.2).

Table 6.2. Effect of bio-organics and a pesticide on herb and oil yield of menthol mint and root-knot indices of *Meloidogyne incognita*

Treatment	Herbage yield (t/ha)	% oil content	Oil yield (kg/ha)	Root-knot indices
Untreated control	1.466	1.8	70.4	3.66
Carbofuran	1.880 (+28.4)	2.3	114.2 (+62.2)	2.33 (-34.7)
Mustard cake	1.998 (+36.3)	2.7	143.6 (+104.0)	2.00 (-45.4)

Neem cake	2.100 (+43.2)	2.7	151.0 (+114.5)	2.00 (-45.4)
Glomus aggregatum	1.758 (+19.9)	2.4	112.2 (+59.4)	2.66 (-27.3)
Trichoderma harzianum	1.750 (+19.4)	2.5	116.4 (+65.3)	2.66 (-27.3)
CD (P=0.05)	0.284	0.5	40.4	0.1

(iii) Chemical Methods: Application of carbofuran at 3 kg a.i./ha increased plant growth and oil yield of *M. arvensis* and reduced the root-knot nematode population.

(iv) Biological Methods: Application of arbuscular mycorrhizal fungi such as *G. aggregatum, G. fasciculatum* and *G. mosseae* improved plant growth, enhanced herbage and oil yield and effectively inhibited root-knot nematode infection (Pandey *et al.*, 1997) (Table 6.3).

Table 6.3. Effect of arbuscular mycorrhizal fungi on the productivity of mint infected with *Meloidogyne incognita*

Treatment	Fresh herbage weight (g)	% Oil yield	% Mycorrhizal colonization	Root-knot index
Untreated - Inoculated	285.0	0.38	---	3.6
Glomus aggregatum (Ga)	322.0	0.40	42.3	2.3
G. fasciculatum (Gf)	333.0	0.41	58.2	2.0
G. mosseae (Gm)	388.0	0.48	63.5	1.6
Ga + Gf + Gm	360.0	0.46	78.5	1.3

(v) Host Resistance: The lowest root-knot (*M. incognita*) infection level was found on Kalka. Spear mint cvs. Neera and Arka were resistant to *M. incognita* (Pandey, 2003). Peppermint cv. Kukrail; Spearmint cvs. Arka, Neera and Neer-Kalka; and Bergamot mint cv. Kiran were found resistant to *M. incognita* (Pandey and Patra, 2001).

(vi) Integrated Methods: Application of carbofuran and neem cake in combination improved the growth of Japanese mint and oil yield and reduced root-knot nematode population very effectively.

The maximum reduction in *M. incognita* population was recorded in beds treated with mustard cake along with *T. harzianum* and produced significantly higher herbage and oil yields (Pandey, 2005a).

6.1.2. Lesion Nematodes, *Pratylenchus minyus, P. thornei, P. penetrans*

Pathogenicity of *P. penetrans* on *M. piperita* has been demonstrated and reported to cause yield losses up to 34%. *M. arvensis* favours the high population build up of the nematode followed by *M. cardiaca, M. piperita* and *M. citrata.*

6.1.2.1. Symptoms

High population of *P. thornei* was found to be associated with different mints and caused wilting and chlorosis of the plants (Haseeb, 1992). The lesion nematode causes light brown to dark coloured lesions on the suckers and roots. In severe cases whole underground portion becomes black and rotting of cortical portion takes place, first due to the nematode infection and later because of the invasion of secondary pathogens on necrotic cells.

6.1.2.2. Interaction with Other Pathogens

P. penetrans and *P. minyus* increased the incidence and severity of *Verticillium albo-atrum* and *V. dahliae* f. sp. *menthae* wilt on mentha.

Bergeson (1963) reported that when *P. penetrans* infected *M. piperita* plants were inoculated with *V. albo-utrum,* diagnostic symptoms of wilt appeared approximately two weeks earlier than the plants inoculated with the fungus in the absence of nematode.

Presence of *P. miniyus* increased both incidence and severity of wilt (*V. dahliae* f. sp. *menthae*) disease, reduced the period for the development of wilt by 2 to 3 weeks, increased the reproduction of nematode and reduced the dry weight of peppermint plants up to 68% when both the pathogens were present (Faulkner and Skotland, 1965).

6.1.2.3. Management Methods

(i) Cultural Methods: Application of neem and mustard cakes at 1.5 t/ha reduced the lesion nematode population (*P. thornei*), improved the growth and oil yield of *M. piperita, M. citrata* and *M. spicata* (Shukla and Haseeb, 1996).

(ii) Chemical Methods: Mint plants treated with phenamiphos at 5.6 kg per ha gave maximum reduction of *P. scribneri* population (Rhoades, 1984). Ingham *et al.* (1988) found that ethoprophos (Mocap) at 6 kg/ha was highly effective in reducing the population of *P. penetrans* both in soil and root tissue of *M. piperita.* However, oxamyl (Vydate) at 1 kg/ha reduced the population of *Pratylenchus* spp. on the crop significantly.

(iii) Host Resistance: Menthol mint cvs. Kosi and Kalka were resistant to *P. thornei* and had least *M. incognita* infection. Bergamot mint cv. Kiran and Spear mint cv. Neer-Kalka were having least *P. thornei* infection.

6.1.3. The Reniform Nematode, *Rotylenchulus reniformis*

Haseeb (1992) reported *R. reniformis* on *M. arvensis*, *M. cardiaca*, *M. citrata*, *M. piperita* and *M. spicata*. *M. piperita* and *M. spicata* are good hosts of *R. reniformis*. The reniform nematode was found to reproduce well on *M. arvensis* and to cause significant reduction in herb and oil yields.

6.1.3.1. Symptoms

R. reniformis was observed to cause stunted growth of mint plants, withering of branches with chlorotic leaves at Hessaraghatta near Bangalore (Khan and Parvatha Reddy, 1992).

6.2. BASIL

Among several species of basils, sweet basil (*Ocimum basilicum*) is considered the most important crop for its high quality of essential oil. Major components of its oil are linalool (43-50%), menthyl chavicol (18-33%), eugenol and isoeugenol (5-6%) and the minor constituents are alpha and beta pinene, camphor, geraniol, etc. The oil of sweet basil owes its importance to its extensive use in condimentary products, cosmetics, toiletry, perfumery and confectionary industries, particularly in European countries.

Root-knot nematodes are recognized as the major limiting factors in successful production of basil crop.

6.2.1. The Root-knot Nematode, *Meloidogyne incognita*

Recently, *M. incognita* and *M. javanica* have been found to be the major problems in the cultivation of *O. basilicum*, *O. canum*, *O. sanctum*, *O. gratissimum* and *O. kilmandescharicum* (Haseeb *et al.*, 1988; Haseeb and Pandey, 1987). Balasubramanian and Rangaswami (1964) reported *O. sanctum* as a host of *M. javanica*. Krishnamurthy and Elias (1967) reported that *O. basilicum* was severely affected by *M. incognita*.

6.2.1.1. Management Methods

(i) Cultural Methods: Haseeb *et al.* (1988) evaluated neem oil cake, aldicarb, bavistin, mentha oil cake and carbofuran against *M. incognita* on *O. basilicum* and observed neem oil cake as the best for plant growth and oil yield. Application of mahua cake also increased the oil yield (Haseeb *et al.*, 1988).

6.3. JASMINE

Jasmine (*Jasminum* spp.) is one of the leading traditional flowers of India. Its flowers are used for making garlands, adorning hairs of women, in religious and ceremonial functions, and for producing perfumery oil. It is commercially cultivated in Karnataka, Tamil Nadu, Andhra Pradesh, Maharashtra, Gujarat, Haryana, Punjab, West Bengal, Rajasthan, Assam and Uttar Pradesh. Apart from internal trade, fresh flowers of jasmine are exported to Malaysia, Singapore and Sri Lanka.

The root-knot and the burrowing nematodes are major problems in jasmine cultivation. Bajaj (1989) recorded *Tylenchulus semipenetrans* on *J. sambac* from Haryana.

6.3.1. The Root-knot Nematode, *Meloidogyne incognita*

M. incognita was reported on *J. sambac* and *J. flexile* from Tamil Nadu by Rajendran and Rajendran (1979).

6.3.1.1. Symptoms

The roots exhibited very small swellings and enlarged rootlets. The minute galls were more in *J. sambac*. Pale coloured leaves and dieback symptoms were associated (Rajendran and Rajendran, 1979).

6.3.1.2. Management Methods

(i) Cultural Methods: The planting material should be raised in nematode-free or solarized nursery beds. Treatment of nursery beds with neem or pongamia cakes at 2 kg/sq. m. helps in reducing the nematode population.

(ii) Chemical Methods: Application of phorate at 4 g a.i. per plant during May and September months and incorporation of 20 kg FYM increased flower yield of jasmine by 50% and reduced the root-knot nematode population by 70% (Sundarababu, 1992).

Nursery beds should be treated with carbofuran 3G or phorate 10G at 25 g /sq. m. and main field with carbofuran 3G or phorate 10 G @ 2.5 kg a.i./ha before planting.

(iii) Integrated Methods: Cuttings or planting material should be raised in nematode-free or treated nursery beds. Application of neem cake enriched with *T. harzianum* at 1 t/ha is recommended.

Nursery beds should be treated with neem or pongamia cakes enriched with mycorrhizal spores at 1 kg/sq. m.

6.3.2. The burrowing nematode, *Radopholus similis*

6.3.2.1. Symptoms

Khan and Parvatha Reddy (1989) recorded a high population of *R. similis* from stunted *J. pubescens* plants in Bangalore. Roots of infected plants exhibited yellow to brown lesions. In some cases blackening also occurs. Infected plants become severely stunted. Branches become dry with chlorotic leaves which ultimately drop.

6.4. PATCHOULI

Patchouli (*Pogostemon patchouli*) is a highly aromatic bushy under-shrub. It is cultivated for its highly fragrant leaves which contain a very sweet smelling oil of lasting sticky odour. The crop is grown in small pockets in Karnataka, Kerala and Tamil Nadu. It has a very characteristic aroma and blends well with other essential oils. The oil is used in very low concentration (2 ppm) in scenting soaps, cosmetics, after-shave lotions, detergents and many fancy products. The oil is also used to flavour foods, beverages, candy and baked products. In combination with sandal wood oil, it is used in blending tobacco and making incense sticks.

Root-knot, lesion and spiral nematodes are recognized as the major limiting factors in successful production of patchouli crop.

6.4.1. Root-knot Nematodes, *Meloidogyne* spp.

6.4.1.1. Economic Importance

It was responsible for 47.0 and 86.7 per cent loss in shoot weight and shade dried leaf yield, respectively (Prasad and Reddy, 1984).

6.4.1.2. Symptoms

In patchouli, the root-knot nematode causes chlorotic leaves, premature drying and shedding of leaves (Prasad and Reddy, 1984).

6.4.1.3. Management Methods

(*i*) *Cultural Methods:* Neem cake at 4 t/ha controlled the root-knot nematodes (Sarwar *et al.*, 1982).

(*ii*) *Chemical Methods:* Aldicarb at 1.5 kg a.i. per ha was effective in controlling phytonematodes infecting patchouli.

(*iii*) *Biological Methods:* Application of 150 kg N and 50 kg *Glomus fasciculatum*/ha enhanced fresh herbage (297.2 g/plant compared to 212.7 in control), dry herbage (92.0 g/plant compared to 64.0 in control) and oil yield (110.4 kg/ha compared to 28.0 kg/ha in control); increased mycorrhizal colonization (90.5% compared to 65.5% in control); reduced *M. incognita* population (145/200 ml soil compared to 395/200 ml soil in control) and recorded the highest cost: benefit ratio (1: 3.67) (Sumathi *et al.*, 2006).

(iv) Host Resistance: Planting of cultivated patchouli scion grafted on wild patchouli rootstock was effective for the management of root-knot nematodes (Krishna Prasad, 1978).

6.4.2. The Lesion Nematode, *Pratylenchus brachyurus*

P. brachyurus was a major problem on patchouli in Annamalai hills of Tamil Nadu and Hessaraghatta in Karnataka (Khan and Parvatha Reddy, 1992).

6.4.2.1. Symptoms

The infected plants invariably exhibited wilting symptoms.

6.4.3. The Spiral Nematode, *Helicotylenchus dihystera*

6.4.3.1. Symptoms

H. dihystera causes heavy damage to patchouli roots and is responsible for decline.

6.4.3.2. Management Methods

(i) Chemical Methods: Aldicarb, carbofuran and phorate each at 2 to 3 kg a.i. per ha significantly reduced *H. dihystera* and increased leaf yield of patchouli.

6.4.4. Root-knot Nematode and Root Rot Disease Complex

6.4.4.1. Management Methods

(i) Integrated Methods: Combined mortality due to root rot and root-knot nematode could be minimized by the application of *Trichoderma harzianum* + karanj cake at 5 t/ha.

6.5. DAVANA

Davana (*Artemisia pallens*) is the most important aromatic herb for its high value essential oil used in the perfumery and cosmetic industries. Its foliage and floral tops produce a viscous essential oil, emitting a delicate, persistently fruity fragrance which is used in floral decorations, bouquets and cosmetics. It is also used in flavouring of cakes, pastries, beverages and tobacco products. India is one of the major exporters of davana oil to the rest of the world (mostly to USA). It is grown in 2,000 ha mainly in Karnataka and to a lesser extent in Tamil Nadu and Maharashtra.

Root-knot nematodes are recognized as the major limiting factors in successful production of davana crop.

6.5.1. The Root-knot Nematode, *Meloidogyne incognita*

The root-knot nematode is a major problem in the cultivation of davana. *M. incognita* is responsible for more than 30% reduction in oil yield of davana (Haseeb and Pandey, 1990).

6.5.1.1. Symptoms

The main symptoms were chlorotic and stunted plants with less number of flower buds (which are the major source of essential oil) showing patchy appearance in the field. Their roots were severely galled by root-knot nematodes (Fig. 6.1). One larva/2 g soil has been found as economic threshold level of *M. incognita* on this crop.

Fig. 6.1. Root-knot nematode on davana. Left – Healthy roots; Right – Infected roots (Courtesy: Haseeb, 1992).

6.5.1.2. Management Methods

(i) Cultural Methods: Soil application of neem cake at 1g N/kg soil significantly suppressed root-knot nematode population and increased oil yield (Haseeb and Butool, 1991). Better efficacy of neem cake over the other organic materials viz., mahua cake, castor cake and nematicides like aldicarb and carbofuran has been demonstrated on *M. incognita* infecting *A. pallens* (Pandey, 1994b) (Table 6.4). Increase in nematicidal efficacy of neem cake and FYM for 90 to 120 days in comparison to 60 days for fenamiphos has also been noticed (Anita and Vadivelu, 1997).

Table 6.4. Effect of oil cakes and pesticides on plant growth and yield of davana infected with *Meloidogyne incognita*

Treatment	Fresh weight (g)	% Oil content	Root-knot index
Control	52.5	0.25	3.6
Mahua cake	54.9	0.27	1.4
Castor cake	73.4	0.33	2.4
Neem cake	87.5	0.36	1.4
Carbofuran	75.0	0.33	2.4
Aldicarb	83.3	0.35	1.4
Bavistin	53.0	0.27	2.6

A considerable enhancement of herbage biomass, flower buds and oil yields were noticed when davana plants were treated with the distillation waste of palmarosa, citronella, curry leaf, menthol mint and vermicompost of pyrethrum, marigold and menthol mint. Least nematode infections were recorded in distillation waste of curry leaf, palmarosa, lemon grass and vermicompost of pyrethrum, marigold and menthol mint (Pandey, 2005b) (Table 6.5).

Table 6.5. Effect of distillation waste and vermicompost on growth and root-knot development of davana infected with *Meloidogyne incognita*

Distillation waste / vermicompost	Fresh wt. of herbage & flower buds (g)	Total nematode population (soil + root)	Repro-duction factor	Root-knot index
Untreated-Uninoculated control	40.0	---	---	---
Untreated-Inoculated control	24.7	3480	3.48	4.00
Carbofuran control	36.8	1800	1.80	2.66
Lemon grass (Distillation waste)	41.8	1260	1.26	2.00
Palmarosa (Distillation waste)	48.8	1100	1.10	1.66
Citronella (Distillation waste)	44.2	1040	1.04	1.66
Menthol mint (Distillation waste)	42.7	1640	1.64	2.33
Curry leaf (Distillation waste)	58.5	1020	1.02	1.00
Geranium (Distillation waste)	32.3	1680	1.68	2.66
Patchouli (Distillation waste)	31.3	1800	1.80	2.66
Marigold (Distillation waste)	41.6	1600	1.60	2.33
Menthol mint (Vermicompost)	51.3	1260	1.26	2.00
Citronella (Vermicompost)	51.0	1380	1.38	2.00

Chrysanthemum (Vermicompost)	44.0	1240	1.24	1.66
Marigold (Vermicompost)	37.9	1040	1.04	1.66
CD (P=0.05)	2.176	---	---	0.118

(ii) Biological Methods: A considerable enhancement of herbage biomass, flower bud and oil yields were noticed when davana plants were treated with the bioagent, *Trichoderma harzianum* (Pandey, 2005b) (Table 6.6).

Table 6.6. Effect of vermicompost and bio-agents on growth and root-knot development of davana infected with *Meloidogyne incognita*

Bio-agents	Fresh wt. of herbage & flower buds (g)	Total nematode population (soil + root)	Reprod-uction factor	Root-knot index
Untreated-Uninoculated control	40.5	---	---	---
Untreated-Inoculated control	27.3	3480	3.48	4.00
Carbofuran control	39.0	1760	1.76	2.66
Trichoderma harzianum	56.7	1480	1.48	2.00
Glomus aggregatum	29.5	2080	2.08	2.66
G. fasciculatum	30.4	1880	1.88	2.66
CD (P=0.05)	2.174	---	---	0.137

6.6. SCENTED GERANIUM

Rose or scented geranium (*Pelargonium graveolens*) is an aromatic perennial herb. It is grown in cooler, sub-tropical climate of Mysore and Bangalore (Karnataka), and extended to Hyderabad (Andhra Pradesh). Its oil has refreshingly delicate long-losting rosy odour with fruity under-note, containing 66-78% primary alcohols (rhodinol). It is used in manufacturing of perfumes, creams, talcum powder and body lotion. It is stable in alkaline medium and is therefore used for scenting soaps.

The root-knot nematode is recognized as the major limiting factor in the successful production of scented geranium crop.

6.6.1. Root-knot Nematodes, *Meloidogyne* spp.

M. hapla was found to be widely distributed on geranium in Nilgiri Hills of Tamil Nadu (Khan and Parvatha Reddy, 1992).

6.6.1.1. Management Methods

(i) Chemical Methods: Aldicarb at 1 g a.i. per plant gave best control of *M. incognita* and increased shoot height, number of branches and leaf yield in gernanium.

6.7. CHAMOMILE

Chamomile (*Matricaria recutita*), an important medicinal and aromatic herb, is widely cultivated for its flowers which are used as medicinal tea (which acts as mild sedative and digestive and provides relief in cough and cold) in Europe and the USA. Their extracts are being used in antiflogistic creams and ointments. The oil of chamomile is utilized in alcoholic and nonalcoholic beverages, ice creams, candies, baked foods as well as in high-grade perfumery throughout the world. It is traditionally employed in manufacturing of pain-relieving balms. It is used in scenting of shampoos, pomades and face creams. It is grown in small area in north-central parts of Uttar Pradesh and Himachal Pradesh.

The introduction of chamomile as a commercial crop brought into the limelight the threat posed by root-knot nematode, *M. incognita* for its cultivation (Pandey *et al.*, 1999).

6.7.1. The Root-knot Nematode, *Meloidogyne incognita*

M. incognita causes considerable reduction in growth, flower buds and essential oil yield of chamomile (Pandey *et al.*, 1999).

6.7.1.1. Management Methods

(*i*) *Chemical Methods:* Maximum reduction in root-knot severity and nematode population occurred with chemical activators such as 2-Chloro-salicylic acid, O-Acetylsalicylic acid and 2-Chloronicotinic acid (Pandey and Kalra, 2005) (Table 6.7).

Table 6.7. Effect of chemical activators on root-knot nematodes, plant growth and flower yield of chamonile

Treatment	Total nema popn. (soil + roots)	Reproduction factor (Rf = Pf/Pi)	Root-knot index (RKI)	Dry weight of plant (g)	Flower yield (g)
Control	3018	3.01	4.00	4.1	3.3
Isonicotinamide	2498	2.49	3.00	3.0	3.3
2-Chloronicotinic acid	1710	1.71	2.00	6.3	6.1
5-Nitrosalicylic acid	1818	1.81	2.00	7.7	6.3
4-Chlorosalicylic acid	1412	1.41	1.33	9.6	6.7
DL-2 Aminobutyric acid	1777	1.77	2.00	6.7	6.1
2 Aminobutyric acid	2020	2.02	2.33	4.3	3.7
0-Acetyl salicylic acid	1640	1.64	1.66	10.6	7.4

4 – Amino salicylic acid	2176	2.17	2.66	6.7	6.2
Salicylic acid	2010	2.01	2.00	8.6	6.5
Carbofuran	1920	1.92	2.00	4.9	6.0
CD (P = 0.05)	321.83	---	0.021	---	0.01

(ii) Integrated Methods: Integration of *G. mosseae* and neem cake reduced the severity of root-knot disease and enhanced the growth, biomass and flower yield by 12 to 37% of chamomile.

6.8. CITRONELLA

Citronella (*Cymbopogon winterianus*) is a large, perennial, stem less, aromatic grass. Its oil from leaves has a strong characteristic lemonic odour extensively used in scenting soaps, cosmetics, producing deodorants and mosquito repellent products. The oil is used as starting material to produce pure citronellol, geraniol and related upstream high-value aroma chemicals used in perfumery industry. It is cultivated in 6,000-10,000 ha land in high rainfall tracts of Assam, Arunachal Pradesh, Meghalaya, Nagaland and Manipur. It is also cultivated on small scale in Karnataka, coastal Andhra Pradesh and Orissa.

Root-knot and lesion nematodes are recognized as the major limiting factors in successful production of citronella crop.

6.8.1. Lesion Nematodes, *Pratylenchus* sp.

6.8.1.1. Symptoms

Very high populations of *Pratylenchus* sp. were encountered from the rhizosphere of apparently diseased plants of citronella which were stunted, chlorotic with less number of tillers (Haseeb, 1992).

6.8.2. The Root-knot Nematode, *Meloidogyne incognita*

The root-knot nematodes were consistently isolated from the rhizosphere of citronella. Pandey (1994) reported that *M. incognita* was responsible for 50% loss in oil yield of citronella.

Chapter 7

PLANTATION CROPS

Plantation crops serve a variety of human needs such as food, oil, industrial raw materials, beverages and confectionary items. They generate huge employment opportunities directly or indirectly to several million people in their production, processing, marketing and international trade sectors. Plantation crops are a large group of crops. These are grown over an area of 3.102 million ha with production of 13.161 million tonnes in India (Table 7.1). While the major plantation crops include coconut, areca nut, oil palm, cashew, coffee, tea and rubber, minor plantation crops are cocoa and betel vine. They play an important role in export as well as domestic requirement and employment generation and poverty alleviation particularly in rural sector. India has emerged as the largest producer of areca nut and holds third position in coconut production. The annual income generated is more than Rs.5,52,590 million and export earning is approximately Rs.93,350 millions.

Table 7.1. Area, production and productivity of plantation crops in India (2004-05) (Chamber, 2006)

Plantation crops	Area ('000 ha)	Production ('000 tonnes)	Productivity (tonnes/ha)
Coconut	1933.7	12178.2	6.3
Areca nut	365.0	439.2	1.2
Total	3102.0	13161.0	4.2

7.1. COCONUT

The coconut palm (*Cocos nucifera*), originated in South East Asia, is now cultivated widely throughout the tropics of the world. The largest producers of coconut are the Philippines, Indonesia, India and Sri Lanka. In all the coconut producing countries, it is an important source of export earning despite being a regular constituent of the food. Coconut is an important small holder's plantation crop. The total area under coconut cultivation in India is 1.933 million ha with an annual production of 12.178 million tonnes and the average productivity of 6.3 tonnes/ha is the highest in the world. It provides proteins, fats and some vitamins. Coir and coir products exported from India earned Rs.322 million during the above period. It is supporting the livelihood of 10 million people across the country. The four Southern States – Kerala, Tamil Nadu, Karnataka and Andhra Pradesh together accounts for 90% of the total area and production. Coconut is predominantly grown under rainfed conditions in Kerala and parts of coastal Karnataka, Maharashtra and Tamil Nadu. In rest of the country, it is mainly grown under irrigated conditions.

Of the several nematode species reported on coconut, the important ones are the burrowing nematode, *Radopholus similis* and the red ring nematode, *Rhadinaphelenchus cocophilus*.

7.1.1. The Burrowing Nematode, *Radopholus similis*

The burrowing nematode was reported from coconut palms in Kerala, India by Weischer (1967). *R. similis* is the most important nematode pest of coconut and is responsible for considerable amount of root rotting (Koshy *et al.*, 1978). It causes 30% yield loss in coconut (Koshy and Geetha, 1992). The threshold inoculum density required to cause significant reduction in various growth parameters of coconut is 100 nematodes/625 ml sandy loam soil over a period of 5 years under field conditions (Koshy, 1986).

7.1.1.1. Distribution

R. similis occurs in most tropical and subtropical areas of the world and has been reported from coconut palms in Florida, Jamaica, Sri Lanka and India. Surveys conducted in coconut plantations in South India recorded 24% incidence of *R. similis* (Sosamma, 1984).

7.1.1.2. Symptoms

The burrowing nematode infested coconut palms exhibit general decline symptoms like yellowing, button shedding, reduction in number and size of leaves and leaflets, delay in flowering and reduced yield which are non-specific. Symptoms on the root are more specific. *R. similis* on infestation produces isolated elongate orange coloured lesions on tender and semi-hard roots. Consequent to nematode parasitization and multiplication, these lesions enlarge and coalesce to cause extensive rotting of roots. Tender roots on heavy infestation become spongy in texture. On semi-hard orange coloured roots, surface cracks are commonly seen. As high as 4,000 nematodes were recovered from one gram (one inch length) of main roots. The drastic reduction in the number and mass of tertiary feeder roots on parasitization by the nematode limits plant growth (Koshy and Sosamma, 1987).

7.1.1.3. Life Cycle

R. similis is a migratory endoparasite capable of spending its entire life cycle within the roots. Except the fourth stage male juveniles and adult males, all other juvenile stages and adult females are found to be infective. The nematode completes its life cycle (J2 to J2) with in 25 days at a temperature range of 25°-28°C (Koshy and Sosamma, 1977).

7.1.1.4. Host Range

Large number of crops like banana, black pepper, ginger, turmeric, sweet potato, groundnut, sugarcane, betel vine, etc. are known hosts of *R. similis*.

7.1.1.5. Histopathology

Burrowing nematodes penetrate the absorbing regions behind the root cap (covered by very delicate epidermis) by the lysis of cells. Such entry points are 1-2 cells in diameter and surrounded by sclerenchymatous cells to a depth of 10-15 cells. The outer cortex harbour maximum number of cavities and nematodes. In the early stages of infestation, the roots contain independent cavities separated by several cells. Consequent to nematode multiplication and lysis of cell walls and cytoplasm, adjacent cavities merge with each other leading to almost complete destruction of the cortex. Longitudinal section of the roots showed nematodes in all stages of development (Koshy and Sosamma, 1987).

7.1.1.6. Survival and Spread

Under field conditions, *R. similis* survives for 6 months in moist soil (27°-36°C) and one month in dry soil (29°-39°C). In roots of stumps of felled coconut palms, the nematode survives for up to 6 months (Sosamma and Koshy, 1986). The infested coconut roots yielded maximum number of *R. similis* during October to November and minimum during March to July. A mean soil temperature below 25°C and light rain fall coupled with availability of tender fleshy roots are the factors favourable for *R. similis* multiplication (Koshy and Sosamma, 1978).

Infested coconut seedlings help in the dissemination of the nematode to distant places. Apart from coconut, infested planting materials of intercrops such as areca nut, banana, pepper, ginger, turmeric also serve as sources of inoculum.

7.1.1.7. Management Methods

(i) Cultural Methods: Avoiding banana as a shade crop in coconut nurseries and use of nematode-free planting material of coconut and other intercrops are the other management approaches.

Organic Amendments: Application of oil cakes, FYM and green foliage to the basins, growing of intercrops like cocoa that enriches the soil with sizeable quantities of shed foliage which help in the build up of beneficial organisms which may inhibit nematode multiplication.

Application of 50 kg cow dung/FYM, 2 kg neem/marotti cake and 25 kg green manure to the coconut basins in November is quite effective in bringing down the nematode population. Thirty per cent increase in yield has been recorded by application of *Hydnocarpus* sp. oil cake at 8 kg/palm/year in June-July and October-November to the burrowing nematode infested coconut palms (Koshy, 1986).

Green Manure Cropping: *Crotalaria juncea* may be grown as a green manure crop in the basins and interspaces and its incorporation helps in the inhibition of nematode multiplication.

(ii) Chemical Methods

Nursery Bed Treatment: In infested coconut nurseries, complete control of *R. similis* can be obtained by soil application of phenamiphos or phorate at 2.5 kg a.i./ha during September, December and May (Koshy and Sosamma, 1979).

Various chemicals tried for the management of *R. similis* on coconut under nursery conditions is presented in Table 7.2.

Table 7.2. Effect of various chemicals for the management of *R. similis* in coconut under nursery conditions

Chemical	Dosage	Duration & type of treatment	Effectiveness
Fensulfothion	50 kg a.i./ha	Thrice a year (in Sept., Dec. & March)	Reduced nematode infestation
Phenamiphos	25 kg a.i./ha	Thrice a year (in Sept., Dec. & May)	Controlled nematodes in nursery seedlings
Phorate	25 kg a.i./ha	Thrice a year (in Sept., Dec. & March)	Controlled nematodes in nursery seedlings

Bare Root-dip Treatment: Bare root dip of coconut seedlings in 1000 ppm of DBCP for 15 min. eliminated *R. similis* infection.

Main Field Treatment: Application of phorate at 10g a.i./palm in June-July and in September-October is effective in reducing the population of burrowing nematodes. Maximum increase in yield was obtained with the application of phenamiphos at 10g a.i./palm.

Various chemicals tried for the management of *R. similis* on coconut under field conditions is presented in Table 7.3.

Table 7.3. Effect of various chemicals for the management of *R. similis* in coconut under field conditions

Chemical	Dosage	Duration & type of treatment	Effectiveness
Aldicarb	0.5-1.0 kg a.i./ha	Applied once	Reduced nematode infestation
	10-50 g a.i./palm	Twice at 3 monthly intervals	Reduced nematode infestation
Phenamiphos or Phorate	10 g a.i./palm	Twice a year in June-July & October-November	Increased yield by 30% & reduced nema popn.

(iii) Biological Methods: Introduction of biocontrol agents like *Paecilomyces lilacinus* and AMF to nursery potting mixture helps in the reduction of nematode population and at the same time impart resistance/tolerance to nematode infestation (Koshy *et al.*, 1993). Among the AMF, *Acaulospora bireticulata* was found to be more effective in reducing the nematode population. Soil application of *Cylindrocarpon effusum* reduced the rate of multiplication of the nematode and damage to coconut seedlings (Sosamma and Koshy, 1978; Koshy and Sosamma, 1987).

(iv) Host Resistance: The dwarf cvs. Kenthali and Klappawangi and the hybrids such as Java Giant x Kulasekharam Dwarf Yellow, Java x Malayan Dwarf Yellow and San Ramon x Gangabondam recorded the least nematode multiplication and lesion indices (Sosamma *et al.*, 1980). The cv. Philippines Ordinary was found to be least susceptible to *R. similis* with lesion index of 1.22 (Sudha, 1998).

(v) Integrated Methods: The following measures are suggested for an integrated management schedule for *R. similis* infestation on coconut palms:

- Application of cow dung, FYM, oil cakes and green manure to the basins. *Crotalaria juncea* may be cultivated in the basins and interspaces and used as a green manure.

- Application of phorate at 10g a.i./palm twice yearly.

- Avoiding banana as a shade crop in coconut nurseries.

- Use of nematode-free planting material of coconut and other intercrops.

- Use of less susceptible/tolerant cvs. such as Kenthali and Klappawangi or their hybrids in infested areas.

Thirty per cent increase in yield and 5 to 10% decrease in disease indices of palms affected with root wilt disease was recorded by the application of *Hydnocarpus* oil cake and phorate granules at 10g a.i./palm in June-July and in October- November.

7.1.2. The Red Ring Nematode, *Rhadinaphelenchus cocophilus*

Three 'red ring' disease of coconut caused by *R. cocophilus* is very destructive disease on some islands of Caribbean which threatens the existence of this crop. The extensive losses caused by this nematode are a major catastrophe to natives of these areas. The nematode is transmitted by *Rhynchophorus palmarum*. The disease was first reported as occurring in Trinidad by Hart in 1905. It has not been reported from India so far.

At present, most Latin American and Caribbean countries show losses ranging from less than 1% to more than 20% of 1 to 10 year old trees. Thirty five per cent mortality of young coconut palms has been reported in Trinidad and 80% loss in single plantation of Tobago (Esser, 1969).

7.1.2.1. Distribution

The disease is reported to occur in 20 countries seriously affecting their economy. At present *R. cocophilus* has been reported from the West Indian Islands of Trinidad, Tobago, Grenada and St. Vincent and from Latin America, Dominican Republic, Venezuela, Guyana, Surinam, French Guyana, Colombia, Ecuador, Peru, Mexico, Brazil, Panama, Nicaragua, Guatemala, Costa Rica, Honduras, Belize and El Salvador.

7.1.2.2. Host Range

Though *R. cocophilus* is primarily found on coconut and oil palm, it has also been found to occur naturally in the date palm and cabbage palm (Hagley, 1962).

7.1.2.3. Symptoms

The disease is present most commonly in young coconut trees (3-10 years old) with maximum incidence in trees of 4-7 years old. Chlorosis first appears at the tips of the oldest leaves and spreads towards their bases. The brown lower leaves may break across the petiole or the lower part of the rachis or they may become partly dislodged at the base and hang down (Fig. 7.1). Premature shedding of the nuts may occur simultaneously with the development of the leaf symptoms or slightly before. As a result of the severe damage caused by the palm weevil internally, the crown often topples over about 4 to 6 weeks after the appearance of the first symptoms. The trunk then remains standing in the field for several months and finally it decays (Griffith and Koshy, 1989).

The most characteristic symptom of red ring disease is the internal lesions. In the beginning, about 3-5 cm beneath the stem surface, scattered reddish dots of about 1 mm diameter appear, which later coalesce to form an orange red ring of about 3 cm in width (Fig. 7.1). The ring extends the whole length of the stem and roots and in petioles it assumes a crescent-like shape. Large number of juveniles are seen in the centre of the discoloured areas and adults in the periphery. Shedding of green nuts of all sizes takes place.

Fig. 7.1. Red ring nematode (*Rhadinaphelenchus cocophilus*) on coconut. A-Infected plant. B-Cross section of trunk showing red ring symptoms (Courtesy: Sasser, 1971).

R. cocophilus causes little leaf disease of coconut and oil palm in Surinam and Guyana. A thermostable phytotoxin was produced due to breakdown of coconut tissue. Water uptake of infested coconut palms was less due to occlusion of xylem vessels. In coconut roots, the nematodes attack cortical tissues.

7.1.2.4. The Vector

The vector of the red ring nematode is the palm weevil (*Rhynchophorus palmarum*). The adult palm weevil acts as the carrier. The female weevil makes punctures in the softer tissue of the stem of the coconut palm with the help of its proboscis to lay its eggs. Normally the punctures are made in the soft axils to a depth of 5-8 cm and the eggs are placed deep in the puncture.

The nematodes are injected into the tissues of the coconut palm when the insect deposits its eggs. Larvae of the palm weevil while feeding in diseased trees get infested with the nematodes and the adult beetles which develop from them in turn become the carriers.

7.1.2.5. Life Cycle

Blair (1966) carried out preliminary studies on the life-history of *R. cocophilus* in immature nuts of coconut palm. When the nuts were inoculated with eggs containing active larvae, adult nematodes and eggs were recovered 9-10 days later. The nematodes completed one life cycle in 9-10 days which is probably one of the shortest life cycles reported for plant parasitic nematodes.

7.1.2.6. Survival

R. cocophilus can survive for a maximum of 15 days in soil. The third stage juvenile is the most persistent form.

7.1.2.7. Management Methods

(i) Cultural Methods: 'Cut and burn' the infested trees is one of the oldest and most effective practices used to prevent further spread of the disease.

(ii) Chemical Methods: More recent methods involve the destruction of palms using arboricides, herbicides or phytotoxic nematicides applied to holes made in tree-trunks. Hoyle (1968) recommended the use of cacodylic acid ('Silvisar') to poison trees. The chemical prevented weevil breeding, killed the diseased trees within two weeks, but had no nematicidal effect. Victoria *et al.* (1970) used sodium and potassium arsenate to kill both diseased trees and nematodes.

Adult weevils are attracted to anaerobic fermentation products such as ethyl alcohol, n-butyl alcohol and to certain volatile esters extracted from diseased palm tissue (Griffith, 1987). This behaviour can be exploited to attract weevils to poisoned baits at 25/ha.

The trap or guard baskets protect plantations by attracting and killing the palm weevils which may enter the plantations from nearby diseased trees. The guard baskets are made of 2 cm mesh wire. They are cylindrical, 1 m high and 0.3 m in diameter. These baskets are filled with chunks of fresh tissue from diseased coconut trees to attract the beetle. The guard baskets are sprayed completely with about 4.5 litres of methomyl solution and distributed on the ground in the plantations at 2.5 baskets/ha of young coconut trees. This procedure is especially recommended in the dry season when the weevils are most active in the cool nights. Guard baskets remain for about two weeks, after which the tissue and insecticide in the basket should be burnt. Fresh tissue should be placed in the basket and treated as previously described.

(iii) Biological Methods: Griffith (1977) reported the pathogenicity of a bacterium, *Micrococcus agilis* to the palm weevil and suggested its potential use as a biocontrol agent. The rhabditid nematodes of the genera *Steinernema* and *Heterorhabditis* can heavily parasitize the weevils by maintaining a high population of nematodes in the environment (Griffith and Koshy, 1989).

7.2. ARECA NUT

Areca nut (*Areca catechu*), otherwise known as the betel nut, has its origin in the humid regions of Asia and Malay Islands. It is an important cash crop of India. India is the largest producer of areca nut in the world covering an area of 3.65 lakh ha with annual production of 4.392 lakh tonnes. The major areca nut producing states include Kerala, Karnataka, West Bengal and Andaman and Nicobar islands. It is masticatory and the ripe fruits are sometimes used as an antihelminthic and astringent in Europe. It has a toning effect and stimulates the nerves. Areca nut industry forms the economic backbone of nearly 6 million people in India and for many of them it is the sole means of livelihood. The extract after boiling the nuts contains tannins which are used for tanning leather.

Although a number of nematodes have been reported from the rhizosphere of arecanut, the endoparasite reported as an important pest is only the burrowing nematode, *Radopholus similis*.

7.2.1. The Burrowing Nematode, *Radopholus similis*

R. similis was first reported from soil around areca nut roots in Mysore, Karnataka, India by Kumar *et al.* (1971). The threshold inoculum level causing significant damage to growth of areca nut was found to be 100 nematodes per seedling or one nematode in 800 g laterite soil (Koshy, 1986). Ten-fold increase in yield was recorded by treatment with nematicides (Sundraraju and Koshy, 1986).

7.2.1.1. Symptoms

The burrowing nematode infested areca palms exhibit non-specific above-ground symptoms like general yellowing and visible reduction in growth, vigour and yield. The characteristic symptom of *R. similis* infestation is the appearance of lesions and rotting of roots. The nematode produces small, elongate, orange-coloured lesions on the young, succulent creamy-white to light orange coloured portion of the main and lateral roots. Later these lesions coalesce and cause extensive rotting. The thick primary roots arising from the bole of the palm exhibit large, oval, sunken dark lesions. In areca nut, the tips of lateral and tertiary roots on infestation becomes black.

7.2.1.2. Life Cycle

At a temperature range of 21-31°C, *R. similis* takes 25-30 days to complete one life cycle (J2 to J2) on areca nut seedlings.

7.2.1.3. Ecology

Maximum population density of *R. similis* in roots was recorded during October-November and minimum during March-June.

7.2.1.4. Histopathology

The burrowing nematodes are found in inter and intracellular positions in the cortex. They do not enter the stellar region (Sundraraju and Koshy, 1988a).

7.2.1.5. Management Methods

(i) Cultural Methods: Use of nematode-free planting materials of areca nut and other inter/mixed crops and avoiding *R. similis* susceptible inter/mixed crops like black pepper and banana in infested areas helps in reducing the nematode population.

Complete control of *R. similis* was obtained on areca nut palms affected by yellow leaf disease (YLD) after four years of continuous application of neem cake at 1.5 kg/palm thrice a year (June, September, January). Significant increase in yield and decrease in disease indices of YLD-affected palms were recorded only in the fifth year.

(ii) Chemical Methods: Application of aldicarb or fensulfothion at 1g a.i./seedling thrice a year for three years controlled *R. similis* infestation in the seedling stage. In adult palms, reduction in nematode population and 10-fold increase in yield was obtained by application of fensulfothion at 50g a.i./palm in May-June, September-October and December-January for five years (Sundararaju and Koshy, 1986).

Complete control of *R. similis* was obtained on areca nut palms affected by yellow leaf disease (YLD) after four years of continuous application of fensulphothion at 50g a.i./palm or aldicarb 10g a.i./palm or DBCP at 10ml a.i./palm thrice a year (June, September, January). Significant increase in yield and decrease in disease indices of YLD-affected palms were recorded only in the fifth year.

Various chemicals tried for the management of *R. similis* on areca nut under nursery and field conditions is presented in Table 7.3.

Table 7.3. Effect of various chemicals for the management of *R. similis* in areca nut under nursery and field conditions

Chemical	Dosage	Duration & type of treatment	Effectiveness
i) Nursery level			
Aldicarb or Fensulfothion	1 g a.i./seedling	Thrice a year (in Sept., Dec. & March) for 3 years	Controlled soil and root population
ii) Field level			
Aldicarb	1 g a.i./palm	Thrice a year (in June, September & January)	Reduced nema popn. & increased the yield
DBCP	10 ml a.i./palm	Thrice a year (in June, September & January)	Reduced nema popn. & increased the yield
Fensulfothion	50 g a.i./palm	Thrice a year (in June, September & January)	Reduced nema popn. & increased the yield

(iii) Host Resistance: The areca nut hybrid VTL-11 x VTL-17 has been found to be highly resistant to *R. similis*. The cvs. Indonesia-6, Mahuva B and Andaman-5 are tolerant to *R. similis* and hence can be recommended for infested areas (Koshy *et al.*, 1979; Sundararaju and Koshy, 1982). The cvs. Indonesia-6 and Singapore (VTL-17) are known to yield 50% more nuts over local South Kanara variety.

(iv) Integrated Methods

- Use of nematode-free planting material of areca nut and other inter/mixed crops.

- Avoiding *R. similis* susceptible inter/mixed crops like black pepper and banana in infested areas.

- Use of tolerant/resistant varieties.

- Application of 5-10 kg of green manure/palm preferably *Glyricidia* or *Crotolaria*.

- Application of 1 kg neem oil cake/palm/year.

- Application of phorate at 3g a.i./plant in the root zone of areca nut, banana and black pepper in June-July and in September-October in areca nut based farming systems.

Integration of neem oil cake application at 1 kg/plant with phorate at 15 g a.i./plant gave effective control of *R. similis* in areca nut. In areca nut + banana + black pepper cropping system, integration of phorate at 15 g a.i./plant with neem oil cake at 500 g/plant was found most effective in reducing the nematode population (Sudha and Sundararaju, 1998).

7.3. COFFEE

Coffee (*Coffea arabica, C. canephora*) is an important commercial plantation crop mostly grown on the hilly slopes of the traditional coffee tracts in South India (Karnataka, Kerala, Tamil Nadu and Andhra Pradesh) and non-traditional Orissa, Maharashtra, Madhya Pradesh and North-Eastern Council States. The coffee industry is well established and flourishing enterprise. Besides being one of the major foreign exchange earner to a tune of around 2000 million rupees annually. Coffee industry supports more than four lakh people to earn their livelihood. The two main species of coffee in cultivation are *Coffea Arabica* (Arabica) and *Coffea canephora* (Robusta). *C. arabica* produces finer seeds which are more valued and grows best at higher elevations (1000 and 1500 m MSL). *C. canephora* is sturdier and better adapted to the warm climate prevailing in the lower elevations (500 and 1000 m MSL). Robusta, in general, is more tolerant to many diseases and pests including nematodes, as compared to Arabica. Coffee in India is grown under a canopy of shade trees of various species along with other plantation crops like black pepper, cardamom, banana, cocoa, etc. The present area under coffee in India is above 100,000 hectares, nearly three-fourth of which is under the fine quality Arabica species with most of the remaining area under Robusta.

Lesion, root-knot and reniform nematodes are recognized as the major limiting factors in successful production of Arabica coffee crop.

7.3.1. The Lesion Nematode, *Pratylenchus coffeae*

The lesion nematode is the most important pest of coffee in India. The annual loss of 20 million rupees was assessed in coffee due to this nematode in an area of about 1,000 ha in Karnataka alone (Van Berkum and Seshadri, 1970).

P. coffeae destroyed about 95% of *Coffea arabica* plantations in Java, which was subsequently found in other regions, including India and Central and South America.

7.3.1.1. Distribution

P. coffeae is proved to be highly destructive to Arabica coffee in Karnataka, Kerala and Tamil Nadu. Over 3,000 hectares have been infected with this nematode in coffee growing tracts of South India (Kumar, 1984).

7.3.1.2. Symptoms

In nursery, *P. coffeae* causes death of coffee plants.

The diseased trees are stunted, showing thin stems, and as a result, complete destruction of secondary roots and rootlets (Fig. 7.2). In some trees, severe injury caused by *P. coffeae* results in a corky region at the base of the trunk. Complete wilting of the foliage is also a typical symptom during dry seasons. Total defoliation occurs under severe attacks.

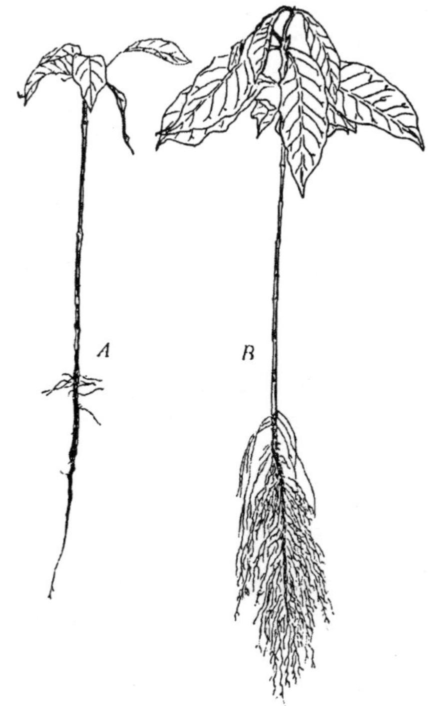

Fig. 7.2. Lesion nematode on coffee. Left – Infected; Right – Healthy
(Courtesy: Thorne, 1961).

The infected plants show loss of primary roots. The lesion nematode causes wounds in roots through which other pathogenic organisms, such as fungi and bacteria, enter the root tissues. The interaction of these agents results in the formation of lesions that finally destroys the root tissues. The root system of seedlings infected by *P. coffeae* is poor and the plants can be pulled out of the ground easily. Leaves and stems may show symptoms of mineral deficiencies.

7.3.1.3. Life Cycle

Lesion nematodes are typically migrant parasites of plant roots. The nematode penetrate slightly behind the elongating zone in the piliferous region. Adults and juveniles of various ages constantly migrate into and out of the roots. The first molt in the life cycle of *P. coffeae* takes place within the egg while the remaining three occur outside the egg. Eggs hatch in 6-8 days at 28-30°C. Within 6 more days, the second-stage juvenile forms. Development to the third stage occurs at 21 days and to the fourth at 28 days. Adult males emerge at 29-32 days.

7.3.1.4. Host Range

P. coffeae is known to attack coffee, banana, citrus, grapevine, apples, strawberry, abaca, yam, cocoa, rubber, potato, sweet potato, peas, beans, cabbage, cauliflower, tomato, chrysanthemum, caladium, dahlia, marigold, peanut, areca nut, cardamom, etc.

7.3.1.5. Survival and Dispersal

P. coffeae is capable of surviving for about 18 months in moist soil under laboratory conditions but failed to exist beyond 12 months under field conditions.

The nematode is disseminated through working implements and running water especially across the slope. Nematodes enter the nursery through jungle soil and breed there.

7.3.1.6. Biological Races

Existence of as many as four biological races in *P. coffeae* i.e., coffee, cardamom, pavetts and bamboo races has been recognized and coffee race is specific to coffee (Kumar, 1988).

7.3.1.7. Management Methods

(i) Cultural Methods: Nematode control in coffee is achieved through cultural practices like maintaining heavy shade, providing mulch to individual plants, application of FYM and giving frequent irrigation or providing nutrients through foliar spray. Soil application of coffee pulp at 18 kg/sq.m. reduced the nematode population and increased the plant growth.

Lordello (1968) reported that replanting of old nematode-infested plantations preceded by fallow period of at least 2 years gave effective control of the lesion nematode.

(ii) Chemical Methods: In field trials with young trees of Arabica coffee, phenamiphos at 0.5 to 1.0 g/plant gave good control of *P. coffeae* and remained effective for 90 days after application. Repeated applications were necessary to maintain continuous control (Kumar, 1982).

(iii) Integrated Methods: It is advisable to change nursery site once every 7-8 years. Jungle soil and FYM should be dried and well sieved before use. The nursery site should also be dug and exposed to sun. The nematode-free plants should be used for planting. Affected plants should be pulled out and burnt. The planting site should be dug and the soil exposed to the sun for at least one summer season. Care should be taken to keep the site free from weeds. The site should be planted with either Robusta or Arabica-Robusta grafted plants. Grafting of Arabica scion on Robusta rootstock in 40 day-old 'topee stage' plants by using wedge-cleft method can be adopted (Kumar, 1988) (Fig. 7.3).

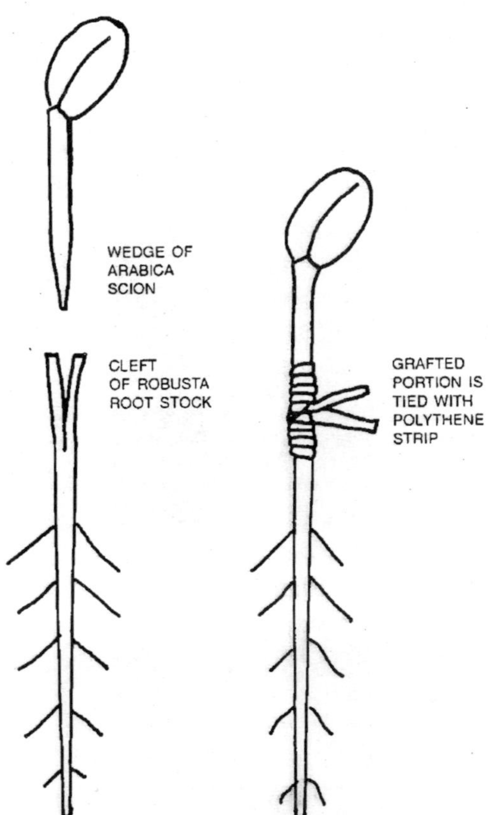

Fig. 7.3. Grafting of Arabica coffee scion on Robusta rootstock
(Courtesy: Indian Coffee).

7.3.2. Root-knot Nematodes, *Meloidogyne exigua, M. coffeicola*

Srinivsan and D'Souza (1965) reported *M. exigua* on coffee from South India. *M. exigua* is a pest of *Coffea arabica* but not of *C. canephora* (robusta coffee) which is very resistant (Schieber and Sosa, 1960).

7.3.2.1. Economic Importance

A 5% loss in yield due to *M. coffeicola* damage has been reported in Hawaiin coffee plantations. The yield of non-infected plants was more than that of artificially inoculated plants (104 and 50 kg, respectively). Losses due to *Meloidogyne* spp. in Brazil are estimated at 15% of that country's production

7.3.2.2. Symptoms

Affected trees look unthrifty and have chlorotic, wilting and dying leaves. *M. exigua* produces small elongated galls that are located primarily at the root tips (Fig. 7.4). In addition, roots may show cracks, necrosis and sloughing of cortical tissue. The effectiveness of root system is greatly diminished, with rootlets and root hairs practically eliminated. Trees may die, particularly during periods of drought and cold temperatures. Affected plants lack vigour and are unable to withstand adverse conditions.

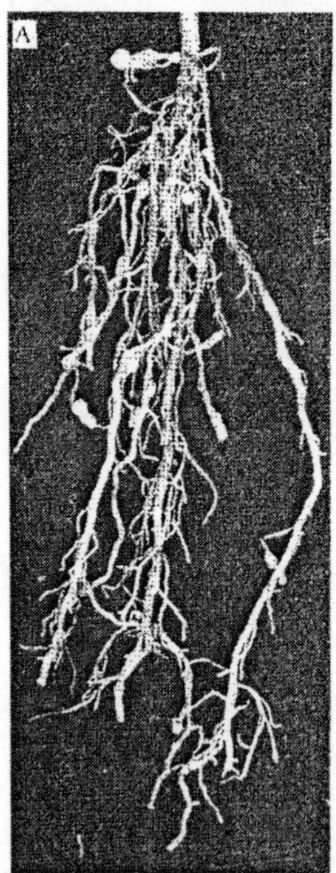

Fig. 7.4. Arabica coffee roots infected with *Meloidogyne exigua*
(Courtesy: Lordello, 1972).

7.3.2.3. Life Cycle

M. exigua takes 32-42 days at 25-30°C to complete the life cycle. The egg masses of *M. exigua* are mostly located under the epidermis of coffee roots.

7.3.2.4. Hosts

M. exigua is known to attack coffee, tea and sweet pepper

7.3.2.5. Survival and Spread

M. exigua does not survive in the soil, in the absence of the host, for more than six months after eradication of infested plants. The nematode spreads through the infested seedlings used for transplanting in the main field.

7.3.2.6. Interaction with Other Pathogens

The disease complex involving *Rhizoctonia solani* and *M. exigua* caused more root necrosis and defoliation when both pathogens were inoculated. Extensive fungal colonization within the coffee root systems was observed.

7.3.2.7. Management Methods

(i) Cultural Methods: At the crop renovation time, the infested coffee area may be turned to pasture. Most graminicolous plants are non-hosts to *M. exigua* and *M. coffeicola*. The decision of growing forage crops and turning back again the area to coffee will certainly reduce drastically the population of *Meloidogyne* species in 6-12 months.

Rotation of coffee with cotton, soybean and maize in *M. exigua* infested areas for one year significantly reduced the nematode population, after which the farmer can return to coffee cultivation. Coffee can also be rotated with groundnut, castor (both immune), *Styzolobium deeringiatum* and *Crotalaria spectabilis* (both resistant).

(ii) Chemical Methods: Application of aldicarb, carbofuran and phenamiphos each at 1.6 to 6.0g a.i./plant, in one or two applications during the year gives effective control of root-knot nematodes. The first application should be in the beginning of the rainy season followed by the second 3 months later.

(iii) Host Resistance: Coffea canephora (line 2258 from Costa Rica) showed resistance to *M. exigua* and *M. incognita*. Planting of *Coffea arabica* var. Mundo Novo or Catuai Vermelho grafted on line 2258 in field showed good growth and production in Brazil.

7.3.3. The Reniform Nematode, *Rotylenchulus reniformis*

The reniform nematode has been observed on coffee in India, Puerto Rico, Brazil and Hawaii.

7.3.3.1. Symptoms

R. reniformis infection of coffee trees results in an almost complete absence of feeder roots, a poorly developed tap root, and the yellowing and wilting of above-ground parts. In soils heavily infested with reniform nematode, Arabica coffee cannot be grown.

7.3.3.2. Life Cycle

R. reniformis is a sedentary ectoparasitic nematode. Only females feed on plant tissues; males are non-parasitic. These nematodes reproduce by cross fertilization. Females lay about 120 eggs approximately 9 days after entering the root. The eggs are deposited in a gelatinous matrix from which second stage juveniles hatch within 8 days. Juveniles of both sexes develop rapidly in the soil where 3 molts occur. Apparently, individual nematodes do not feed during juvenile development.

After penetrating the root, the young female feeds, enlarges progressively, the body eventually assuming a kidney shape, hence the common name, reniform nematode. The posterior portion of the body usually protrudes from the root surface and can easily be seen in properly stained material.

7.3.3.3. Management Methods

(i) Cultural Methods: In Puerto Rico, crop rotation with Pangola grass and sugarcane have effectively controlled *R. reniformis*. Rotation with sugarcane has also provided effective control in Brazil.

(ii) Host Resistance: Coffeae canephora var. robusta may have some resistance to infection by *R. reniformis*.

7.3.4. Ring Nematodes, *Hemicriconemoides* spp.

7.3.4.1. Symptoms

Association of large number of the ectoparasitic nematodes like *Hemicriconemoides coffeae, H. cocophilus* and *H. gaddi* with Arabica and Robusta coffee has been reported to cause "Crinkle Leaf" disorder of coffee in South India (Kumar, 1985).

7.3.4.2. Management Methods

(i) Chemical Methods: Both phenamiphos (Nemacur 5G at 25 g/plant) and carbofuran (Furadan 3G at 30 g/plant) were effective in reducing the population of *H. gaddi*. The treated plants put forth more new and healthy vegetative growth.

7.4. TEA

India is the largest producer tea (*Camellia sinensis*) in the world. It accounts for 20% of the total area under tea in the world, 28% of world production, 22% of global tea consumption and 15% of the total global tea exports. The total area under tea is 5,07,200 ha. The tea industry provides direct employment for one million workers of which sizeable number are women. The largest producers of tea are Assam (50.7%), West Bengal (22.1%), Tamil Nadu (15.9%) and Kerala (8.3%).

More than half of the one million hectares of tea cultivation is in India and Sri Lanka, on areas ranging from 225,000 to 350,000 hectares. Between them, these two major producers account for more than three-quarters of the world's export trade of 600,000 metric tonnes of this commodity. This apparently disproportionate share in the export trade is because in these two countries, domestic consumption is far outstripped by total production. In India, tea cultivation is restricted to the North Eastern and Southern regions.

Root-knot, lesion and burrowing nematodes are recognized as the major limiting factors in successful production of tea crop.

7.4.1. Root-knot Nematodes, *Meloidogyne incognita* (in nurseries), *M. brevicauda* (in field)

Barber (1901) reported *Meloidogyne* sp. in tea seedlings from south India. Seshadri and Kumarswami (1964) reported *M. brevicauda* on tea in Tamil Nadu. Root-knot nematodes are the most frequently encountered in the tea growing regions of India. Most of the species of *Meloidogyne* except *M. brevicauda* are associated only with nursery plants.

7.4.1.1. Symptoms

(i) On Young Tea: The first report of root-knot nematode in young tea was from South India, where large number of seedlings were found infected (Barber, 1901). *Meloidogyne incognita, M. javanica* and *M. hapla* have been found associated with tea in India.

Young nursery plants, both seedlings and vegetatively propagated clonal plants, are severely damaged by root-knot nematodes. Seedling plants in which both the tap-root and lateral rots are severely attacked suffer greater damage than do clonal tea plants of equivalent age, probably because seedling tea plants possess less than half the root bulk of clonal plants of similar age. A marked increase in resistance is observed between 8 and 15 months.

(ii) On Mature Tea: M. brevicauda is the only root-knot nematode that is a serious pest of mature tea. Affected bushes are unthrifty, have severely

galled main roots and lack feeder roots. *M. brevicauda* occurs in southern India at altitudes of 3000 to 3500 ft. in the Wynaad area (Venkata Ram, 1963). It can attack both seedlings and mature tea.

Root-knot nematodes (*M. brevicauda*) cause severe galling on the roots of mature tea bushes with resultant decline (Fig. 7.5). Tea plants of all ages are susceptible to infection by this nematode which has been recorded in only South India and Sri Lanka.

Fig. 7.5. Heavy root galling of tea with *Meloidogyne brevicauda* (Sivapalan, 1972).

The infected bushes grow more slowly and the leaves tend to become smaller, dull and yellow in appearance. The roots show the characteristic swelling and pitting that is typical of *M. brevicauda*.

The effect of infection on the growth of tea bushes is most evident during the period of recovery from pruning. Healthy bushes are usually harvested about 16 weeks after pruning, but the harvesting of severely infected bushes is necessarily delayed up to a further 6 months. The production of new branches by infected bushes is very restricted, and as a consequence the cropping capacity of the bush is reduced. The most severely affected bushes fail to recover from pruning.

7.4.1.2. Life Cycle

The average size of mature female of *M. brevicauda* is about 5 or 6 times that of a mature female of *M. incognita*. Despite this massive size, the females are often observed to be empty, with only a few eggs. The mean hatch per egg mass is around 10, whilst in the other common species this is in the order of 200 to 600 juveniles/egg mass. The rest of the life cycle is very similar to the other species of *Meloidogyne*.

7.4.1.3. Management Methods

(i) Management in Nurseries

(a) Physical Methods: Soil solarization in nurseries is adequate to manage root-knot and other nematodes as the plants develop complete resistance by about 9-15 months (Gnanapragasam *et al.*, 1989).

(b) Cultural Methods: Every effort should be made to raise the nurseries in nematode-free soil. Incorporation of organic matter into nematode infested soil is known to suppress the incidence of root-knot nematodes on tea.

The cultivation of marigold as a pre-plant crop considerably reduces the population of *Meloidogyne* spp. and *Pratylenchus loosi*. The reduction in the nematode population is significantly greater than when the land is left fallow. The interplanting of marigold in tea has resulted in a 7% increase in yield compared to 10% increase that followed soil fumigation with DBCP.

The rehabilitation crops such as Mana grass (*Cymbopogon confertiflorus*) or Guatemala grass (*Tripsacum laxum*) (which help to improve the soil) are resistant to *Meloidogyne* spp. The soil population of parasitic nematodes decline rapidly when these grasses are grown for one or more years before replanting tea.

(c) Chemical Methods: In field trials, aldicarb at 0.5 and 1.0g and carbofuran at 1g/tea seedling were found effective for the control of *M. incognita* and also caused significant increase in plant height (Basu and Gope, 1985).

(d) Integrated Methods: Combination of neem cake at 5 g/seedling + carbofuran 3G at 1.5 g/seedling improved plant growth characters and reduced number of galls, egg masses and eggs/egg mass (Kalita and Bora, 2006).

(ii) Management Under Field Conditions

(a) Cultural Methods: The first step in the management of nematodes in replanted areas is by ensuring a very thorough uprooting of all old tea roots (up to pencil thickness). In Sri Lanka, such uprooted old tea lands are routinely reconditioned by planting deep-rooted grass species including Guatemala grass and Mana grass which are non-hosts to parasitic nematodes and bring down the residual nematode population (Kerr and Vythilingam, 1966).

Shade trees and green manure crops help to provide shade to the tea plants, aid in N fixation and provide regular supplies of mulch in the form of leaf litter and loppings which help in reducing the nematode population during their decomposition. The cultivation of shade trees and green manure crops susceptible to tea nematodes should be discouraged in tea plantations.

Plants such as marigolds, *Tagetes erecta* and *T. patula*; *Vetiveria zizanoides* and *Eragrostis curvula* have been found to help reduce nematode populations.

Large inputs of organic matter including cow dung and well decomposed plant residues have been reported to suppress nematode populations. In India, neem cake has been recommended to manage root-knot nematodes in the tea field (Muraleedharan, 1991). In Sri Lanka, castor oil cake has been proved to be the most effective one for suppressing nematodes, followed by mahua, neem and karanj cakes (Gnanapragasam, 1991).

(b) Chemical Methods: The recommended nematicides for the management of root-knot nematodes in mature tea include phenamiphos (Nemacur 5%G) and carbofuran (Furadan or Curaterr 3%G) at 10g/plant applied after pruning (Gnanapragasam, 1987a).

7.4.2. The Lesion Nematode, *Pratylenchus coffeae, P. loosi*

The most destructive and most intensively investigated nematode pest of tea is the root lesion nematode *P. loosi*, which is equally destructive to both young and mature tea plants and is a problem in nurseries and in mature tea fields. Crop loss in tea infested with *P. loosi* has been estimated to average between 225-350 kg made tea/ha/year (Sivapalan, 1972). Sasser (1989) estimated the average crop loss on a worldwide scale to be 8.2%.

7.4.2.1. Symptoms

The infested bushes show unthrifty tea with weak maintenance foliage. The leaves appear pale yellowish and dull mainly due to nutritional imbalances caused by altered rate of uptake of essential nutrients. The heavily infested plants show tendency for premature flowering and fruiting.

The roots of infested plants show typical marked reduction in the feeder root system. The feeder roots appear brownish in colour and dark brown necrotic patches appear on both the feeder roots, as well as the storage roots. Under greenhouse conditions, the damage threshold level of *P. loosi* to tea was estimated to be 40 nematodes/100 g soil at 24°C in a clayey loam soil (Gnanapragasam and Manuelpillai, 1984).

7.4.2.2. Host Range

The presence of crops such as sorghum, *Tephrosia vogelii*, *Sesbania* spp. and *Acacia* spp., as well as certain weeds, increase the incidence of this nematode in tea fields.

7.4.2.3. *Survival and Spread*

P. loosi is known to survive in host-free soils in the lesions of the larger old storage roots of tea that are left uncleared, following uprooting of old tea fields, for as long as three years.

One of the most important means of spread of *P. loosi* amongst tea areas, is by the dissemination of infested plants to fields from contaminated nurseries. Spread of nematodes could also occurs through: (i) movement of infested soil and water; (ii) uprooting of old tea fields; (iii) use of contaminated irrigation water in nurseries.

7.4.2.4. *Management Methods*

(i) Cultural Methods: The incorporation of large quantities of green manure such as loppings from dadaps, *Tephrosias* and marigolds, resulted in population decline of *P. loosi*. Mulching with Guatemala grass at 20 to 30 t/ha gave effective control of *P. coffeae* on tea. The use of specific potassium-enriched fertilizers to both young and mature tea plants is known to retard population build up of *P. loosi* and at the same time helps the plants to better tolerate nematode parasitism (Gnanapragasam, 1982).

The cultivation of marigold (*Tagetes erecta* and *T. patula*) as a pre-plant crop considerably reduces the population of *P. loosi* and *Meloidogyne* spp. The interplanting of marigold in tea has resulted in a 7% increase in yield. *P. loosi* is attracted to marigold roots, but upon entry the nematodes die within a week. Roots of marigold in the vegetative stage are effective in this way, but roots of plants in the full bloom or post-bloom are less effective. Consequently, it is preferable to maintain the plants in a non-flowering state by periodic lopping. Growing of grasses like Guatemala (*Tripsacum laxum*), Mana (*Cymbopogon confertiflorus*) and *Eragrostis curvula* helps to bring down the population of *P. loosi*.

The incorporation of oil cakes such as neem, coconut as well as decomposed tea waste, help to curtail pathogenicity caused by *P. loosi*. Increasing the dosage of K fertilizer brought about a decline in the population level of this nematode. Application of N in the form of urea brought about a significant suppression in nematode population.

7.4.3. The Burrowing Nematode, *Radopholus similis*

R. similis has been reported attacking tea in Java. Tea is a good host of the burrowing nematode. All stages of the life cycle can be found in tea roots. *R. similis* is known to cause damage to tea at elevations ranging from 200 to 1000 meters.

7.4.3.1. Symptoms

R. similis forms brownish lesions mostly on feeder roots and the infested roots appear dried up. Under greenhouse conditions, the damage threshold level of *R. similis* to tea was estimated to be 28 nematodes/100 g soil at 24°C in a clayey loam soil (Gnanapragasam and Herath, 1989).

7.4.3.2. Host Range

R. similis also infects banana and black pepper. The latter is commonly intercropped along with tea in the mid-elevation tea areas of Sri Lanka. Guatemala grass which is often used to recondition old tea fields prior to replanting, was also found to be a host for *R. similis*.

7.4.3.3. Survival and Spread

The method of spread of *R. similis* in tea is very much similar to that of *P. loosi*. However, the survival rate in host-free soil is much shorter for *R. similis*.

7.4.3.4. Management Methods

(i) Cultural Methods: Eragrostis curvula, marigold and *Vetiveria conyzoides* appear to suppress soil populations of *R. similis*.

(ii) Host Resistance: In Sri Lanka, tea clone CH 13 was found resistant to *R. similis* (Gnanapragasam, 1983).

7.4.4. The Reniform Nematode, *Rotylenchulus reniformis*

R. reniformis has been found to be responsible for large scale casualties in young tea fields. Damage by this nematode species is found in nursery plants and newly planted young tea fields.

7.4.4.1. Symptoms

Thorne (1961) reported that in tea nurseries in Java, the young tea plants were killed by the attack of *R. reniformis* on the tap root, but not when a strong tap root had first been established.

7.4.4.2. Host Range

Perennial crops that are sometimes intercropped with tea, including black pepper, coffee and cloves, as well as grass cover crops, including Guatemala and Mana that are planted in uprooted tea fields for a process of reconditioning, are good hosts of *R. reniformis*.

7.5. COCOA

Cocoa (*Theobroma cacao*) is an important commercial crop grown the world over. It is native of Amazon Valley of South America. In India, it is being cultivated in Kerala, Karnataka and Tamil Nadu. Kerala account for about 76% of the area and 78% of the total production in the country.

Root-knot nematodes are recognized as the major limiting factors in successful production of cocoa crop.

7.5.1. Root-knot Nematodes, *Meloidogyne* spp.

Root-knot nematodes are the most important nematodes of cocoa due to their pathogenicity and wide distribution in cocoa producing regions. *Meloidogyne* spp. have been found in cocoa since 1990 (Sosamma *et al.*, 1980a), and they have been reported from Zaire, Sao, Java, Cotterel, Ghana, Malawi, Ivory Coast, Nigeria, Venezuela, Brazil and India (Sosamma *et al.*, 1980b). *M. incognita* seems to be the most frequently found in cocoa. *M. incognita* and *M. javanica* have caused losses in cocoa areas around the world including yield decrease, sudden death of trees and growth retardation of seedlings in nursery.

7.5.1.1. Symptoms

M. incognita causes dieback, stunting, wilting, yellowing of leaves and small leaves. On roots tiny galls and females with egg masses can be observed. The leaf tips and margins first turn brown and become dried. This spreads to the entire leaves which are eventually shed. The infested plant exhibit the syndrome of sudden death.

7.5.1.2. Survival and Spread

The commonly used shade plants such as banana may become a source of inoculum in the cocoa plantation (Sosamma *et al.*, 1980). Nursery soil infested with the nematodes will allow the production of infested seedlings which will spread nematodes into plantations. Run off water may also spread nematodes.

7.5.1.3. Management Methods

(i) Cultural Methods: The most efficient control strategies include production of seedlings free from major pathogenic nematodes and to cultivate in soils or areas from which the nematodes are absent. Soil to be used in the nursery should be collected from areas that are not infested by root-knot and *Pratylenchus* spp. Care must be taken with the selection of shade plants, avoiding trees susceptible to root-knot or lesion nematodes such as *Leucaena glauca* and banana (Sosamma *et al.*, 1980a).

(ii) Chemical Methods: Application of granular nematicides such as phenamiphos (Nemacur 10G), aldicarb (Temik 10G) at 50 mg of commercial product/plant (Sharma and Ferraz, 1977) helps in reducing the nematode population. Tarjan *et al.* (1973) reported increased yield of 11 to 96% after field application of ethoprophos (Mocap) or fensulfothion (Terracur) at 34 kg of the commercial product/ha and phenamiphos (Nemacur) at 22 kg of the commercial product/ha.

7.6. OIL PALM

Oil palm (*Elaeis guineensis*) is the highest oil-yielding plant (4-6 tonnes oil/ha) among perennial oil-yielding crops, producing palm oil and palm kernel oil. These are used for culinary as well as industrial purposes. It occupies an area of 44,413 ha in Andhra Pradesh (80% area), Karnataka, Assam, Gujarat, Goa, Kerala, Maharashtra, Orissa, Tamil Nadu, Tripura and West Bengal.

The red ring nematode is recognized as the major limiting factor in successful production of oil palm crop.

7.6.1. The Red Ring Nematode, *Rhadinaphelenchus cocophilus*

The red ring disease caused by *R. cocophilus* is the most important disease of oil palm. *R. cocophilus* causes economic loss in oil palm in South Africa. Malaguti (1953) observed that in a 7 year-old oil palm grove suffering from little leaf, many trees had died.

7.6.1.1. Symptoms

Diseased palms in young groves are initially found clustered which gradually expand. In older groves, such palms have a random distribution. Little leaf is the prominent symptom in diseased palms which have erect, short and deformed leaves. The top part of the main vein bore suberized patches. The pinnae are shorter, wavy and necrotic at the tips and yellow patches are present on the petioles and leaf bases. The diseased palm has a brownish discolouration. *R. cocophilus* is found in discoloured tissue and thrived more in the petioles of young folded leaves than in stems and roots. A notable feature is that the band of necrotic tissue is always very small. In diseased oil palms, an average of less than 500 nematodes is found in one gram of infected tissue.

7.6.1.2. Management Methods

Same as in coconut.

7.7. BETEL VINE

Betel vine (*Piper betle*) is a perennial, dioecious, evergreen creeper grown in India. It is a chewing stimulant of brain, lungs and heart. The leaves contain B and C vitamins. It is an important cash crop in Andhra Pradesh, Assam, Bihar, Karnataka, Kerala, Madhya Pradesh, Maharashtra, Orissa, Tamil Nadu, Tripura, Uttar Pradesh and West Bengal with an annual turnover of about Rs.700 crores. Its leaves are exported to Pakistan, Bangladesh, Indonesia, Malaysia, Burma and Thailand.

Root-knot nematodes are recognized as the major limiting factors in successful production of betel vine crop.

7.7.1. The Root-knot Nematode, *Meloidogyne incognita*

M. incognita was responsible for 21-38% loss in leaf yield of betel vine (Saikia, 1992; Jonathan *et al.*, 1990).

7.7.1.1. Symptoms

Blackening and drooping of the growing tip and yellowing of leaves occurs due to root-knot nematode infestation. Galls of varying sizes and shapes develop in roots leading to quick root decay.

7.7.1.2. Interaction with Other Pathogens

Association of *M. incognita* with severe wilt symptoms of betel vine was reported from India (Mammen, 1974) and *M. incognita* is known to predispose betel vine to root-rot caused by *Phytophthora palmivora* (Sivakumar *et al.*, 1987).

7.7.1.3. Management Methods

(i) Physical Methods: White polyethylene mulching for 15 days on beds prepared for planting betel vine revealed high reduction in nematode populations due to increased soil temperature of 44.1°C (Sivakumar and Marimuthu, 1987).

(ii) Cultural Methods: Nematode-free planting material is a cardinal principle in the management of betel vine nematodes. Summer ploughing, summer ploughing and solarization or rice cropping have been beneficial. Application of FYM at 25 t/ha or neem cake at 1 t/ha before planting or crop rotation with rice helps to reduce the nematode population in soil. Considerable reduction in nematode populations in the soil and number of galls on

roots has been reported after application of 50-75 kg K$_2$O/ha (Jagdale *et al.*, 1985a).

Crop Rotation: A pre-rice crop reduces populations of *M. incognita, Rotylenchulus reniformis* and *Helicotylenchus incisus* in the subsequent betel vine crop. Crop rotation with non-host crops like marigold, sesame and groundnut also suppress root-knot nematode populations. Support plants such as *Erythrina indica, Moringa oleifera* and *Bombax malbaricum* which are not susceptible to root-knot nematodes can be used in betel vine gardens.

Intercropping: In pots having 5 seedlings of *Tagetes erecta* around a betel vine plant, the population of *M. incognita* was reduced by 41.4% and root galling by 54% (Medhana *et al.*, 1985).

Botanicals: Green leaf manuring with African marigold, neem or *Xanthium* have also been quite effective against root-knot, lesion and other nematodes.

Soil application of neem oil cake at 1 tonne/ha and sawdust at 2 tonnes/ha can reduce nematode populations, number of galls and increase the number of leaves harvested significantly (Jagdale *et al.*, 1985b). Significant reduction (60%) in the nematode population has been observed in beds amended with chopped and shade dried leaves of *Calotropis gigantea* at 2.5 t/ha followed by neem oil cake and poultry manure (44.4 and 40.9%, respectively). Beds amended with *C. gigantea* leaves yielded 14.2 kg/4840 leaves and with neem oil cake yielded 12.1 kg/4220 leaves. Soil amendment with sawdust at 2 t/ha + NPK (3638 leaves) and neem cake at 2 t/ha is effective in reducing nematode numbers and increasing yields (Sivakumar and Marimuthu, 1986b).

(iii) Chemical Methods: Application of aldicarb and carbofuran at 0.75 kg a.i./ha reduces nematode populations by 71 and 55%, respectively, resulting in increased yields. The granules degraded to non-detectable levels 41 days after application (Sivakumar *et al.*, 1987). Aldicarb, carbofuran and benfurocarb applied at 1.5, 3.0 and 5.0 kg a.i./ha, respectively, in furrows on either side of the rows can reduce *M. incognita* populations in soil and galling on the roots significantly (Dethe and Pawar, 1987).

(iv) Biological Methods: Three applications of *Paecilomyces lilacinus* grown on neem cake at 10 g/vine at 60 days interval gave a significant reduction in nematode infestation and increased vine growth and leaf yield under field conditions (Jonathan *et al.*, 2000). Nakat *et al.* (1995) reported that the application of *P. lilacinus* at 8 g/kg soil as the best treatment for the management of root-knot nematodes and increasing the leaf yield by 38.3 to 56.4% in betel vine.

(v) Host Resistance: The cvs. Kakair, Bangla, Karapaku, Gachipan, Aswani pan and Berhampuri are reported to be tolerant to root-knot (Anon,

1987). The cv. Kuljedu had the lowest root-knot index and number of egg masses per plant (Jagdale *et al.*, 1985a; Sivakumar *et al.*, 1987).

(vi) Integrated Methods: Trichoderma viride mixed with mustard oil cake applied in three split doses reduced the nematode infestation and enhanced the leaf yield.

Because of the residue problems on leaves, it is preferable to manage root-knot nematode infestation on betel vine by adopting the following non-chemical measures:

- Crop rotation wherever possible.
- Use of resistant/tolerant cultivars.
- Use of non-living standards or nematode-resistant live standards for supports.
- Soil solarization by mulching the land with clear polythene (100 gauge) before planting.
- Application of organic amendments such as leaves of neem and *Calotropis*, and sawdust at 2 t/ha.
- Supply of nitrogen through neem oil cake at 2 t/ha.

7.7.2. The Burrowing Nematode, *Radopholus similis*

The burrowing nematode has been reported to cause yellow/slow wilt disease of betel vine in India.

7.7.2.1. Symptoms

The symptoms produced on betel vine are akin to the symptoms caused by *R. similis* on black pepper vines (Eapen *et al.*, 1987).

7.7.2.2. Management Methods

The integrated management schedules suggested for the control of nematodes on black pepper, other than application of nematicides, can be largely adopted with modification to suit the local conditions for controlling *R. similis* on betel vine.

7.7.3. The Reniform Nematode, *Rotylenchulus reniformis*

Acharya and Padhi (1987) found *R. reniformis* to be pathogenic to betel vine. At inoculum levels of 1,000 and 20,000 nematodes/cutting, the reduction in number of leaves was 20 and 60%, respectively.

7.7.4. Root-knot Nematode, *Meloidogyne incognita* and Foot Rot, *Phytophthora capsici* Disease Complex

7.7.4.1. Management Methods

(i) Integrated Methods

- Summer ploughing and exposing the field to sunlight during May prior to sowing of *Sesbania* sp. (live standard) minimizes the initial load of inoculum of both the nematode and the fungus.

- Selection of healthy seed vines from nematode-free and disease-free mother plants for planting.

- Dipping seed vines in 0.25% Bordeaux mixture solution for 5 min before planting.

- Application of FYM at 30 t/ha to promote multiplication of antagonistic microbes which in turn kills nematodes.

- Spot application of neem cake enriched with *P. lilacinus* at 3 t/ha in 3 split doses -1st dose at 45 DAP and remaining 2 splits at 45 days interval during North-East monsoon season (Oct-Dec).

- Rotation of betel vine crop with rice.

Combined application of *Pseudomonas* sp. (Pfbv 22) and *Bacillus* sp. (Bbv 57) gave significant reduction in nematode infestation (gall index, no. of egg laying females and soil population), wilt disease incidence and increase in leaf yield. The treatment also enhanced the biochemical markers responsible for induced systemic resistance such as peroxidase, polyphenol oxidase and phenylalanine ammonia lyase (Jonathan *et al.*, 2004).

Table 7.4. Biological control of plant parasitic nematodes in plantation crops

Crop	Nematode	Effective bioagent/s	Reference
Coconut	*R. similis*	*G. mosseae, G. fasciculatum, G. macrocorpus, G. versiforme, Acaulospora bireticulata, Sclerocystis rubiformis, Scutellospora nigra*	Koshy *et al.*, 1999
Areca nut	*R. similis*	*P. lilacinus*	Sudha *et al.*, 2000
Betel vine	*M. incognita*	*P. lilacinus*	Jonathan *et al.*, 1995
			Nakat *et al.*, 1995
			Hazarika *et al.*, 1998
	M. incognita & R. reniformis	*P. lilacinus*	Jonathan *et al.*, 1995
	R. similis	*P. lilacinus*	Sosamma *et al.*, 1994

Chapter 8

SPICE CROPS

From time immemorial, India is regarded by the rest of the world as the "magic land of spices". Spices are natural plant products used to improve flavour, aroma, taste and colour of food products, beverages, liquors, pharmaceutical, cosmetic and perfumery products. No other country in the world has such a diverse variety of spice crops as India. Indian spices are also noted for their excellent aroma, flavour and pungency not so easily matched by any other country. Even in minute quantities, spices are a real delight to all our sensory organs, making the food dishes more palatable. Usage of spices in food processing industries is increasing quite fast in all the countries. Hence, the world demand for spices in recent years is continuously on the rise.

India is the largest producer, consumer and exporter of spices in the world. The annual production of spices is more than 5 million tonnes valued at US$ 1,500 million. India's share of the world spices trade is estimated at 45 to 50 % by volume and 25% by value. The present annual production of spices in the country is 5.113 million tonnes from over 5.155 million hectares (Table 8.1). About 8.5% of India's export earnings from agricultural and allied products come from spices. The important spice crops include black pepper, cardamom, ginger, garlic, turmeric and chilli. The major spice-growing states are Rajasthan, Andhra Pradesh, Gujarat, Madhya Pradesh, Karnataka, Orissa, Tamil Nadu, Kerala, Uttar Pradesh, Maharashtra and West Bengal.

Table 8.1. Area, production and productivity of spices in India (2004-05) (Chamber, 2006)

Spice crop	Area ('000 ha)	Production ('000 tonnes)	Productivity (tonnes/ha)
Black pepper	223.1	70.6	1.3
Turmeric	149.4	529.0	3.5
Ginger	85.1	305.9	3.6
Coriander	282.5	172.3	0.6
Total	5155.0	5113.0	1.0

8.1. BLACK PEPPER

Black pepper (*Piper nigrum*), known as the "King of spices", has its origin in the South-Western hills of India. It is a branching and climbing perennial shrub belonging to the family Piperaceae and is grown in the hot and humid parts of the world. Pepper is loved for its spicy factor, piperine. It is used in various culinary seasonings, as a preservative for meat and fish, in medicine and perfumery (pepper oil). In India, it is a common practice to use live trees as standards for climbing pepper. Pepper is the most important spice produced in the country and accounts for 50-60% of the total export earnings. It is being cultivated in 0.223 million hectares producing 0.070 million tonnes of black pepper with an average yield of 1.3 tonnes per hectare.

Although several nematodes are reported on black pepper, the two most important pathogens recorded are the burrowing nematode, *Radopholus similis* and the root-knot nematode, *Meloidogyne* spp.

8.1.1. The Burrowing Nematode, *Radopholus similis*

Van der Vecht (1950) first reported the association of the burrowing nematode with the "Yellows disease" of black pepper and also proved the pathogenicity of *R. similis*. "Yellows disease" was reported to have wiped out 22 million pepper vines within 20 years in Bangka Island in Indonesia. *R. similis* is found associated with "Slow-wilt" disease of black pepper in Kerala and Karnataka, India (Mohandas and Ramana, 1987; Ramana *et al.*, 1987a).

A population level of 250 nematodes/g of root was constantly recorded with slow wilt affected pepper vines in Kerala (Ramana, 1986). Menon (1949) reported mortality up to 10% of the vines due to the disease. Mohandas and Ramana (1991) reported that *R. similis* was responsible for 59% loss in yield of black pepper.

8.1.1.1. Distribution

R. similis was reported on black pepper from Indonesia (Van der Vecht, 1950), India (Venkitesan, 1972), Malaysia, Thailand (Sher *et al.*, 1969; Reddy, 1977) and Sri Lanka (Gnanapragasam *et al.*, 1985).

8.1.1.2. Symptoms

Primary symptoms of slow-wilt (yellows) disease is the appearance of pale yellow drooping leaves on the vines whose number gradually increases and finally the entire foliage becomes yellow. Yellowing is followed by shedding of leaves, cessation of growth and die-back symptoms. Within 3-5 years of the initiation of the primary symptoms, all the leaves are shed and the death of the vine takes place. Young and old plants are susceptible. In bearing vines, shedding of the spikes is another major symptom. At the base of the affected vines, large numbers of shed spikes can be seen. In plantations, barren standards that have lost the vines or standards supporting dead vines are of common occurrence. The foliar symptoms become more pronounced when the soil moisture is depleted.

The symptoms on the roots are the presence of characteristic orange to purple coloured lesions on the thin, white feeding roots which are not clearly seen on old brown roots. The main roots remain devoid of fine feeder roots and extensive root rotting takes place. Subsequently, extensive necrosis of the larger lateral roots takes place (Fig. 8.1).

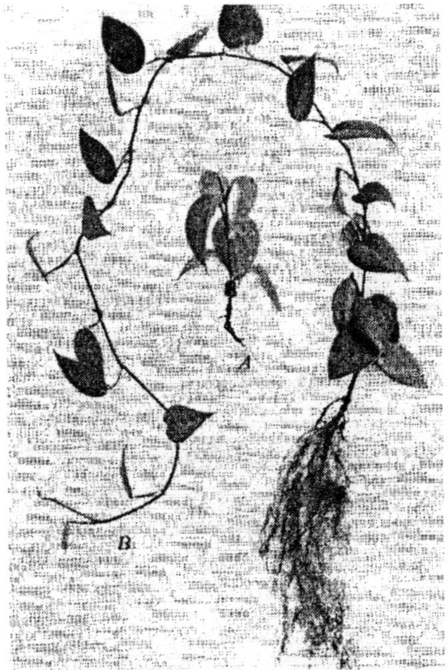

Fig. 8.1. Burrowing nematode on black pepper. A-Infected; B-Healthy
(Thorne, 1961).

8.1.1.3. Life Cycle

The nematode completes its life cycle within 25-30 days at temperature range of 21-31°C. The black pepper isolate of *R. similis* can be cultured axenically on carrot discs (Koshy, 1986).

8.1.1.4. Host Range

Crops like coconut and areca nut which are commonly used as live standards are good hosts of *R. similis*. Intercrops like banana, ginger, turmeric and cardamom are also susceptible to *R. similis* infestation.

8.1.1.5. Histopathology

R. similis penetrates the roots within 24 hours of inoculation and the cells around the penetration site become brown (Venkitesan and Setty, 1977). Nematodes do not enter stellar portions of the root and plugging of the xylem vessels with a gum like substance has been reported. Nematodes are found in inter and intracellular positions within the cortex.

8.1.1.6. Interaction with Other Pathogens

Yellows disease in Indonesia is caused by a nematode-fungus complex involving *R. similis* and *Fusarium* spp. and possibly other fungi. *R. similis* predisposed black pepper seedlings to attack by a weakly pathogenic isolate of *Fusarium solani* causing severe root damage.

8.1.1.7. Management Methods

(i) Cultural Methods: Use of healthy and nematode-free planting material can ensure better establishment, survival and higher yield. Soil solarization of nursery sites or potting mixture and random rotation of nursery sites are reported to be effective in black pepper. Mulching (with Guatemala grass) has an ameliorating effect on the symptoms of slow wilt/pepper yellows (*R. similis*). Planting nematode-free rooted cuttings, uprooting of affected vines and replanting after a period of 9-12 months, use of non-living standards/supports, avoiding *R. similis* susceptible intercrops like banana, ginger, turmeric and exclusion of susceptible trees as standards for trailing the vines reduces the nematode population. Application of organic amendments like neem cake (200 g), green foliage (3-5 kg) or FYM (1 kg per vine) and earthing up in September-October are recommended. Growing marigold as a trap crop and uprooting the trap crop at flowering stage and burning reduce the incidence of nematodes. *R. similis* was not encountered when black pepper was intercropped with coffee (Venkitesan, 1976).

Application of fertilizers at 400 kg N, 180 kg P, 480 kg K, 425 kg Ca and 112 kg Mg in combination with a mulch gave effective control of yellows disease.

(ii) Chemical Methods: Under Indian conditions, aldicarb/carbofuran/phorate each at 3g a.i./vine applied in May-June and again in September-October resulted in reduction of the nematode population and remission of foliar yellowing. Among the above three nematicides, phorate proved to be the best (Ramana, 1986; Mohandas and Ramana, 1987a). Aldicarb sulphone at 8kg a.i./ha gave best control of *R. similis* followed by fensulfothion (Venkitesan, 1976).

Various chemicals tried for the management of *R. similis* on black pepper under nursery and field conditions is presented in Table 8.2.

Table 8.2. Effect of various chemicals for the management of *R. similis* in black pepper under nursery and field conditions

Chemical	Dosage	Duration & type of treatment	Effectiveness
i) Nursery level			
Aldicarb sulfone	4-8 kg a.i./ha	At nursery stage	Eliminated infection
Fensulfothion	4-8 kg a.i./ha	At nursery stage	Eliminated infection
ii) Field level			
DBCP	40 litres/ha	Field application	Effective in controlling slow wilt disease
Phorate	3 kg a.i./ha	Field application	Effective in controlling slow wilt disease
	3 g a.i./vine	Twice a year	Reduced nema popn. & foliar yellowing

(iii) Biological Methods: Pasteuria penetrans, Paecilomyces lilacinus and *Pochonia chlamydosporia* are effective in suppressing the root-knot and burrowing nematodes. Five isolates of rhizobacteria (IISR-552, 528, 658, 853 and 859) were good growth promoters (26.7 to 55.6%) having dual nematicidal action (suppressing both *M. incognita* and *R. similis*).

(iv) Host Resistance: A local variety at Peringamala, Kerala, India was found not to be invaded by *R. similis*

(v) Integrated Methods: Integrated methods of nematode management that can be suggested are:

- Planting nematode-free rooted cuttings.
- Use of non-living standards and excluding *R. similis* susceptible trees as standards for trailing pepper vines and avoiding infested planting materials of intercrops such as banana, ginger and turmeric.
- Uprooting of affected vines and replanting after 9-12 months with nematode-free rooted cuttings of high yielding tolerant variety and protecting them with phorate at 1g a.i./vine at planting itself.
- Application of phorate at 3g a.i./vine in May-June and September-October. If the vine is trailed on a susceptible tree like areca nut, phorate has to be applied at 6g a.i./standard.
- Application of 5 kg green manure (*Glyricidia, Crotolaria*), 1 kg FYM, along with the recommended dosages of NPK, neem oil cake at 200g/vine and earthing up to 50 cm radius.

- Avoiding susceptible intercrops such as banana, ginger, turmeric etc. or application of phorate at 3g a.i./banana plant twice a year and 1g a.i./ ginger/turmeric plant at planting.

8.1.2. Root-knot Nematodes, *Meloidogyne incognita, M. javanica*

Butler (1906) reported root-knot nematodes from black pepper in Wynad, Kerala, India. *M. incognita* and *M. javanica* have been reported from India. The root-knot infestation is a serious problem in Government nurseries in Kerala. Up to 91% root-knot infestation was reported from Para, Brazil (Ichinhoe, 1975) and Kerala, India (Ramana *et al.*, 1987a).

An initial inoculum level of 10 J_2 per rooted cutting was found to reduce growth by 15%, while at 1,00,000 J_2 level 50% reduction in growth was observed over one year period (Koshy *et al.*, 1979a). More than 50% death of transplants occurs in field planting of infected cuttings. *M. incognita* was responsible for 46% loss in yield of black pepper (Mohandas and Ramana, 1991).

8.1.2.1. Distribution

The root-knot nematodes have been reported from India, Brazil, Sarawak, Borneo, Cochin-China, Malaysia, Brunei, Kampuchea, Indonesia, Philippines, Thailand and Vietnam.

8.1.2.2. Symptoms

Prominent symptoms of root-knot infestation on black pepper are unthrifty growth and yellowing of leaves. Inter-veinal yellowing of the foliage is also noticeable. The leaves exhibit dense yellowish discolouration of the interveinal areas making the leaf veins prominent with deep green colour. Heavy galling of the root system is also present.

8.1.2.3. Host Range

Among commercially used standards, *Oroxylon indicum, Erythrina lithosperma, Ceiba pentandra, Bombax malabaricum* are highly susceptible to root-knot nematodes (Koshy *et al.*, 1977).

8.1.2.4. Life Cycle

The eggs are laid in a gelatinous matrix. Its contents undergo a series of determinate cleavage giving rise to the first juvenile stage which moults within the egg and gives rise to second stage juvenile which hatches out. The J2 juveniles penetrate host plants above the meristematic zone and orient parallel to the stele. It starts feeding on the pericycle cells for several weeks, gradually swelling. A series of 3 moults occur in quick succession. Reproduction may be parthenogenetic or bisexual with a female laying 400-500 eggs. Egg laying is completed within a week. The life cycle under optimum conditions takes 3-4 weeks.

8.1.2.5. Interaction with Other Pathogens

M. incognita and *Fusarium solani* were found associated with black pepper vines. Both the organisms together were found to do more harm than either of them alone. Increased susceptibility of *M. incognita* and *M. javanica* infested pepper to *Phytophthora* infection has been reported.

8.1.2.6. Management Methods

(*i*) *Cultural Methods*: Live standards for black pepper such as *Artocarpus heterophyllus, A. hirsutus, Ailanthes malabarica, Mesopsis emini, Peltophorum pterocarpum, Swietenia macrophylla, Tamarindus indica, Garuga pinnata* and *Macaranga peltata* are resistant to *M. incognita.*

Growing of non-host cover crops like Siratro (*Macroptilium atropurpureus*) in the interspaces and mulching with Guatemala grass (*Imperata cylindrica*) are recommended to reduce the root-knot nematode (*M. incognita*) population on black pepper in the Amazonian region (Ichinohe, 1980).

Application of neem cake reduces nematode populations followed by increase in yield.

(*ii*) *Chemical Methods:* Fumigation of nursery potting mixture with methyl bromide is effective in checking the infestation.

Green berry yields can be doubled by four applications of carbofuran incorporated into mound soil at 114 g/vine per application in black pepper fields infested with *M. incognita* and *M. javanica*. Application of aldicarb at 1.25g a.i./vine or carbofuran at 10g a.i./vine twice a year, including at planting around cuttings, can reduce populations of *M. incognita* on black pepper. Phenamiphos at 1g a.i./vine followed by carbofuran and ethoprophos was effective in controlling nematodes.

(*iii*) *Biological Methods*

Antagonistic Bacteria: : Root-knot nematodes in pepper can be managed by the application of *Bacillus macerans* at 1.2×10^8 cells/vine just before the monsoon period.

Antagonistic Fungi: Application of *Paecilomyces lilacinus* and *Pochonia chlamydosporia* has potential to control nematodes which reduced the rate of infection to 15 and 12%, respectively. Soil application of *P. chlamydosporia* reduced nematode population and increased the yield (5.14 kg/vine) with cost: benefit ratio of 1: 7.12.

Arbuscular Mycorrhizal Fungi: AMF such as *Glomus fasciculatum, G. etunicatum* and *Acaulospora laevis* reduced root-knot nematodes and enhanced the growth of black pepper. Pre-inoculation of *A. laevis* in black pepper cv. Panniyur-1 reduced the root-knot index (0.75) followed by *G. mosseae/ Gigaspora margarita* (1.25) compared to control (3.5) (Anandraj *et al.,* 1991).

Effective strains of AMF (*Glomus* sp. and *Gigaspora* sp.) isolated from black pepper rhizosphere were very effective in suppression of root-knot nematodes.

(iv) Host Resistance: The root-knot nematode infestation was less in cvs. Veliakaniakadan, Karimunda, Kalluvalli, Balankotta, Narayakodi and Padapan. A cultivated germplasm 'Ottaplakal-1' (CLT-P-812) showed resistant reaction to *M. incognita.* Its yield (4.7 kg green/vine) was on par with the popular varieties – Paniyur-1 and Karimunda. It has an average yield of 7526 kg green/ha at fifth year with a potential yield of 17,280 kg green/ha. This variety also has high oleoresin content (13.8%) and was released as 'Pournami' and recommended for cultivation in all black pepper areas having root-knot nematode problem.

(v) Integrated Methods: Black pepper vines combinedly inoculated with *P. lilacinus* and *P. penetrans* had put out 23-112 per cent increase in plant growth over control and were very effective in the management of root-knot nematodes (Sosamma and Koshy, 1995).

Integration of neem cake application at 1 kg/vine along with phorate/ carbofuran at 3g a.i./vine during May-June and again during September-October gave effective control of nematodes infesting black pepper.

Application of aldicarb at 1g a.i./vine twice a year (May/June and October/November) integrated with fertilizers (N-100g, P- 40g, K-140g/vine) in two equal splits, earthing up to 50 cm radius at the base of the vines and mulching the base of the vines with leaves reduces foliar yellowing by 83% and *M. incognita* population by 33-38% (Venkitesan and Jacob, 1985).

8.1.3. Root-knot nematode, *Meloidogyne incognita* and Foot Rot, *Phytophthora capsicii* Disease Complex

Increased susceptibility of *M. incognita* and *M. javanica* infested cultivars of black pepper to *Phytophthora* infestation has been reported.

8.1.3.1. Management Methods

(i) Integrated Methods: Integrated management of foot rot (*P. capsicii*) and nematodes (*M. incognita* and *R. similis*) on black pepper was achieved by (i) mixing AMF and *Trichoderma harzianum* in solarized nursery mixture to

raise healthy and robust seedlings, (ii) application of *T. harzianum* and FYM in planting pit, (iii) field application of neem cake at 1 kg/vine mixed with 50 g of *T. harzianum* during August.

8.2. CARDAMOM

Cardamom (*Elettaria cardamomum*), known as the "Queen of Spices" has its origin in the evergreen rain forests of Western hills of South India. It is a perennial plant with an underground stem (rhizome) and aerial shoots and valued for their fruits (capsules). The fruits are small, trilocular capsules containing 15-20 seeds. India and Guatemala are the main producers and exporters of cardamom. Other cardamom growing countries are Tanzania, Sri Lanka, El Salvador, Vietnam, Laos, Kampuchea and Papua New Guinea. It is used for flavouring various food preparations, confectionery, beverages, liquors and medicines. The area under cultivation in India during 1987-88 was 1,05,000 ha, with an annual production of 3,200 metric tonnes. The export earning during 1987-88 was Rs. 34 million.

The most important nematode problem is the root-knot nematode, *Meloidogyne* spp. The lesion nematode, *Pratylenchus coffeae* and the burrowing nematode, *Radopholus similis* are also known to cause root rotting (D'souza *et al.*, 1970; Kumar *et al.*, 1971; Khan and Nanjappa, 1972; Viswanathan *et al.*, 1974; Sundararaju *et al.*, 1979b).

8.2.1. Root-knot Nematodes, *Meloidogyne* spp.

M. incognita and *M. javanica* have been reported as widely occurring in the cardamom nurseries and plantations in Kerala, Karnataka and Tamil Nadu (Kumar *et al.*, 1971; Koshy *et al.*, 1976; Ali, 1982, 1986). An yield loss of 32-47% due to root-knot infestation has been reported (Ali, 1984, 1986). An initial population of 100 nematodes/plant causes discernible damage to cardamom (Eapen, 1987).

8.2.1.1. Host Range

Many annual weeds and common shade trees like *Erythrina indica* and *E. lithosperma* present in cardamom plantations are susceptible to root-knot nematodes.

8.2.1.2. Symptoms

Heavy infestation of root-knot on mature plants in a plantation causes stunting, reduced tillering, yellowing, premature drying of leaf tips and margins, narrowing of leaf blades, delay in flowering, immature fruit drop and reduction in yield. Infested roots are characterized by excessive branching. Galling of roots is not conspicuous on mature plants.

In primary nursery, the second stage juveniles infest the radicle and plemule as a result of which even 50% of the germinating seeds do not emerge. At the two leaf stage itself the infested seedlings show marginal yellowing and drying of leaves and severe galling of roots. Up to 40% of such seedlings do not establish in the secondary nursery. The infested secondary nursery plants exhibit stunting, yellowing, poor tillering, drying of leaf tips and margins and heavy galling of roots (Ali and Koshy, 1982).

8.2.1.3. Survival and Spread

The nematodes are spread through infested seedlings and rhizomes used for propagation.

8.2.1.4. Interaction with Other Pathogens

In nurseries, presence of *M. incognita* increases incidence of rhizome rot and damping off diseases caused by *Rhizoctonia solani* (Ali, 1986; Eapen, 1987).

In cardamom, *M. incognita* is the predisposing factor for *R. solani* infection, damping off and rhizome rot, prevalent in nurseries (Ali and Venugopal, 1992, 1993).

8.2.1.5. Management Methods

(i) Physical Methods: Soil solarization for 40-45 days using 400 gauge transparent LDPE was proved ideal pre-sowing treatment for the management of nematodes and other soil-borne diseases in cardamom nurseries. Mean soil temperature during the solarization period increased by 8.7°C in solarized beds and the germination was enhanced by 25.5%, while weed growth was suppressed by 82%. *Pythium vexans* was totally eliminated, while populations of *Rhizoctonia solani, Phyllosticta elettaria* and nematodes were suppressed to varying levels. Solarization was also found to stimulate the growth of cardamom seedlings (Eapen, 1995).

(ii) Cultural Methods: It is advisable to change nursery sites every year.

Application of neem cake at 0.5 and 1.0 kg/plant twice a year reduces nematode populations followed by increase in yield (Ali, 1984). Soil application of crushed neem seed can take care of nematode problem.

(iii) Chemical Methods: Disinfestation of nursery beds with methyl bromide at 500 g/10m^2 is effective in controlling root-knot infestation in both primary and secondary nurseries.

Application of aldicarb at 5 kg a.i./ha three times in a year, resulted in increased growth and vigour of seedlings both in primary and secondary nurseries (Koshy *et al.*, 1979; Ali, 1986). Increase in yield of *M. incognita* infested cardamom plants from 47-88% was recorded by the application of aldicarb/carbofuran/phorate at 5g and 10g a.i./plant (Ali, 1984).

(iv) Biological Methods: AMF such as *Glomus fasciculatum* and *Gigaspora margarita* were recorded to be effective in reducing the root-knot nematode population.

P. lilacinus suppressed *Meloidogyne* spp. population by 58.3-86.9 per cent in cardamom nursery. The number of quality seedlings for transplanting was greater in beds treated with this biocontrol agent (Eapen and Venugopal, 1995).

T. harzianum also suppressed *Meloidogyne* spp. by 58.3-68.9 per cent in cardamom nursery and increased the number of quality seedlings for transplanting (Eapen and Venugopal, 1995).

(v) Integrated Methods: Soil solarization of nursery beds and application of bioagents such as *P. lilacinus* or *Trichoderma* spp. improved growth of cardamom seedlings by suppressing root-knot nematode population (Eapen and Venugopal, 1995). Application of *T. harzianum* multiplied on decomposed coffee husks (7 day old) at the time of sowing at 2.5 kg/bed (4.5 m X 1.0 m) and repeated after 3 months is recommended for the control of root-knot nematodes and damping off in nurseries.

Integration of the following practices can help in the successful management of root-knot nematodes:

- Changing nursery sites frequently.
- Disinfecting nursery beds.
- Introduction of biocontrol agents at nursery level.
- Control of susceptible weeds.
- Exclusion of susceptible shade trees.
- Destruction of infested crop residues.
- Mulching with dead leaves.
- Application of neem cake.

8.2.2. Root-knot Nematode, *Meloidogyne* sp. and Rhizome Rot, *Rhizoctonia solani* Disease Complex

M. incognita was found to predispose cardamom seedlings to *R. solani* infection, which causes damping off and rhizome rot in the primary nursery.

8.2.2.1. Management Methods

(i) Chemical Methods: When both fungicide (metalaxyl) and nematicide (carbofuran) were applied for the control of rhizome rot disease and nematodes, the mortality of seedlings was least.

(ii) Integrated Methods: P. lilacinus in combination with *Trichoderma* spp. suppressed *Meloidogyne* spp. and rhizome rot disease (*R. solani*) complex when incorporated in solarized cardamom nursery beds (Eapen and Venugopal, 1995). Application of carbofuran along with neem cake was found to reduce immature fruit-drop and to increase capsule yield.

Soil solarization alone enhanced the germination by 25.5% and suppressed weed growth by 82.0%. Solarization also enhanced the growth and vigour of cardamom seedlings. The disease complex was suppressed by incorporation of *P. lilacinus/T. harzianum* and phorate into the solarized nursery beds. This approach is being adopted on a large scale for the production of nematode-free cardamom seedlings.

8.3. GINGER

Ginger (*Zingiber officinale*), an herbaceous perennial, belonging to the family Zingiberaceae, is valued for its rhizome. The country of its origin is not definitely known but is presumed to be either from India or China. India is the largest producer (3,05,900 metric tonnes) and exporter of dry ginger in the world (export earning of Rs.191 million). The other ginger producing countries are Jamaica, Sierra Leone, Nigeria, Southern China, Japan, Taiwan and Australia. In India, It is being cultivated in 85,100 hectares producing 0.306 million tonnes of ginger with an average yield of 3.6 tonnes per hectare.

Although a large number of nematode species have been reported from ginger, the root-knot nematode, *Meloidogyne* spp., the burrowing nematode, *Radopholus similis* and the lesion nematode, *Pratylenchus coffeae* are the most important.

8.3.1. Root-knot Nematodes, *Meloidogyne* spp.

M. incognita, M. javanica, M. arenaria and *M. hapla* have been reported to be associated with ginger from various countries. Under pot conditions, an initial inoculum level of 10,000 nematodes per plant over a period of six months caused 74% reduction in rhizome weight. A population level of one juvenile/30 g of soil was found to cause significant reduction in yield (Sukumaran and Sundararaju, 1986). Kaur (1987) estimated 41-59% yield loss in ginger when the crop was raised using apparently healthy rhizomes in nematode infested fields.

8.3.1.1. Symptoms

Heavily infested plants exhibit stunting and chlorotic leaves with marginal necrosis. The root-knot nematodes cause galling and rotting of roots and underground rhizomes. Infested rhizomes show brown, water-soaked lesions in the outer tissues, particularly in the angles between shoots. The J2 of *M. incognita* invade the rhizome through the axils of leaf sheaths in the shoot apex. In fibrous roots, penetration occurs in the area of differentiation and in fleshy roots. The entire length of root is invaded. In the rhizomes and fleshy roots extensive internal lesions develop.

8.3.1.2. Life Cycle

In fleshy and fibrous roots the nematode develops to maturity in 24 days, while in rhizomes it takes 43 days at a soil temperature of 30°C (Huang, 1966).

8.3.1.3. Spread

Infested rhizomes help in long distance dissemination of nematodes.

8.3.1.4. Histopathology

In all infested tissues, except rhizome meristems, abnormal xylem and hyperplastic parenchyma are observed. Giant cells are found in all infected tissues. Infection sites in differentiated rhizomes and fleshy roots, wound cork and lignified wall thickening of endodermis and pericycle are formed (Huang, 1966; Shah and Raju, 1977).

8.3.1.5. Interaction with Other Pathogens

Bacterial wilt of ginger, caused by *R. solanacearum* is influenced by *M. incognita* (Samuel and Mathew, 1983).

M. incognita had synergistic effect in rhizome rot/soft rot (*Pythium aphanidermatum*) of ginger, particularly when nematodes were inoculated to the plants prior to fungal inoculation. The damage to the crop was more and the onset of the disease was early when *M. incognita* was inoculated 40 days prior to fungal inoculation (Ramana *et al.*, 1998).

8.3.1.6. Management Methods

(i) Physical Methods: Summer ploughing at an interval of 15 days and use of hot water treated ginger rhizomes (at 50°C for 10 min.) as seed proved promising in reducing the nematode population and in increasing the yield up to 65%.

(ii) Cultural Methods: Ginger rhizomes used for planting should be collected from nematode-free fields.

Organic Amendments: Application of well decomposed FYM/compost at 25-30 t/ha or neem cake at 2 t/ha and mulching with green leaves at 10-12 t/ha at planting and repeating the mulching during the growth period helps in reducing nematode multiplication (Kaur, 1987).

Two applications of neem cake at 1 t/ha as basal and 45 DAP significantly reduced nematode population and increased the yield.

Crop Rotation: Crop rotation with cereals and millets should be practiced, preferably with rice at least once in 3 years. For ginger cultivation in Fiji, a ginger-taro-fallow rotation has been recommended. Intercropping of ginger with maize, capsicum and mulching with FYM are effective in reducing the nematode population.

(iii) Chemical Methods: Kaur (1987) found that presowing dip of ginger rhizomes in 0.04% phenamiphos (Nemacur) or 0.18% Mancozeb for 30 min. helps in reduction of nematode population resulting thereby in higher crop yields. In Himachal Pradesh, application of phenamiphos at 3 kg a.i./ha resulted in 70-144% increase in yield of ginger in fields infected with *M. incognita* (Kaur, 1987).

Two applications of carbofuran at 1 kg a.i./ha as basal and 45 DAP significantly reduced nematode population and increased the yield.

(iv) Integrated Methods: P. lilacinus along with *T. harzianum/P. chlamydosporia* was effective in suppression of root-knot nematodes under field conditions.

Application of neem cake at 1 t/ha at planting, followed by carbofuran at 1 kg a.i./ha at 45 DAP gave effective control of nematodes associated with ginger.

The following integrated nematode management measures are suggested:

(a) Production of nematode-free planting material by:

- Selecting an area where ginger has not been grown in the previous season and has no history of severe nematode infestation.

- Preparation of land and fumigation of soil with DD or EDB at 330 litres/ha in August (at a depth of 20 cm in rows 30 cm apart). The time interval between fumigation and planting should be at least two weeks.

- Selection of nematode-free planting material and treatment in hot water at 40°C for 20 min. It is followed by cooling the rhizomes before cutting and dipping in benomyl. Seed should be planted within one week of hot water treatment.

- Growing under saw dust mulch. If saw dust is not available, phenamiphos granules should be sprinkled over the soil between the plants at 11 kg/ha in mid-November and again in mid-January.

(b) Fumigation of land two or more weeks before planting.

8.3.2. The Burrowing Nematode, *Radopholus similis*

R. similis infestation in ginger was first reported in India by Charles and Kuriyan (1979). The coconut isolate of *R. similis* in Kerala reproduced well on ginger (Koshy and Sosamma, 1975, 1977). *R. similis* has been reported from more than 50% of the total area in Fiji causing yield reduction up to 40%. An initial inoculum level of 10,000 nematodes/plant causes 74% reduction in rhizome weight (Sundararaju *et al.*, 1979b).

8.3.2.1. Symptoms

R. similis infestation causes stunting, reduced vigour and tillering of infested plants and they tend to mature and dry out faster than healthy plants. The top most leaves become chlorotic with scorched tips. Lesions on the rhizomes results in small, shallow, sunken, water-soaked areas (Sundararaju *et al.*, 1979a). Intra-cellular migration of nematodes lead to large infection channels or galleries within the rhizomes.

8.3.2.2. Spread

The perpetuation and dissemination of the nematode takes place through infested rhizomes used for planting.

8.3.2.3. Management Methods

The methods suggested for the control of root-knot nematodes were found effective in reducing the burrowing nematode infestation also.

8.3.3. The Lesion Nematode, *Pratylenchus coffeae*

The association of *P. coffeae* on ginger has been reported from Kerala (Charles and Kurian, 1979) and Himachal Pradesh. An initial inoculum level of even 10 nematodes proved highly pathogenic to 15 day-old ginger plants in sterilized soil (Kaur, 1987).

8.3.3.1. Management Methods

(i) Chemical Methods: In Himachal Pradesh, application of fenamiphos at 3 kg a.i./ha resulted in 70-144% increase in yield of ginger in fields infested with *P. coffeae* (Kaur, 1987).

8.4. TUEMERIC

Turmeric (*Curcuma longa*) is best known as an important spice crop. It is believed to have its origin in South-East Asia. Turmeric is valued for its rhizomes, often processed and used commercially. It is mainly cultivated in India and to a small extent in China, Indonesia, Peru and Jamaica. It is an important source of yellow dye and is also used as a colouring matter in food industries, confectionery and drugs. In India, it is cultivated in 1,49,400 ha with annual production of 5,29,000 tonnes during 2004-05. India's export earning from turmeric was Rs.92 million annually.

Of the many plant parasitic nematodes reported in association with turmeric, the root-knot nematode, *Meloidogyne* spp., the burrowing nematode, *Radopholus similis* and the lesion nematode, *Pratylenchus coffeae* are important.

8.4.1. Root-knot Nematodes, *Meloidogyne* spp.

M. incognita and *M. javanica* have been reported on turmeric of which *M. incognita* is more important. In pot experiments, 100,000 nematodes/plant resulted in 76.6% reduction in the rhizome weight after 6 months (Sukumaran *et al.*, 1986).

8.4.1.1. Symptoms

Turmeric plants infected with *M. incognita* exhibit stunting, yellowing, reduced tillering and marginal and tip drying of leaves. Galling and rotting of roots can also be noticed. Infected rhizomes have brown, water soaked areas in the outer tissues and lose their bright yellow colour (Mani *et al.*, 1987). High populations of *M. incognita* in field cause stunting, yellowing and withering of plants in large patches. Premature death of plants takes place leaving a poor crop stand.

8.4.1.2. Management Methods

(i) Physical Methods: Use of nematode-free rhizomes for fresh planting, washing seed material to free of soil and drying in shade before planting helps to reduce the inoculum. Hot water treatment of rhizomes at 50-55°C for 10 min or 45°C for 30 min eliminates nematode infection. Soil solarization of beds for 40 days during summer reduces nematode population in soil.

Rabbing was most effective for the management of root-knot nematodes on turmeric which gave maximum plant height and rhizome yield with lowest root-knot index (Patel *et al.*, 2001).

Summer ploughing at an interval of 15 days and use of hot water treated turmeric rhizomes as seed proved promising in reducing the nematode population and in increasing the yield up to 65%.

(ii) Cultural Methods: Crop rotation with cereals and millets should be practiced, preferably with rice at least once in 3 years. Deep summer ploughing with furrow turner ploughs should be practiced. Application of well decomposed FYM/compost at 25-30 t/ha or neem cake at 2 t/ha and mulching with green leaves at 10-12 t/ha at planting and repeating the mulching during the growth period helps in reducing nematode multiplication (Kaur, 1987).

(iii) Chemical Methods: Application of phenamiphos at 2.5 kg a.i./ha one day before planting gave yield increase of 59-187% (Patel *et al.*, 1982). Aldicarb and carbofuran at 1 kg a.i./ha increased yield by 71% and 68%, respectively over control with a cost: benefit ratio of 1: 6 in aldicarb and 1: 2 in carbofuran treatments (Gunasekharan *et al.*, 1987). Carbofuran at 4 kg a.i./ha applied in rows to a four-month old turmeric crop has resulted in 81.6% reduction in root-knot nematode population (Mani *et al.*, 1987).

(iv) Biological Methods: Complete suppression of nematodes could be achieved by the application of *Fusarium oxysporum* (isolate IISR-11) at 50 g/bed of 3 sq. m. at the time of planting. *Paecilomyces lilacinus, Pochonia chlamydosporia* and *Aspergillus nidulans* suppressed root-knot nematode population.

Turmeric rhizome treatment with *P. fluorescens* at 10 g/kg increased plant growth characters (58 and 31% increase in pseudo stem height and number of tillers over control) and decreased nematode population (40%) and gall indices (25%) (Seenivasan *et al.*, 2001).

(v) Host Resistance: Turmeric cvs. Kodur, Chayapasupu, Duggirala, Guntur-1, Guntur-9, Rajampet, Sugandham and Appalapadu are resistant to *M. incognita* (Gunasekharan *et al.*, 1987).

(vi) Integrated Methods: Soil application of *T. harzianum* + neem cake at 1 t/ha is recommended.

8.4.2. The Burrowing Nematode, *Radopholus similis*

Koshy and Sosamma (1975) reported turmeric as a host of *R. similis*. An initial inoculum level of 10 nematodes/plant resulted in 35% reduction in rhizome weight after four months. A reduction of 65 and 76% in rhizome weight was observed after 4 and 8 months, respectively with an inoculum level of 100,000 nematodes (Sosamma *et al.*, 1979).

8.4.2.1. Symptoms

Drying of leaf tips and margins is conspicuous. Infested rhizomes have shallow, water-soaked brownish areas on the surface. The golden yellow colour of healthy rhizomes change to yolk-yellow colour on infestation. Infested plants age and dry faster than healthy plants. Rotting of roots takes place and the decayed roots remain devoid of cortex and stellar portions (Sosamma *et al.*, 1979).

8.4.2.2. Survival and Spread

The nematodes are spread through infested planting material. Populations of *R. similis* from coconut are known to infest turmeric and the use of turmeric as an intercrop in *R. similis* infested coconut and areca nut based farming systems should be avoided.

8.4.2.3. Management Methods

(i) *Cultural Methods:* Avoiding use of turmeric as an intercrop in *R. similis* infested coconut and areca nut based farming systems and use of only clean, nematode-free rhizomes for fresh planting helps in reducing the nematode population.

8.4.3. The Lesion Nematode, *Pratylenchus coffeae*

8.4.3.1. Symptoms

P. coffeae has been reported to cause discolouration and rotting of mature rhizomes of wild turmeric, *Curcuma aromatica*. Affected rhizomes show dark brown necrotic lesions internally. The rhizomes in advanced stages of infection become red to dark brown in colour. Rhizomes become less turgid and wrinkled with dry rot symptoms. Fingers are more affected than mother rhizomes (Sarma *et al.*, 1974).

8.5. CORIANDER

Coriander (*Coriandrum sativum*) is used as a common flavouring substance. The stems, leaves and fruits have a pleasant aroma. The whole young plant is used in preparing chutney. Its leaves are used for flavouring curries, sauces and soups. Dry seeds are extensively used in preparation of curry powder, pickling spices, sauces and seasonings. This plant contains vitamin A. In medicines, its seeds are used as a carminative, refrigerant and diuretic. The decoction is used in indigestion, cold and other diseases. The oil from seeds is used in sweet meats, coco eatables, liquor and for agreeable smell and taste. The oil is also useful as an insecticide. It is being cultivated in 0.282 million hectares producing 0.172 million tonnes of coriander seeds with an average yield of 0.6 tonnes per hectare. In India, coriander is cultivated in Andhra Pradesh, Madhya Pradesh, Karnataka, Tamil Nadu and Uttar Pradesh.

The root-knot nematode is an important limiting factor in successful production of coriander crop.

8.5.1. The Root-knot Nematode, *Meloidogyne incognita*

The root-knot nematode has been reported to cause damage to coriander. Midha and Trivedi (1991) reported that *M. incognita* was responsible for 52% loss in seed yield of coriander.

8.5.1.1. Management Methods

(i) **Biological Methods:** *P. lilacinus* was effective in increasing the plant growth parameters and in reducing root galling and egg mass production due to *M. incognita* in coriander (Sobita Devi and Pandey, 2001).

Table 8.3. Biological control of plant parasitic nematodes in spice crops.

Crop	Nematode	Effective bioagent/s	Reference
Black pepper	*M. incognita*	*P. lilacinus* *P. fluorescens*	Sosamma & Koshy, 1995
		P. penetrans	Eapen *et al.*, 1997
		G. mosseae	Sosamma & Koshy, 1995 Sivaprasad & Sheela, 1999
		G. mosseae, G. fasciculatum, Acaulospora laevis, Gigaspora margarita	Anandraj *et al.*, 1991
Cardamom	*Meloidogyne* spp.	*P. lilacinus, Trichoderma* spp.	Eapen & Venugopal, 1995
Turmeric	*M. incognita*	*P. fluorescens*	Seenivasan *et al.*, 2001
Coriander	*M. incognita*	*P. lilacinus*	Midha *et al.*, 2001

Chapter 9

TUBER CROPS

Tuber crops form an important part of the diet in many areas of the tropics and subtropics. They are most important after cereals and grain legumes. They are mostly grown for the carbohydrates stored in underground storage organs viz., roots or rhizomes. In addition, the leaves or leaf stalks of some of the tuber crops are rich in protein, vitamins and minerals and are used for either human consumption or cattle feed. Tuber crops are a major source of food and calories in many tropical countries. The annual tropical root crop production is in the range of 170 million metric tonnes (Table 9.1). Root and tuber crops are predominantly grown in the upland or rain fed areas, mainly by resource poor small and marginal farmers. The major tuber crops include cassava (tapioca), sweet potato, colocasia, amorphophyllus, yams and coleus. Cassava and sweet potato are grown to a larger extent in southern and eastern India. Their cultivation has spread to Maharashtra, Gujarat and North-eastern states as well. The Salem belt in Tamil Nadu and the Samalkot belt in Andhra Pradesh are known for cassava as an industrial crop. Yams and Colocasia are popular in Andhra Pradesh, Tamil Nadu, West Bengal, Uttar Pradesh and Orissa states.

Table 9.1. Area, production and productivity of tuber crops in India (2004-05) (Chamber, 2006)

Tuber crop	Area ('000 ha)	Production ('000 tonnes)	Productivity (tonnes/ha)
Sweet potato	136.5	1211.0	8.9
Cassava (Tapioca)	281.3	7900.8	28.1

Tuber crops, especially cassava can be successfully substituted as alternative source of feed for livestock in both tropical and subtropical countries. Cassava has also been used as a substrate for the production of ethyl alcohol. The tropical root crops, in general, have a great potential in meeting basic food and energy needs of the developing world. Tuber crops are very efficient in solar energy transfer, having a clear superiority over cereals in biological efficiency and are comparatively free from severe pests and diseases. Cassava and sweet potatoes account for about 30% of the total production of root crops from developing countries. The potential of root crops as nutritionally rich sources of β-carotene, anti-oxidants, dietary fibre and minerals like calcium have begun to be recognized as a result of the multifarious research programmes worldwide.

9.1. YAM

Yam (*Dioscorea* spp.) belongs to the family Dioscoreaceae, which is widespread throughout the tropics and subtropics. Normally yams are consumed as boiled, baked or fried vegetables. They contain 18-20% starch with mucilaginous substance. Yam also contains alkaloids, tannins and steroids which have pharmaceutical value. *Dioscorea alata* (greater yam), *D. rotundata* (white or white Guinea yam), *D. cayensis* (yellow Guinea yam) and *D. esculenta* (lesser yam) produce edible tubers. *D. alata* was first cultivated in S.E. Asia and is widely grown in Asia and the Pacific, is less favoured in Africa, but much grown in the Caribbean. *D. rotundata* is the most important species in W. Africa, where it is originated. *D. cayensis* is a native of W. Africa, where it is extensively cultivated. *D. esculenta* originated in Indo-China and is extensively grown in S.E. Asia and elsewhere in the tropics.

In India, greater yam and lesser yam are under cultivation. Madhya Pradesh, North-Eastern states, West Bengal, Bihar, Orissa, Uttar Pradesh, Kerala, Tamil Nadu, Gujarat and Maharashtra are major yam-growing states.

The most important nematode problem is the dry rot nematode. Lesion and root-knot nematodes are also known to cause root rotting in yam crop.

9.1.1. The Yam Nematode, *Scutellonema bradys*

The most important nematode problem on yams is the dry rot of tubers caused by *S. bradys*. Nadakal and Thomas (1967) reported the occurrence of dry rot of *Dioscorea alata* and the association of *S. bradys* from Kerala. *Dioscorea alata, D. cayenensis, D. esculenta* and *D. rotundata* are good hosts of the yam nematode. Sesame and cowpea support high root populations and melon can increase soil populations.

9.1.1.1. Economic Importance

Dry rot of yams alone causes a marked reduction in the quality, marketable value and edible portions of tubers, and these reductions are more severe in stored yams. When dry rot is followed by wet rot in stored yams, losses of whole tubers can be as high as 80-100%. Weight differences between healthy and diseased tubers harvested from the field have been estimated to be 20-30%.

9.1.1.2. Symptoms

S. bradys causes a disease of tubers commonly known as 'dry rot'. In early stages of infection, small yellow lesions can be observed just below the skin; these become dark brown and eventually form a continuous dark, dry rot layer just beneath the surface in heavily infested tubers. Outwardly the infected tubers have cracks in the skin, which are often deep, can be malformed, or the skin flakes off exposing dark brown, dry rot tissue underneath. *S. bradys* reproduces and builds up large populations in stored yam tubers and causes severe damage during storage.

Nematode damage to the tubers results in considerable reduction in the edible portion and marketable value of the tubers. Moisture loss is also significantly greater from diseased tubers during storage.

9.1.1.3. Life Cycle

S. bradys is an endoparasite of plant roots and yam tubers. The nematodes attack the primary roots growing from the rhizomatous head of the tuber and can enter the developing tubers through their growing points. The life-cycle is simple. Eggs are laid in soil or plant tissues (roots and tubers) where they hatch and the juveniles develop into adults by subsequent molting. All stages seem to be infective.

9.1.1.4. Histopathology

The nematodes were found in the intercellular spaces of the tissues lying within the periderm where extensive cell destruction was noticed.

9.1.1.5. Survival and Spread

Sizeable populations of the nematode are maintained in the absence of yams probably on other host plants. Yams are propagated from whole tubers or pieces of tuber which are the principal means of spread of *S. bradys*. Infested seed tubers rather than soil are probably the main source of nematode inoculum in yam fields.

9.1.1.6. Interaction with Other Pathogens

The more extensive internal decay of tubers known as "wet rot", 'soft rot' or "watery rot" is associated with fungal (*Botryodiplodia theobromae* and

Fusarium spp.) and bacterial (*Erwinia* sp.) pathogens. This general decay of tubers, which is a serious problem in stored yams, is increased when tubers are wounded or damaged. The damage caused by nematodes can predispose the tubers to invasion by decay organisms resulting in complete rotting of the tubers.

9.1.1.7. *Management Methods*

(i) Physical Methods: Treatment of yam tubers at 50-55°C for 40 min. has given over a 98% kill of *S. bradys* without damage to *D. rotundata* tubers.

(ii) Cultural Methods: To check the spread of *S. bradys* in yams, Decker *et al.* (1967) suggested use of healthy seed tubers. The bottom or distal portions of the tubers have the least nematodes and can be selected for planting. Mixing cow dung in yam mounds before planting at 1.5 kg/mound (1886.3 kg/ha) can increase yields of tubers and significantly decrease nematode numbers.

Nematodes will decline in the absence of host plants and a fallow period of several months has been suggested to control *S. bradys* in *D. alata* tubers.

Soil populations of *S. bradys* will be reduced if a non-host or poor-host crop, such as, peanut, chilli pepper, tobacco, Indian spinach, cotton, maize or sorghum are grown prior to yams.

NPK fertilizers can reduce *S. bradys* populations in tubers of *D. alata* to a very low level. N alone can increase both populations of *S. bradys* and the percentage of infested tubers of *D. rotundata*, whereas P alone can reduce percentage of infested tubers. These results support observations by farmers in certain yam growing areas of Nigeria that yams fertilized with N alone do not store well, but yams fertilized with mixtures that contain P store longer.

(iii) Chemical Methods: Bare root-dip treatment of diseased yams in fensulfothion (Dasanit), thionazin (Nemafos) and DBCP (Nemagon) at 1250 ppm for 15 min. and 625 ppm for 30-60 min was found effective in eliminating the nematode infection (Ayala and Acosta, 1971). Significant increases in yield have been obtained by soaking tuber pieces of *D. alata* infected with *S. bradys* for 30 min. in 1000 ppm aqueous solutions of carbofuran and oxamyl.

Application of aldicarb, oxamyl, carbofuran and miral or isazophos each at 2 kg a.i./ha two weeks after planting reduced soil populations of *S. bradys* to very low levels with remarkable yield increases recorded.

(iv) Host Resistance: Yam cv. Florido of *Dioscorea alata* did not seem to be susceptible to nematode attack (Ayala and Acosta, 1971).

9.1.2. The Lesion Nematode, *Pratylenchus coffeae*

P. coffeae also attack yams and cause dry rot symptoms.

9.1.2.1. Symptoms

Symptoms of damage to yam tubers by *P. coffeae* are similar to those caused by *S. bradys*. The nematodes are restricted to the outer layers of tuber tissue, cause severe necrosis or dry rot, cracking of the skin, and are associated with complete deterioration of tubers.

9.1.2.2. Life Cycle

P. coffeae is a migratory endoparasite of yam roots and tubers. The life cycle is completed in 3-4 weeks on yams.

9.1.2.3. Host Range

P. coffeae is a parasite of *D. alata, D. cayenensis, D. rotundata, D. bulbifera* and *D. trifida*. In addition to yams, *P. coffeae* has an enormous host range covering almost all plant families.

9.1.2.4. Management Methods

(i) Physical Methods: Immersion of *D. rotundata* tubers in hot water maintained at 46-52°C for 15-30 min. has given good control of *P. coffeae*.

(ii) Cultural Methods: Using plant material which is free of nematodes is an effective means of controlling or reducing damage by *P. coffeae*. Collection of central or distal tuber pieces, which generally contain least *P. coffeae*, are recommended for propagative material.

(iii) Chemical Methods: Immersion of *D. rotundata* tubers in DBCP and fensulfothion at 1250 ppm for 15-30 min. effectively reduced *P. coffeae* present to a depth of 6 mm in the tuber tissues.

Aldicarb as a single application at planting at 5.4 kg a.i./ha can give 72% control of *P. coffeae* and significantly increase high quality tuber yields of *D. rotundata*. Significant increases in yield of *D. rotundata* have also been obtained by a combination of foliar and seed tuber treatments with oxamyl.

(iv) Host Resistance: D. alata cv. Florido is not susceptible to attack by *P. coffeae* (or *S. bradys*). *D. esculenta* is possibly less susceptible to *P. coffeae* because of its different growth habit.

9.1.3. Root-Knot Nematodes, *Meloidogyne* spp.

The root-knot nematode is another important nematode occurring on yams and is present world wide. Yams have been found infested by *M. incognita, M. javanica, M. arenaria* and *M. hapla. M. javanica* populations of 30,000 nematodes/plant can reduce yields of *D. opposita* by over 50%. It is estimated that there is a reduction of 39-52% in the price of galled tubers compared to healthy ones.

9.1.3.1. Symptoms

Yam seedlings infested with root-knot nematodes can be severely stunted and chlorotic; very young seedlings may be killed by severe infestation. The roots swell in size at the site of infection producing the characteristic and typical knots on the roots. Tuber formation was severely affected, often infested tubers are smaller and most of the tuber forming roots does not develop. Rotting of tubers has been reported.

9.1.3.2. Life Cycle

The life cycle of *M. incognita* in *D. rotundata* or *D. alata* tubers is 35 days. In *D. alata*, most nematodes are concentrated to a depth of 2 mm with none beyond the 8 mm depth. In *D. rotundata* they are concentrated at depths between 4 to 6 mm with few at 14 mm.

9.1.3.3. Host Range

Susceptible yam hosts of *M. incognita* are *D. alata, D. cayenensis, D. esculenta, D. bulbifera, D. composita, D. floribunda, D. trifida, D. praehensilis, D. spiculiflora* and *D. rotundata*. Hosts of *M. javanica* are *D. alata* and *D. rotundata*.

9.1.3.4. Survival and Spread

Where *Meloidogyne* juveniles and/or eggs survive in stored tubers, they will be spread in propagative material. However, *Meloidogyne* spp. have extremely wide host ranges and damaging populations will come from field soil having survived on other weed hosts, or be introduced into yam fields on infested seedlings of other crops.

9.1.3.5. Management Methods

(i) Physical Methods: Hot water treatment of corms at 51°C for 30 min. eliminates root-knot nematode infection.

(ii) Cultural Methods: The root-knot nematode can be controlled by growing sweet potato cv. Shree Bhadra as a trap crop.

(iii) Chemical Methods: Application of carbofuran at 3 kg a.i./ha gave effective control of root-knot nematodes. Granular oxamyl at 3 or 6 kg a.i./ha applied at planting and at 3, 4-week intervals can control *M. javanica* on *D. rotundata*. In the presence of both *M. javanica* and *Pratylenchus brachyurus*,

tuber yields can be increased by over 40% when granular oxamyl at 3 kg a.i./ha applied at planting is combined with subsequent applications of calcium nitrate or ammonium sulphate incorporated at 3, 4-week intervals, each at 60 kg N. These treatments also reduce the incidence of rot in stored yams associated with nematodes.

(iv) Host Resistance: The only yam species consistently found to be resistant to attack by *M. incognita* is the cluster yam, *D. dumentorum. D. alata* cv. Obunenyi is reported to be resistant to *M. incognita* and *D. cayenensis* can be resistant to *M. incognita* and *M. javanica.* Yam cvs. Sree Latha and Sree Kirthi were found resistant to root-knot nematodes.

9.2. SWEET POTATO

Sweet potato (*Ipomea batatas*) is an economically important member of the family Convolvulaceae and probably originated in Central America, where it has long been cultivated. It is now grown extensively throughout the tropics and subtropics. It is one of the life sustaining crops rich in carbohydrates that stand between man and starvation. Sweet potato ranks fourth and sixth on the list of dry matter production per hectare and edible energy production per hectare, respectively. It is also used for making starch, syrup and alcohol. It occupies a prime place in terms of calories production/unit area and time. It is grown as a starchy food crop throughout the tropical, sub-tropical and frost-free temperate climatic zone in the world. It is being cultivated in 0.136 million hectares producing 1.211 million tonnes of sweet potatoes with an average yield of 8.9 tonnes per hectare. India ranks sixth in area but the productivity is very low (8.9 t/ha), lower than the world productivity level. Orissa, Uttar Pradesh, West Bengal, Karnataka, Maharashtra, Bihar and Madhya Pradesh are the leading states in sweet potato cultivation.

The root-knot and the reniform nematodes are the two most widely occurring and important parasites of sweet potato throughout the world. *Radopholus similis* caused significant reduction in root weight, tuber weight and tuber length even at a low inoculum level of 100 nematodes/plant (Koshy and Jasy, 1990).

9.2.1. Root-Knot Nematodes, *Meloidogyne* spp.

M. incognita is the major nematode species damaging sweet potato, but the crop is also attacked by *M. javanica, M. hapla* and *M. arenaria.* In India, *M. incognita* and *M. javanica* are reported on sweet potato (Ray *et al.,* 1990). Gapasin and Valdez (1979) reported tuber reduction to the extent of 47.7 and 50.6% with an initial inoculum level of 20,000 eggs of *M. incognita* and *M. javanica,* respectively.

9.2.1.1. Symptoms

Nematode infection in primary roots cause swelling (knotting) of the entire root, and heavy infection can inhibit apical growth. The most obvious symptoms of damage on enlarged roots are longitudinal cracking and general rough appearance of the skin. In many cases lenticels are noticeably enlarged. Both the roots and tops of infected plants are reduced. In fibrous roots, the nematode infestation produces small swellings or knots which are visible in most cases to the naked eye. Infected feeder roots are usually shorter and have fewer secondary roots and root hairs. Symptoms such as reduction in vine growth, yellowing or abnormally abundant production of flowers could also be due to infestation by the nematode.

9.2.1.2. Life Cycle

The nematode may complete several generations during the cultivation of this crop. Feeder and storage roots are attacked at the same rate. All races of *M. incognita* can attack sweet potatoes at varying degrees.

9.2.1.3. Survival and Spread

Meloidogyne juveniles and/or eggs survive in storage roots and can be disseminated in root, but not stem, propagative material. Irrigation water and unclean farm tools and machinery can aid dissemination of the nematodes. Nematodes can survive on many alternate weed hosts.

9.2.1.4. Ecology

Meloidogyne species seem to do well in light, friable, sandy loam soil which happens to predominate and constitute the major portion of the world's sweet potato growing areas. *M. incognita* requires warm temperature for completion of its life cycle. During a normal growing season it can undergo 4-5 generations. Therefore, it is capable of increasing its population to a level of economic threshold in a short period.

9.2.1.5. Interaction with Other Pathogens

M. incognita interacts with *Fusarium* spp. and *Ralstonia solanacearum* causing severe wilting and premature death. Although there are several *Fusarium* resistant cultivars, their resistance is broken in the presence of *M. incognita*.

9.2.1.6. Management Methods

(i) Physical Methods: Hot water treatment at 45°C for 30 hr, 46.7°C for 65 min or 50°C for 3-5 min eliminates root-knot nematode infection without

seriously impairing the viability of roots. Hot air treatment of roots, 5-8 cm in diameter, at 50°C for 4-8 hr eliminates most root-knot nematodes without significantly affecting their keeping qualities.

(ii) Cultural Methods: The clean cuttings of seed roots from the apex are less prone to root-knot damage when planted out than cuttings from the basal end.

Groundnut or maize in rotation with sweet potato decreases numbers of *M. incognita* to low levels.

(iii) Chemical Methods: Application of carbofuran granules at 6.7 kg a.i./ha, 1 to 4 days before planting; fensulfothion/ethoprophos granules each at 3.4 kg a.i./ha in a 30 cm band 2 weeks before planting have shown to control root-knot nematodes and increase the quality and yield of sweet potato.

Total yield and number of marketable roots can also be increased by treating nursery beds with ethoprophos and aldicarb granules at 1g/30 cm^2.

Bare root-dip treatment of the propagating material in a solution of oxamyl or side dressing with nematicides at the time of planting will allow the establishment of the crop by providing early protection against nematodes.

(iv) Host Resistance: Sweet potato cvs. Heartogold, Nemagold, Redmar, Jersey types and Kyushu No. 52 were reported to be resistant to root-knot nematodes (Giamalava *et al.*, 1960). The cv. Jasper is moderately resistant, while cv. Painter shows some tolerance. Mohandas and Palaniswami (1990a) reported the availability of high degree of resistance in high yielding released varieties of sweet potato, *viz.*, Sree Vardhini and Sree Nandni. Mohandas and Ramakrishnan (2002) reported that sweet potato cvs. Sree Vardhini, Sree Nandni, Sree Bhadra and Kanjangad were resistant to *M. incognita*.

(v) Integrated Methods: Control of root-knot damage to sweet potatoes involves integration of at least 3 methods:

- The selection of seed roots that are either free of nematodes or freed of nematodes by hot water treatment.
- Planting of the seed roots into beds of sand or coarse textured soil that is either nematode-free or, if infested, is pre-plant fumigated.
- Transplanting of the clean 'slips' into nematode-free soil or pre-plant fumigated soil.

9.2.2. The Reniform Nematode, *Rotylenchulus reniformis*

R. reniformis is another important nematode species reported from sweet potato from a large number of countries. Gapasin and Valdez (1979) reported

60.6% yield reduction with an initial inoculum level of 5,000 larvae of *R. reniformis*.

9.2.2.1. Symptoms

The reniform nematode is associated with leaf yellowing and reduction in yield and size of swollen tubers. The most serious injury occurs when nematode population levels are high at planting time and there is a seasonal increase of the nematodes. Infestation by *R. reniformis* cause cracking of storage roots.

9.2.2.2. Interaction with Other Pathogens

R. reniformis interact with *Fusarium* spp. in development of disease complex.

9.2.2.3. Management Methods

(i) Physical Methods: Hot water treatment of sweet potatoes at 50°C for 3 to 5 minutes eliminated the reniform nematode (Martin, 1970).

(ii) Chemical Methods: Good control of *R. reniformis* and increase in sweet potato grade have been achieved by application of granules in 30 cm band of fensulfothion/aldicarb each at 5.6 kg a.i./ha or ethoprophos at 3.4 to 6.7 kg a.i./ha.

(iii) Host Resistance: The sweet potato variety Goldrush is less susceptible to *R. reniformis* than other varieties, but is highly susceptible to root-knot nematodes. Selection P-104 is reported resistant to cracking.

9.3. CASSAVA

Cassava (*Manihot esculenta*), popularly known as tapioca, is a member of the family Euphorbiaceae, which has long been cultivated in Central America and Brazil and was later carried to Africa (where it is now an important subsistence crop), Asia and the Pacific Islands. It has drought tolerance (mainly due to in-built mechanism to shed or drop the leaves under adverse moisture conditions) and produces more calories/unit area. The tubers contain 25-30% starch, 1.5% protein and 2.1% minerals. It is being cultivated in 0.281 million hectares producing 7.9 million tonnes of cassava with an average yield of 28.1 tonnes per hectare. The Salem belt of Tamil Nadu and Samalkot belt of Andhra Pradesh are known for cassava as an industrial crop. The major cassava producing states include Tamil Nadu, Kerala, Karnataka and Andhra Pradesh.

Important nematodes parasitizing cassava are root-knot and root lesion nematodes. Ray *et al.* (1990) reported *Helicotylenchus multicinctus* and *Tylenchorhynchus mashhodi* from Orissa.

9.3.1. Root-Knot Nematodes, *Meloidogyne* spp.

Meloidogyne incognita, M. javanica and *M. arenaria* have been reported from cassava. Nirula (1963) reported *M. javanica* for the first time from India. Attack by root-knot nematodes was considered serious when cassava was intercropped with other susceptible hosts like brinjal and roselle. Jatala (1988) reported considerable yield reduction in cassava due to *M. incognita* and *M. javanica* infestation. *M. incognita* race 2 reduced fresh storage root weight by 49% and dry top weight by 22%.

9.3.1.1. Symptoms

Root-knot nematodes produce the typical swelling or knotting of cassava roots. *M. incognita* race 2 and *M. javanica* significantly reduced stalk length, stalk weight and storage root weight of cassava after 15.5 months growing period (Caveness, 1981).

9.3.1.2. Management Methods

(i) Cultural Methods: The summer fallowing which is already being practiced by farmers in many parts of cassava growing areas in Kerala due to non-availability of water is an excellent means of controlling nematode population. It is also essential to avoid planting susceptible crops either in the previous season or as intercrops.

Cassareep, a by-product of the cassava industry, was apparently effective in controlling *M. incognita* and *M. javanica* on cassava.

(ii) Host Resistance: Cassava cvs. Sree Sahya and Narayaniyakappa were found resistant to root-knot nematodes (Sreeja *et al.*, 2001). Freitas and Moura (1986) reported that cassava cv. Mandiocol to be resistant to *M. incognita* and *M. javanica*.

9.3.2. The Lesion Nematode, *Pratylenchus brachyurus, P. sefaensis*

Pratylenchus brachyurus and *P. sefaensis* were found attacking cassava. *P. brachyurus* is regarded as a serious pest of cassava causing damage to cortical tissues of roots and also predisposing plants to infection by secondary pathogens. *P. sefaensis* reduced fresh storage root weight by 45% and dry top weight by 21%.

9.3.2.1. Symptoms

The lesion nematode produces root lesions and subsequent rotting of roots. Affected plants are stunted with yellowing of leaves, die-back of twigs and produce small or no tubers. Leaves also show nutrient deficiency symptoms (Mohandas *et al.*, 1990).

9.3.2.2. Management Methods

(i) Chemical Methods: A 7.2% increase in tuber yield was recorded after application of DBCP at 50 l/ha.

(ii) Host Resistance: A number of cassava varieties (Atitogen, Ba Pou II, Agba Tiega, Agba Boquia, Sodjievi) show some resistance to the lesion nematode.

9.4. COLOCASIA

Colocasia (*Colocasia esculenta*), also known as cocoyam, taro, dasheen and eddoe, is a member of the family Araceae and occurs wild in S.E. Asia. It is cultivated throughout the humid tropics and is of great importance in the Pacific Islands. It is mostly a staple food or subsistence crop grown commercially in some countries.

The root-knot nematode is an important limiting factor in successful production of colocasia crop.

9.4.1. The Root-knot Nematode, *Meloidogyne incognita, M. javanica*

M. incognita and *M. javanica* are widely reported on colocasia. Nirula (1959) was the first to report root-knot nematode on colocasia from India. *M. javanica* causes severe losses to colocasia in India. *M. incognita* was responsible for 24% loss in tuber yield of colocasia (Anon, 1990b).

9.4.1.1. Symptoms

The foliage of nematode infested plants at first becomes yellow and then turns brown and ultimately dies back. No corms were found in plants attacked in the earlier stages and late stage attack resulted in deformed and galled corms (with blister-like swellings varying in size from 2-15 mm) of little market value. Rotting of corms associated with the nematode, occur during storage.

9.4.1.2. Survival and Spread

Meloidogyne spp. can be carried over from one *Colocasia* crop to next in the wide range of other host crops and weeds. As the nematodes feed and reproduce in corm tissues, they can be spread in corms and cormels if infested material is used for propagation.

9.4.1.3. Management Methods

(i) Physical Methods: Root-knot can be controlled in corms by dipping in hot water at 50°C for 40 min.

(ii) Cultural Methods: Meloidogyne populations could be suppressed when colocasia is grown in very wet or flooded conditions.

Use of nematode-free planting material will prevent spread into the field. Seed corms or cormels should be free of root-knot damage.

(iii) Chemical Methods: Srivastava *et al.* (1971) found that application of DBCP at 3.6 l/ha as soil drench was more effective than aldicarb (2.5 kg a.i./ha), fensulfothion (3.0 kg a.i./ha) and dichlofenthion (2.0 litres a.i./ha) for the control of *M. javanica* in India.

(iv) Host Resistance: Mohandas and Palaniswami (1990b) reported resistance in colocasia cv. Sree Reshmi and very high resistance in C-9 (a popular variety among farmers). In cv. C-9, no nematode could be detected either from root or from tubers. Cv. Dodare was found to be completely resistant to both *M. incognita* and *M. javanica*, while cv. Samra is described as moderately susceptible to these two species.

9.5. AMORPHOPHALLUS

The root-knot and the lesion nematodes are important pests of amorphophallus (*Amorphophallus* spp.).

9.5.1. The Root-knot Nematode, *Meloidogyne incognita*

Amorphophallus spp. are highly susceptible to *M. incognita*. The nematode infestation is very high in Kerala, Andhra Pradesh and Tamil Nadu. The root-knot nematodes have been recorded in high numbers from Tamil Nadu leading to crop failure in many farms.

9.5.1.1. Symptoms

The roots produce typical root-knot symptoms. In corms, the galls appear as irregular projections which harbour adult females and eggs. The area of infestation in tuber was discoloured when the infestation was severe, the infested area dried up resembling dry rot. Infested tubers were deformed and smaller. The degree of infestation was very high in cormels compared with that in corms.

9.5.2. Lesion Nematodes, *Pratylenchus* spp.

The lesion nematodes have been recorded in high numbers from Tamil Nadu leading to crop failure in many farms.

9.5.2.1. Symptoms

The surface of the infested tubers were black in colour with cracks. The blackening extended further deep inside the tubers reducing the edible portion. The infested tubers were smaller in size.

Chapter 10

FUTURE OUTLOOK

In looking to the future, what are some of the things we need to do from a realistic and practical stand point of view?

We should continue research on nematode management to include efforts to obtain nematicides which are more effective, can be used at lower dosage, and cause less pollution; to identify and develop new generation chemicals which may not kill the nematodes but either change their behaviour or delay the development and force them into the inactive stages or alert the plant to defend the nematode attack.

Although valuable improvement has been obtained in development of nematode resistant horticultural crop cultivars, there is a need for breeding varieties with combined resistance to different nematode species and to other organisms such as soil-borne fungi and bacteria which cause wilt and root rot. There is an urgent need to develop resistant cultivars for major crops and at least for the most important nematode species; and knowledge necessary to utilize the benefits and capabilities offered by integrated nematode management systems. The development of transgenic plants with durable nematode resistance keeping pace with the evolution of resistance breaking populations should be considered.

A second area of concern is to advance biological control to its fullest potential, utilizing biological control agents plus organic amendments, sewage, sludge, etc. This would include cultural methods for biocontrol parasites, understand their nature, make them practical, and genetically modify the biocontrol agents to improve their effectiveness. Many fungal and bacterial bioagents have been found effective under *in vitro* conditions. Efforts be made to test them under field conditions. Further, our efforts should be directed towards developing mass production technologies for effective bioagents and easy field application techniques.

In recent years, it has been observed that the combination of fungal bioagents perform better in reducing nematode population and increasing the plant health. Thus the combination of highly toxic fungus, *Aspergillus niger* (killed most of the infective second stage juveniles) and an egg parasite, *Cladosporium oxysporum* (invaded and killed the eggs in egg sac) both at half the doses reduced significantly more *M. incognita* population and exhibited better plant growth than when either of the fungal bioagents in brinjal (Goswami and Singh, 2001).

A third area would be, through surveys and identifications, determine more accurately the numbers and kinds of nematodes which actually exist and pinpoint their distribution. Success of crop rotations and resistant cultivars depend upon knowing what nematodes occur in which areas and fields.

A fourth area would be to develop a better understanding of the biological aspects of all our more important plant nematodes. This would include the molecular genetics of closely related species and races; host-parasite relationships and histochemical details which may provide clues to the nature of plant resistance to nematodes; ecological factors, such as temperature, moisture and soil types which may influence nematode development and survival which in turn affect control measures; and make full use of current biotechnologies as appropriate in this and other nematology research.

With regard to viruses transmitted by soil-borne nematodes, it is advisable to explore the possibility of finding dorylaimid nematodes from the rhizosphere and their role in transmitting soil-borne viruses mainly in crops like fruits, vegetables and ornamentals.

Our future research programmes need to be more based towards development of sustainable and non-chemical methods for the management of nematode pests. There is a need for development of cropping systems based nematode management technologies.

Phytoalexins effectively induce plant resistance mechanisms to nematodes, particularly to sedentary ones. However, the mechanism by which phytoalexins are accumulated in plants is not known. Further research in accurately elucidating the effect of potential phytoalexins on the nematodes is urgently needed.

Once the INM technologies are developed, it is necessary that they should be put to practice under field conditions. The implementation of these programmes will require scientists/trained technicians and efficient extension media. In this direction, national institutes and SAU's can play an important role by holding training courses and carrying out problem oriented researches. Since integrated approach may involve more than one discipline, it is therefore, essential that professional scientists must collaborate in its planning and execution.

We have made much progress in nematology; but we still have a long way to go in the battle against these destructive and insidious plant nematodes. Through our continued dedication and strong research efforts, greater knowledge and better understanding of nematodes will be achieved. This will result in further reductions in losses caused by nematodes and a significant increase of food and fiber in our countries.

Chapter 11

SOURCES OF CRITICAL INPUTS FOR NEMATODE MANAGEMENT

11.1. ARBUSCULAR MYCORRHIZAL FUNGI

Table 11.1. Trade names and source of availability of Arbuscular Mycorrhizal Fungi

Arbuscular Mycorrhizal Fungi	Trade Name	State	Source of Availability
Glomus mosseae	Symbion VAM	Karnataka	Agri Technol Infmn Centre, IIHR, B'lore
		Maharashtra	Bharatiya Agro Industries Foundation, Pune
		Tamil Nadu	T Stanes & Co, Coimbatore
G. fasciculatum	Symbion VAM	Karnataka	Agri Technol Infmn Centre, IIHR, B'lore
		Maharashtra	Bharatiya Agro Industries Foundation, Pune
		Tamil Nadu	T Stanes & Co, Coimbatore
G. aggregatum		Maharashtra	Bharatiya Agro Industries Foundation, Pune
Gigaspora margarita		Maharashtra	Bharatiya Agro Industries Foundation, Pune
Arbuscular mycorrhizal fungi	Multiplex Trishul	Karnataka	Karnataka Agro Chemicals, B'lore

11.2. ANTAGONISTIC FUNGI

Table 11.2. Trade names and sources of availability of antagonistic fungi.

Antagonistic Fungi / Trade Name	State	Source of Availability
Arthrobotrys cladodes var. *macroides*	Karnataka	Project Directorate of Biol Control, Bangalore
A. oligospora	Karnataka	Project Directorate of Biol Control, Bangalore
Aspergillus niger (Kalisena SD, Kalisena SL, Sanjeevni)	Gujarat	Cadila Pharmaceuticals Ltd, Ahmadabad
	New Delhi	Plant Path Divn, Indian Agri Res Inst, New Delhi
Dactylella brochophaga	Karnataka	Project Directorate of Biol Control, Bangalore
Gliocladium catenulatum	Karnataka	Project Directorate of Biol Control, Bangalore
G. deliquescens	Karnataka	Project Directorate of Biol Control, Bangalore
G. roseum	Karnataka	Project Directorate of Biol Control, Bangalore
Pochonia chlamydosporia (Bionema, Biovert, Multiplex Varsha)	Karnataka	Project Directorate of Biol Control, Bangalore Agri Technol Infmn Centre, IIHR, Bangalore Central Sericultural Res &Trng Inst, Mysore Karnataka Agro Chemicals, B'lore
Paecilomyces lilacinus (Bio-Nemator, Biocon, Bioact, PL Plus, Multiplex Niyantran, Abtec Paecilomyces)	Karnataka	Project Directorate of Biol Control, Bangalore Agri Technol Infmn Centre, IIHR, Bangalore
Trichoderma aureoviride	Karnataka	Project Directorate of Biol Control, Bangalore
T. hamatum	Karnataka	Project Directorate of Biol Control, Bangalore
T. harzianum (Binab-T, Ecoderma, F-Stop, Supravit, Tricodex, Trichoderma)	Andhra Pradesh	ADA (BCL), P M Palem ADA (BCL), Kakinada ADA (BCL), Nidadavole ADA (BCL), Ibrahimpatnam ADA (BCL), Ongole ADA (BCL), Nellore DDA (FTC), Nandyal ADA (BCL), Anantapur

	DDA (FTC), Rajendranagar
	ADA (Oilseeds) (BCL), Mahabubanagar
	ADA (BCL), Nalagonda
	ADA (BCL), Mulugu Road
	ADA (BCL), Karimnagar
	ADA (BCL), Adilabad
	Pest Control (India) Pvt Ltd, Hyderabad
Assam	State Biocontrol Laboratory, Guwahati
	Pest Control (India) Pvt Ltd, Guwahati
Bihar & Jarkhand	Pest Control (India) Pvt Ltd, Jamshedpur
Gujarat	Pest Control (India) Pvt Ltd, Vadodara
Haryana	State Biocontrol Laboratory, Sirsa
	Pest Control (India) Pvt Ltd, Chandigarh
Karnataka	Project Directorate of Biol Control, Bangalore
	Agri Technol Infmn Centre, IIHR, Bangalore
	Central IPM Centre, Bangalore
	Central Sericultural Res & Trng Inst, Mysore
	Coffee Research Sub Station, Chethalli
	Central Coffee Res Inst, Entomol Divn, Chikmagalur
	Univ of Agri Sci, Entomol Dept, Dharwad
	Inst of Pulses & Oilseeds Res, Gulbarga
	Regional Res Stn, UAS, Raichur
	Agri Res Stn, UAS, Bijapur
	Kadur Agro Pvt Ltd, Bangalore
	Margo Pvt Ltd, Bangalore
	Pest Control (India) Pvt Ltd, Bangalore
Kerala	Travancore Organic Fertilizers, Kottayam
	VRM Biotech Products, Changanacherry
Madhya Pradesh	Pest Control (India) Pvt Ltd, Bhopal
Maharashtra	Bio Era Technologies, Nagpur
	Hoechst Schering Agr Envt, Mumbai
	KNS Biotech, Nanded
	Nathkrupa Biocontrol Lab, Nagpur
	Om Agro Organics, Yavatmal
	Pest Control (India) Pvt Ltd, Mumbai-400 063
	Pest Control (India) Pvt Ltd, Mumbai-400 059
	Pest Control (India) Pvt Ltd, Thane

	Pest Control (India) Pvt Ltd, New Mumbai
	Pest Control (India) Pvt Ltd, Nashik
	Pest Control (India) Pvt Ltd, Pune-411 033
	Pest Control (India) Pvt Ltd, Pune-411 030
	Sanvardhini Agro Pvt Ltd, Satara
	Soman Biofertilizers, Pune
	Vidyas Biotech Lab, Nagpur
	West Coast Herbochem P Ltd, Mumbai
New Delhi	Biotech Intern Ltd, New Delhi
	Pest Control (India) Pvt Ltd, New Delhi
Orissa	Pest Control (India) Pvt Ltd, Bhubaneswar
Punjab	Pest Control (India) Pvt Ltd, Chandigarh
Rajasthan	Biocontrol Lab, ARS, RAU, Sri Ganganagar
	JDA (Pl Path), Biocontrol Lab, Durgapura
	DDA (Agro), IPM Lab, Ajmeer
	DDA (Agro), IPM Lab, Malikpur
	DDA (Agro), IPM Lab, Rampura
	DDA (Agro), IPM Lab, Cjjatrapura
	DDA (Agro), IPM Lab, Chittorgarh
	DDA (Agro), IPM Lab, Hanumangarh
	DDA (Pl Path), IPM Lab, Banaswara
	Pest Control (India) Pvt Ltd, Jaipur
Tamil Nadu	Centre for Pl Prot Studies, TNAU, Coimbatore
	Bio-control Lab, Papparapatty
	Bio-control Lab, Coimbatore
	Bio-control Lab, Salem
	Bio-control Lab, Villupuram
	Bio-control Lab, Trichy
	Bio-control Lab, Panjupettai
	Bio-control Lab, Kuruppanaichenpalyam
	Bio-control Lab, Vinayagapuram
	Bio-control Lab, Tirunelveli
	Bio-control Lab, Tanjavur
	Bio-control Lab, Namakkal
	Basarass Biocontrol Res, Vallavar
	Greentech Agro Prod P Ltd, Coimbatore
	Jeypee Bio Techs, Virudhunagar
	Pest Control (India) Pvt Ltd, Chennai
	Rajendra Foundn for Agri Res, Chittar

		Shri Durga Agro Services, Coimbatore
		Shrishti Bioprod P Ltd, Coimbatore
		Sun Agro Ind India, Chennai
		TN Co-op Sugar Fedn Ltd, Chengalpattu
		TAP Industries, Chennai
	Uttarakhand	GB Pant Uni Agri & Technol, Pl Path Dept, Pantnagar
		Pest Control (India) Pvt Ltd, Dehradun
	Uttar Pradesh	Crop Health Bioprod Res Centre, Ghaziabad
		Pest Control (India) Pvt Ltd, Bareilly
		Pest Control (India) Pvt Ltd, Lucknow
		Pest Control (India) Pvt Ltd, Kanpur
	West Bengal	Pest Control (India) Pvt Ltd, Kolkata
T. koningii	Karnataka	Project Directorate of Biol Control, Bangalore
T. longibactriatum	Karnataka	Project Directorate of Biol Control, Bangalore
T. polysporum	Karnataka	Project Directorate of Biol Control, Bangalore
T. pseudokoningii (Nursery Guard)	Karnataka	Project Directorate of Biol Control, Bangalore Central Sericultural Res &Trng Inst, Mysore
T. virens (Gligard, WRC-GV)	Karnataka	Project Directorate of Biol Control, Bangalore
T. viride (Bio-Cure, Bio-Cure F, Agroderma, Antogon TV, Bioderma, Dermapak, Ecofit, Monitor, Multiplex Nisarga, Rakshak, Trichosan, Trichoft, Trichoderm, Trichonik, Tricho-X, Trichodermin)	Andhra Pradesh	ADA (BCL), P M Palem ADA (BCL), Kakinada ADA (BCL), Nidadavole ADA (BCL), Ibrahimpatnam ADA (BCL), Ongole ADA (BCL), Nellore DDA (FTC), Nandyal ADA (BCL), Anantapur DDA (FTC), Rajendranagar ADA (Oilseeds) (BCL), Mahabubanagar ADA (BCL), Nalagonda ADA (BCL), Mulugu Road ADA (BCL), Karimnagar ADA (BCL), Adilabad Pest Control (India) Pvt Ltd, Hyderabad
	Assam	State Biocontrol Laboratory, Guwahati Pest Control (India) Pvt Ltd, Guwahati
	Bihar & Jarkhand	Pest Control (India) Pvt Ltd, Jamshedpur

Gujarat	Pest Control (India) Pvt Ltd, Vadodara
Haryana	State Biocontrol Laboratory, Sirsa
	Pest Control (India) Pvt Ltd, Chandigarh
Karnataka	Project Directorate of Biol Control, Bangalore
	Coffee Research Sub Station, Chethalli
	Central Coffee Res Inst, Entomol Divn, Chikmagalur
	Kadur Agro Pvt Ltd, Bangalore
	Karnataka Agro Chemicals, B'lore
	Margo Pvt. Ltd, Bangalore
	Pest Control (India) Pvt Ltd, Bangalore
	Travancore Org Fert Co P Ltd, Bangalore
Kerala	Travancore Organic Fertilizers, Kottayam
	VRM Biotech Products, Changanacherry
Madhya Pradesh	Pest Control (India) Pvt Ltd, Bhopal
Maharashtra	Bio Era Technologies, Nagpur
	Hoechst Schering Agr Envt, Mumbai
	KNS Biotech, Nanded
	Nathkrupa Biocontrol Lab, Nagpur
	Om Agro Organics, Yavatmal
	Pest Control (India) Pvt Ltd, Mumbai-400 063
	Pest Control (India) Pvt Ltd, Mumbai-400 059
	Pest Control (India) Pvt Ltd, Thane
	Pest Control (India) Pvt Ltd, New Mumbai
	Pest Control (India) Pvt Ltd, Nashik
	Pest Control (India) Pvt Ltd, Pune-411 033
	Pest Control (India) Pvt Ltd, Pune-411 030
	Sanvardhini Agro Pvt. Ltd., Satara
	Soman Biofertilizers, Pune
	Vidyas Biotech Lab, Nagpur
	West Coast Herbochem P Ltd, Mumbai
New Delhi	Biotech Intern Ltd, New Delhi
	Pest Control (India) Pvt Ltd, New Delhi
Orissa	Pest Control (India) Pvt Ltd, Bhubaneswar
Punjab	Pest Control (India) Pvt Ltd, Chandigarh
Rajasthan	Biocontrol Lab, ARS, RAU, Sri Ganganagar
	JDA (Pl Path), Biocontrol Lab, Durgapura
	DDA (Agro), IPM Lab, Ajmeer

		DDA (Agro), IPM Lab, Malikpur
		DDA (Agro), IPM Lab, Rampura
		DDA (Agro), IPM Lab, Cjjatrapura
		DDA (Agro), IPM Lab, Chittorgarh
		DDA (Agro), IPM Lab, Hanumangarh
		DDA (Pl Path), IPM Lab, Banaswara
		Pest Control (India) Pvt Ltd, Jaipur
	Tamil Nadu	Centre for Pl Prot Studies, TNAU, Coimbatore
		Bio-control Lab, Papparapatty
		Bio-control Lab, Coimbatore
		Bio-control Lab, Salem
		Bio-control Lab, Villupuram
		Bio-control Lab, Trichy
		Bio-control Lab, Panjupettai
		Bio-control Lab, Kuruppanaichenpalyam
		Bio-control Lab, Vinayagapuram
		Bio-control Lab, Tirunelveli
		Bio-control Lab, Tanjavur
		Bio-control Lab, Namakkal
		Basarass Biocontrol Res, Vallavar
		Greentech Agro Prod P Ltd, Coimbatore
		Jeypee Bio Techs, Virudhunagar
		Pest Control (India) Pvt Ltd, Chennai
		Rajendra Foundn for Agri Res, Chittar
		Shri Durga Agro Services, Coimbatore
		Shrishti Bioprod P Ltd, Coimbatore
		Sun Agro Ind India, Chennai
		TN Co-op Sugar Fedn Ltd, Chengalpattu
		TAP Industries, Chennai
		T Stanes & Co, Coimbatore
	Uttarakhand	GBPant Univ Agri & Technol, Pl Path Dept, Pantnagar
		Pest Control (India) Pvt Ltd, Dehradun
	Uttar Pradesh	Crop Health Bioprod Res Centre, Ghaziabad
		Pest Control (India) Pvt Ltd, Bareilly
		Pest Control (India) Pvt Ltd, Lucknow
		Pest Control (India) Pvt Ltd, Kanpur
	West Bengal	Pest Control (India) Pvt Ltd, Kolkata
Verticillium lecanii	Karnataka	Project Directorate of Biol Control, Bangalore

		Pest Control (India) Pvt Ltd, Bangalore
	Maharashtra	Sio Agro Research Lab, Mumbai
	Tamil Nadu	Sakthi Biocontrol & Biofert Centre, Kallipatti Gobi
V. suchlosporium	Karnataka	Project Directorate of Biol Control, Bangalore

11.3. BACTERIAL BIOPESTICIDES

Table 11.3. Trade names and sources of availability of bacterial biopesticides.

Bacterial Biopesticides/ Trade Name/s	State	Source of Availability
Bacillus sphaericus	New Delhi	Biotech Intern Ltd, New Delhi
	Tamil Nadu	Sinagro Industries India, Chennai
B. subtilis (Kill Dew DP Epic, Kodiac)	Maharashtra	Kumar Krishimitra Bioproducts, Pune
B. thuringiensis var. *kurstaki* (Biolep, Biosap, Delfin, Dipel, Bibot, Spicturin, Halt, Multiplex Bt. K)	Andhra Pradesh	Indian Inst of Chem Technol, Hyderabad
	Karnataka	Karnataka Agro Chemicals, B'lore
	Madhya Pradesh	Indore Biotech Inputs & Res (Pvt) Ltd, Indore
	Maharashtra	Lupin Laboratories (I) P Ltd, Mumbai Sandoz (I) Ltd, Mumbai Wockhardt Ltd, Mumbai
	New Delhi	Biotech Intern Ltd, New Delhi Hindustan Insecticides, New Delhi
	Tamil Nadu	Tuticorn Alkali Chem & Fert Ltd, Chennai
Pseudomonas fluorescens (Bio-Cure B, Bioshield, Multiplex Sparsha, Traco-monas, Sudocel, Dagger-G)	Assam	State Biocontrol Laboratory, Guwahati
	Haryana	State Biocontrol Laboratory, Sirsa
	Karnataka	Agri Technol Infmn Centre, IIHR, Bangalore Karnataka Agro Chemicals, B'lore

	Travancore Org Fert Co P Ltd, Bangalore
Kerala	Plantrich Chem & Fert Ltd, Kottayam
New Delhi	Biotech Intern Ltd, New Delhi
Tamil Nadu	Bio-control Lab, Papparapatty
	Bio-control Lab, Coimbatore
	Bio-control Lab, Salem
	Bio-control Lab, Villupuram
	Bio-control Lab, Trichy
	Bio-control Lab, Panjupettai
	Bio-control Lab, Kuruppanaichenpalyam
	Bio-control Lab, Vinayagapuram
	Bio-control Lab, Tirunelveli
	Bio-control Lab, Tanjavur
	Bio-control Lab, Namakkal
	Basarass Biocontrol Res, Vallavar
	Greentech Agro Prod P Ltd, Coimbatore
	Jeypee Bio Techs, Virudhunagar
	Rajendra Foundn for Agri Res, Chittar
	Sun Agro Ind India, Chennai
	T Stanes & Co, Coimbatore

11.4. COMBINATION PRODUCTS OF BIOCONTROL AGENTS

Table 11.4. Source of availability of Combination Products of Biocontrol Agents.

Combination Products of Biocontrol Agents /Trade Name/s	State	Source of Availability
B. pumilis + P. chlamydosporia (Bacillus Plus-P)	Karnataka	Agri Technol Infmn Centre, IIHR, Bangalore
B. subtilis + P. chlamydosporia (Bacillus Plus-S)	Karnataka	Agri Technol Infmn Centre, IIHR, Bangalore
P. fluorescens + P. chlamydosporia (Pseudomonas Plus)	Karnataka	Agri Technol Infmn Centre, IIHR, Bangalore
Arthrobotrys sp. + *Plilacinus + Verticillium* sp. (Nemastin)	Maharashtra	Kumar Krishimitra Bioproducts, Pune

Aspergillus niger + *T. viride* + *P. lilacinus* (Multiplex Shock)	Karnataka	Karnataka Agro Chemicals, B'lore
P. chlamydosporia + *B. pumilis* (Bacillus Plus-P)	Karnataka	Agri Technol Infmn Centre, IIHR, Bangalore
P. chlamydosporia + *B. subtilis* (Bacillus Plus-S)	Karnataka	Agri Technol Infmn Centre, IIHR, Bangalore
P. chlamydosporia + *P.lilacinus* (Biovert Plus)	Karnataka	Agri Technol Infmn Centre, IIHR, Bangalore
P. chlamydosporia + *P. fluorescens* (Pseudomonas Plus)	Karnataka	Agri Technol Infmn Centre, IIHR, Bangalore
P. chlamydosporia + *T. harzianum* (Trichorich)	Karnataka	Agri Technol Infmn Centre, IIHR, Bangalore
P. chlamydosporia + *T. viride* (Trichovert)	Karnataka	Agri Technol Infmn Centre, IIHR, Bangalore
T. harzianum + *T. viride* (Ecoderma, NIPROT)	Karnataka	Bio Control Res Lab, PCI Pvt Ltd, Bangalore Margo Pvt Ltd, Bangalore
T. harzianum + *T. viride* + *T. virens* (Combat)	Maharashtra	Kumar Krishimitra Bioproducts, Pune

11.5. NEEM-BASED AND OTHER BOTANICALS

Table 11.5. Source of availability of neem – based and other botanicals.

Neem-Based & other Botanicals	State	Source of Availability
Achook	Maharashtra	Bahar Agrochemicals, Ratnagiri Mahagrapes, Pune
Aphidin	Maharashtra	Eco-Max Agrosystems Ltd, Mumbai
Avana	Tamil Nadu	EID Parry (I) Ltd, Bioprod Divn, Chennai
Aza	Andhra Pradesh	SOM-IPM Systems (I) Ltd, Hyderabad
Azadit	Rajasthan	Pesticides India Ltd, Udaipur
Biogold	Tamil Nadu	SPIC, Bio-Tech Divn, Chennai
Bioneem	Maharashtra	Ajay Biotech Lab (P) Ltd, Pune
Bionema	Uttar Pradesh	Bioved Res Soc, Allahabad
Biopest	Uttar Pradesh	Bioved Res Soc, Allahabad
Biosol	Tamil Nadu	AV Thomas Pvt Ltd, Chennai

Eco-Neem	Karnataka	Murkumbi Bio Agro P Ltd, Belgaum Pest Control (India) Pvt Ltd, Bangalore
Eco-Neem Plus	Karnataka	Murkumbi Bio Agro P Ltd, Belgaum Pest Control (India) Pvt Ltd, Bangalore
Ecotin	Karnataka	Monarch Biofert & Res Centre, Bangalore
Field Marshal	Gujarat	Ganesh Biocontrol System, Gondal
Fortuna Aza	Andhra Pradesh	Fortune Bio-tech Limited, Secunderabad
Jai Neem	Haryana	Jai Chemicals, Faridabad
Jawan	Maharashtra	Mc DA Agro Pvt Ltd, Mumbai
Jeevan Crop Protector	Maharashtra	Mc DA Agro Pvt Ltd, Mumbai
Juerken	Tamil Nadu	Madurai Chem & Agro Ind Ltd, Madurai
Limonool	Karnataka	Bio-Multi Tech Pvt Ltd, Bangalore
Margocide-CK	Karnataka	Monofix Agro Products Ltd, Hubli
Margosan-O	Karnataka	Monofix Agro Products Ltd, Hubli
Multineem	Karnataka	Karnataka Agro Chemicals, B'lore
Multiplex Mahaan	Karnataka	Karnataka Agro Chemicals, B'lore
Multiplex Urea Coat	Karnataka	Karnataka Agro Chemicals, B'lore
Neemactin	Maharashtra	Wockhardt Ltd, Mumbai
Neemak	Karnataka	West Coast Herbochem P Ltd, Bangalore
	Maharashtra	West Coast Herbochem P Ltd, Mumbai
Neemarin	New Delhi	Biotech Intern Ltd, New Delhi
Neemark	Karnataka	West Coast Herbochem P Ltd, Bangalore
Neemasol	Tamil Nadu	EID Parry (I) Ltd, Pesticides Divn, Chennai
Neemax	Maharashtra	Eco-Max Agrosystems Ltd, Mumbai
Neemazal-F	Tamil Nadu	EID Parry (I) Ltd, Bioprod Divn, Chennai
Neemazal-T/S	Tamil Nadu	EID Parry (I) Ltd, Bioprod Divn, Chennai
Neemocide	West Bengal	Vinayak Fats & Proteins P Ltd, Kolkata
Neemol	Andhra Pradesh	Ramson Laboratories, Vijayawada
	Tamil Nadu	Agro Links, Trichy
Neemolin	Maharashtra	Khatau Agro Tech, Mumbai
Neemosan	Tamil Nadu	Agronule Industries, Trichy
Neemox	Maharashtra	Prakash Farm Chemicals, Mumbai
Neempourn	Tamil Nadu	Prabhakar Oil Mills, Trichy
Neemta	Gujarat	AJ Chemicals, Ahmedabad

Neethrin	Uttar Pradesh	Amitul Agro Chem Pvt Ltd, Gorakhpur
Neem Base	New Delhi	Indian Agri Res Inst, Agrochem Divn, New Delhi
Neem Based pesticides	Maharashtra	Bahar Agrochemicals, Ratnagiri KNS Biotech, Nanded Om Agro Organics, Yavatmal
	Tamil Nadu	Ecobiocides & Botanicals Pvt Ltd, Theni Jeypee Bio Techs, Virudhunagar
Neem Cake	Uttarakhand	Sri Ram Solvent Extraction Plant, Jaspur
Neem Gold	Tamil Nadu	EID Parry (I) Ltd, Bioprod Divn, Chennai SPIC, Bio-Tech Divn, Chennai
Neem Guard	Maharashtra	Gharda Chemicals Pvt Ltd, Mumbai
Neem Hit	Maharashtra	Skylark Agro Chemicals, Thane
Neem Oil	Karnataka	Agro Extracts Ltd, Bangalore
	Maharashtra	Mc DA Agro Pvt Ltd, Mumbai National Organic Chem Ind Ltd, Mumbai Sio Agro Research Lab, Mumbai
	Uttarakhand	Sri Ram Solvent Extraction Plant, Jaspur
Neem Plus	Maharashtra	BD Keathen & Co, Mumbai
Neem Rich	Maharashtra	National Chem Lab, Pune
Neem Seed	Maharashtra	Cheminova Agro-Tech, Nandubar
Neem Soap	Karnataka	Agri Technol Infmn Centre, IIHR, Bangalore
Neem Top	Tamil Nadu	Sri Krishna Co, Coimbatore
Nim-76	New Delhi	Defence Inst of Physiol, New Delhi
Nimba	New Delhi	Indian Agri Res Inst, Entomol Divn, New Delhi
Nimbasal	New Delhi	Nimba Foods & Agrochem, New Delhi Pesto-Chem India Ltd, New Delhi
Nimbicidine	Tamil Nadu	T Stanes & Co, Coimbatore
Nimbitor	Maharashtra	Zandu Pharmaceutical Works, Mumbai
Nimbosol	Tamil Nadu	Victoria Laboratories, Salem
Nimitox	Karnataka	Rallis India Ltd, Bangalore
Nimlin	Maharashtra	Sunline Agro Chem P Ltd, Dhulia
Nimbo Bas	Tamil Nadu	Basarass Biocontrol Res, Vallavar
Ozoneem	Haryana	Ozone Biotech, Faridabad
Phytowin	Tamil Nadu	Phyto Products Ltd, Thiruthuripoondi

Pongamia oil-based biopesticide	Tamil Nadu	Jeypee Bio Techs, Virudhunagar
Pongamia Soap	Karnataka	Agri Technol Infmn Centre, IIHR, Bangalore
Rakshak	Karnataka	Murkumbi Bio Agro P Ltd, Belgaum Margo Pvt Ltd, Bangalore
RD-9 Repellin	Andhra Pradesh	ITC Ltd, Hyderabad
Replin 555	Tamil Nadu	Micro Chemicals Ltd, Madurai
Shaktiman	Gujarat	Khetiwadi Corner, Vadodara Krishna Biotech Pvt Ltd, Veravel Oceon Agro India Ltd, Baroda
Sukrina	Tamil Nadu	Canster Chemicals Ltd, Chennai
Suneem	Maharashtra	Sunida Exports Ltd, Mumbai
Swaticure	Maharashtra	Swathi Industries Ltd, Mumbai
Tric	Uttar Pradesh	Amitul Agro Chem Pvt Ltd, Gorakhpur
Tricure	Karnataka	Murkumbi Bio Agro P Ltd, Belgaum Margo Pvt Ltd, Bangalore
Vapacide	Andhra Pradesh	Indian Inst of Chem Technol, Hyderabad

11.6. NEMATICIDES

Table 11.6. Globally important nematicides and their manufacturers

Active substance	Chemical group	Trade name	Manufacturer
1,3-Dichloropropene	Halogenated hydrocarbon	Telone	Bayer Crop Science
Dazomet	Methyl isothiocyanate liberator	Basamid	BASF Corporation
Metham sodium	Methyl isothiocyanate liberator	Vapam	Amvac Chemical Corpn.
Ethoprophos	Organophosphorous	Mocap	Bayer Crop Science
Phenamiphos	Organophosphorous	Nemacur	Bayer Crop Science
Fosthiazate	Organophosphorous	Nemathorin	Syngenta
Cadusafos	Organophosphorous	Rugby	FMC Corporation
Aldicarb	Oxime carbamate	Temik	Bayer Crop Science
Oxamyl	Oxime carbamate	Vydate	Du Pont
Carbofuran	Carbamate	Furadan	FMC Corporation

11.7. FULL ADDRESSES OF SOURCES OF CRITICAL INPUTS FOR NEMATODE MANAGEMENT

11.7.1. Andhra Pradesh

Central Government Organizations

Indian Institute of Chemical Technology (IICT), Uppal Road, Hyderabad-500 007.

State Government Organizations

ADA (BCL), P.M. Palem, Visakhapatnam District, Phone: 0891-2781283, Fax: 0891-2504139.

ADA (BCL), Kakinada, East Godhavari District, Phone: 0884-378794, Fax: 0884-2384238.

ADA (BCL), Nidadavole, West Godhavari District, Phone: 08812-222794, Fax: 08812-244014.

ADA (BCL), Ibrahimpatnam, Krishna District, Phone: 08672-2881968, Fax: 08672-252483.

ADA (BCL), Ongole, Prakasham District, Phone: 08592-233126, Fax: 08592-280046.

ADA (BCL), Nellore, Nellore District, Phone: 0861-2326655, Fax: 0862307132.

DDA (FTC), Nandyal, Kurnool District, Phone: 08518-248090, Fax: 08518-277754.

ADA (BCL), Anantapur, Anantapur District, Phone: 08554-231743, Fax: 08554-275984.

DDA (FTC), Rajendranagar, Rangareddy District, Phone: 040-24015188, Fax: 044-24547507.

ADA (Oilseeds) (BCL), Mahabubanagar, Mahabubanagar District, Phone: 08542-242624, Fax: 08542-242398/242203.

ADA (BCL), Nalagonda, Nalagonda District, Phone: 08682-244560, Fax: 08682-232981.

ADA (BCL), ARS Campus, Mulugu Road, Warangal District, Phone: 0870-2543433, Fax: 0870-2511100.

ADA (BCL), Karimnagar, Karimnagar District, Phone: 0878-262161, Fax: 0878-2240343.

ADA (BCL), Adilabad, Adilabad District, Phone: 08732-226454, Fax: 08732-227311.

Private Organizations

Fortune Bio-tech Limited, 14, Ishaq Colony, 108, Bazar Road, Secunderabad-500 051, Phone: 040-819000, 841519, Fax: 040-843945.

ITC Ltd., ILTD Division, IBD 6-3-1110, Amrutha Hall, Begumpet, Hyderabad-500 016.

Pest Control (India) Pvt. Ltd., Flat No. 213, Bhanu Enclave, Sanjeeva Reddy Nagar, Hyderabad-500 038, Phone: 040-2381 3851, 2370 1411, Fax: 040-2370 4305, E-mail: hyderabad@pcil.co.in (Contact person: Mr. S. Srinivasa Reddy, Mobile: 31017824).

Ramson Laboratories, 29-14-53, Prakasam Road, Vijayawada-520 002.

SOM-IPM Systems (India) Ltd., 7-1-80/202, Kemson Apartments, Ameerpet, Hyderabad-500 016, E-mail: somphyto/Hyderabad@dartmaildartnet.com.

11.7.2. Assam

State Government Organizations

State Biocontrol Laboratory, C/O Director of Agriculture, Ulubari, Guwahati-781 007, Phone: 0361-2332215, Fax: 0361-2332796, E-mail: assamagri@sify.com.

Private Organizations

Pest Control (India) Pvt. Ltd., Latasil Lamb Road, Behind Nice Tele Services, Guwahati-781 001, Phone: 0361-263 1230, E-mail: bedanta.baruah@pcil.co.in (Contact person: Mr. Bedanta Baruah, Mobile: 98640 54187).

11.7.3. Bihar & Jharkhand

Private Organizations

Pest Control (India) Pvt. Ltd., "Madhukunj", 3rd Floor, Q Road, Bistupur, Jamshedpur-831 001, Phone: 0657-242 3158, 228 8580, Fax: 0657-242 3158, E-mail: kailash.sharma@pcil.co.in (Contact person: Mr. Kailash Sharma, Mobile: 3105493).

11.7.4. Gujarat

Private Organizations

A.J. Chemicals, Kisan Brothers Pvt. Ltd., New Cloth Market, Ahmedabad-380 002.

Cadila Pharmaceuticals Ltd., Ahmadabad.

Ganesh Biocontrol System, Ganesh Chamber, Gondala Road, Gondal-360 311, Phone: 02825-221610, 228912, E-mail: ganeshbiocontrol@yahoo.com.

Khetiwadi Corner, Shiyabave, Vadodara-390 001.

Krishna Biotech Pvt. Ltd., 8-B, National Highway, Veravel-360 002, Rajkot District.

Oceon Agro India Ltd., Baroda.

Pest Control (India) Pvt. Ltd., 118-119, VIP View, VIP Road, Karelibaug, Vadodara-390 018, Phone: 0265-236 3546, 236 1027, Fax: 0265-236 1027, E-mail: vadodara_branch@pcil.co.in (Contact person: Mr.Y. Ramprasad, Mobile: 3132536).

11.7.5. Haryana

State Government Organizations

State Biocontrol Laboratory, Sirsa, Phone: 01662-226572, Fax: 01662-225713.

Private Organizations

Jai Chemicals, 14/1, Mathura Road, Faridabad-121 003.

Ozone Biotech, Plot No. 6, Site No. 2, Rajdhani Land and Finance, 14/3 Mathura Road, Faridabad-121 001, Phone: 0129-5047602, Fax: 0129-5047604, E-mail: ozonebiotech@rediffmail.com.

Pest Control (India) Pvt. Ltd., SCO No. 9, Cabin No. 5 (Basement), Sector-26, Madhya Marg, Chandigarh-160 019, Phone: 0172-2790230, E-mail: psd_chandigarh@pcil.co.in (Contact person: Mr. Sunil Awasti, Mobile: 0141 312 2470).

11.7.6. Karnataka

ICAR Institutes

Agricultural Technology Information Centre (ATIC), Indian Institute of Horticultural Research (IIHR), Hessaraghatta Lake, Bangalore-560 089.

Project Directorate of Biological Control (PDBC), P.B.No. 2491, H.A. Farm Post, Bellary Road, Bangalore-560 024.

Central Government Organizations

Central Integrated Pest Management Centre, Whitefield, Bangalore.

Central Sericultural Research and Training Institute (CSR&TI), Mysore-570 008.

Coffee Research Sub Station, Coffee Board, Chethalli-571 248, Kodagu District, Phone: 08276-266726, 266292, E-mail: ddcrss@sancharnet.in.

Central Coffee Research Institute (CCRI), Divn. of Plant Pathology, Coffee Research Station-577 117, Chikmagalur District, Phone: 08265-543008, Fax: 08265-543143, E-mail: crsento@yahoo.com.

State Agricultural Universities

Department of Agricultural Entomology, University of Agricultural Sciences (UAS), Krishinagar, Dharwad-580 005.

Institute of Pulses & Oilseeds Research, Aland Road, University of Agricultural Sciences, Gulbarga-581 101.

Agricultural Research Station (ARS), UAS, Bijapur.

Regional Research Station (RRS), UAS, Raichur.

Private Organizations

Agro Extracts Ltd., Plot No. 16, Phase II, Peenya Industrial Area, Bangalore-560 058.

Bio-Control Research Laboratories, Pest Control (India) Pvt. Ltd., No. 36/2, Sriramanahalli, Rajanakunte P.O., Post Box No. 6426, Yelahanka, Bangalore-560 064, Phone: 080-2246 8839/40/41/42, Fax: 080-2846 8838, E-mail: bcrl@pcil.co.in (Contact person: Mr. N. Venkatesh, Mobile: 3188 2116).

Bio-Multi Tech Pvt. Ltd., Ganganagar, Bangalore-560 030.

Kadur Agro & Video Vision Production Pvt. Ltd., R.V. Vidyaniketan, Mylasandra P.O., Bangalore-560 078.

Karnataka Agro Chemicals, 180, Ist Main Road, Mahalakshmi Layout, Bangalore-560 086, Phone: 080-2349 7464, 2349 4406, 2349 7360, Fax: 080-2349 0647, E-mail: multiplex@vsnl.com.

Margo Pvt. Ltd., 344/8, IVth Main, Sadashivnagar, Bangalore-560 080, Phone: 080-23613051/2/3, Fax: 080-23613055, E-mail: margo@vsnl.com, info@margobiocon.com, msm.christopher@margobiocon.com.

Monarch Bio-Fertilizers & Research Centre, 7/32, 47th Cross, Ist Main, 8th Block, Jayanagar, Bangalore-560 082.

Monofix Agro Products Ltd., VG Limbikai Building, Gokul Road, Hubli-580 030.

Murkumbi Bio Agro Pvt. Ltd., BC 105, Havelock Road, Camp, Belgaum-590 001.

Pest Control (India) Pvt. Ltd., B-385, Industrial Shed, KISSDC Ist Stage, Bangalore-560 058, Phone: 080-2372 4707, Fax: 080-2372 4708, E-mail: psd_bangalore@ pcil.co.in (Contact Person: Mr. Hemant Kumar, Mobile: 080 3186 7949).

Rallis India Ltd., Agrochemical Station, Plot No. 21 & 22, Phase II, Peenya Industrial Area, P.B. No. 5813, Bangalore-560 058, E-mail: rallis.research@gems. vsnl.net.in.

Travancore Organic Fertilizers Co. Pvt. Ltd., Biotech Division, 1609, MVR Layout, Kalyan Nagar, Bangalore-560 043.

West Coast Herbochem Pvt. Ltd., 105/B, Industrial Suburb, IInd Stage, Rajajinagar, IIIrd Cross Road, Bangalore-560 022.

11.7.7. Kerala

Private Organizations

Plantrich Chemical & Fertilizers Ltd., Industrial Estate, Manarcad P.O., Kottayam-686 019, Phone: 91481-2371477/877, E-mail: plantrich@bioplantrich.com.

Travancore Organic Fertilizers, Kottayam-686 001.

VRM Biotech Products, Southern Fertilizers & Chemicals, Salim Complex, Changanacherry-686 101, Kottayam.

11.7.8. Madhya Pradesh

Private Organizations

Indore Biotech Inputs & Research (Pvt.) Ltd., 6, Sikh Mohalla Main Road, Opposite Kothari Market, Indore-452 007.

Pest Control (India) Pvt. Ltd., 109, Malaviya Nagar, Bhopal-462 003, Phone: 0755-255 0090, 255 4767, Fax: 0755-255 4767, E-mail: salim.naushad@pcil.co.in (Contact person: Mr. S.M. Naushad, Mobile: 3139019).

11.7.9. Maharashtra

Private Organizations

Ajay Biotech India Ltd., 100/3, Kalpana Apartments, Dr. Katkar Marg, Erandawane, Pune-411 040.

Ajay Biotech Laboratories (Pvt.) Ltd., A-12/5, Kubera Park, Kanchawa, Pune-411 040.

Bahar Agrochemicals, E/24, MIDC, Lote Parashuram Road, Ratnagiri-415 722.

BD Keathen & Co., Bandra (East), Mumbai-440 029.

Bharatiya Agro Industries Foundation, Pune.

Bio Era Technologies, 12, Sri Ram Apartments, Dindayal Nagar, Ring Road, Nagpur-412 022.

Cheminova Agro-Tech, Plot No. 2, Gat N 352, Shahada Dondaicha Road, Mohide Shivar, Nandubar -425 409, Shahada District.

Eco-Max Agrosystems Ltd., Good Value Marketing Company, Mumbai-440 020.

Gharda Chemicals Pvt. Ltd., 48, Hill Road, Mumbai-440 050.

Hoechst Schering Agr. Environment, Hoechst Centre, 54/A, Andheri-Kurla Road, Andheri West, Mumbai-400 093.

Khatau Agro Tech., Khatau House, Moghul Lane, Mahim, Mumbai-440 016.

KNS Biotech, Shivaneri Nagar, Rampur Road, Degloor, Nanded-431 717.

Kumar Krishimitra Bioproducts (I) Pvt. Ltd., 12, Ganeshwadi, Prin K.R. Kanitkar Path, Off F.C. Road, Pune-411 004.

Lupin Laboratories (I) Pvt. Ltd., 259, CST Road, Kalima, Santa Cruz (East), Mumbai-440 098.

Mc DA Agro Pvt. Ltd., Mumbai-440 001.

Nathkrupa Biocontrol Laboratories, 55, Sriram Nagar, Near Wireless Station, Pawan Bhumi Marg, Wardha Road, Nagpur-440 002.

National Organic Chemical Industries Ltd., 951/A, Appasaheb Marathi Marg, Prabhadevi, Mumbai-440 029.

Om Agro Organics, Samarthwadi, Yavatmal-445 002.

Pest Control (India) Pvt. Ltd., 2nd & 5th Floor, Jagdamba House, Next to Annapurna Theatre, P.B.No. 9060, Goregaon (East), Mumbai-400 063, Phone: 022-2686 5550, 5699 0252, Fax: 022-2686 5555, 5699 0256, E-mail: nikhil.chatterjee@pcil.co.in (Contact person: Mr. Nikhil Chatterjee, Mobile: 3898 1560).

Pest Control (India) Pvt. Ltd., 127, Keytuo Industrial Estate, 220, Kondivita Road (off Andheri-Kurla Road), Andheri (East), Mumbai-400 059, Phone: 022-2821 3546/7, 2821 7385, 2832 8481, Fax: 022-2822 1647, E-mail: nanda.khamkar@pcil.co.in (Contact person: Mr. N.B. Khamkar, Mobile: 022 3210 2063).

Pest Control (India) Pvt. Ltd., Krishna Niwas, Ist Floor, Kaduva Lane, Near Collectorate, Thane-400 601, Phone: 022-2534 1131, 2536 1506, Fax: 022-2538 5142, E-mail: siddharth.sarangpani@pcil.co.in (Contact person: Mr. S. Sarangpani, Mobile: 022-3210 1586).

Pest Control (India) Pvt. Ltd., Shop Nos. 11 & 12, Vindhya Commercial Complex, Sector 11, CBD Belapur, New Mumbai-400 614, Phone: 022-2757 2225, 2757 1840, 2757 1837, Fax: 022-2757 1837, E-mail: priyendra.khona@pcil.co.in (Contact person: Mr. Priyendra Khona, Mobile: 022 3210 1587).

Pest Control (India) Pvt. Ltd., 11 & 12, Wastu Park 'A', Ashwin Sector, New CIDCO, Nashik-422 009, Phone: 0253-2393887, 2390807, Fax: 0253-2390807, E-mail: suresh.bramhankar@pcil.co.in (Contact Person: Mr. S.A. Bramhankar, Mobile: 0253 3106357).

Pest Control (India) Pvt. Ltd., Flat No. 5, Sandesh Building, Opp. Gokhale Park, Chinchwad, Pune-411 033, Phone: 020-2745 2250, Fax: 020-2745 2250, E-mail: suneel.joshi@pcil.co.in (Contact Person: Mr.S.P. Joshi, Mobile: 020 3101 0545).

Pest Control (India) Pvt. Ltd., Manashanti Apartments, 504-B, Shaniwar Peth, Pune-411 030, Phone: 020-2448 2756, 2448 2914, Fax: 020-2745 2250, E-mail: pcilpune@vsnl.net (Contact Person: Mr.B.B. Haldar, Mobile: 0202449 4607).

Prakash Farm Chemicals, Gulmohar Cross, Road No. 9, Ville Parle West, Mumbai-440 049.

Sandoz (I) Ltd., Agro Division, Sandoz House, Dr. Annie Besant Road, Worli, Mumbai-400 018.

Sanvardhini Agro Pvt. Ltd., M-14, Addl. MIDC, Satara-415 004.

Sio Agro Research Laboratories, Dhiraj Apartments, Mulund (West), Mumbai-440 080.

Skylark Agro Chemicals, W-92, MIDC, Phase-II, Dombivali (East), Thane-421 204.

Soman Biofertilizers, 594, Sadashiv Peth, 101, Express Towers, Laxmi Road, Pune-411 030.

Sunida Exports Ltd., Bandra (East), Mumbai-440 029.

Sunline Agro Chemicals Pvt. Ltd., P.B. No. 73, Sakri Road, Dhulia.

Swathi Industries Ltd., Morray House, Bandra, Mumbai-440 050.

Vidyas Biotech Laboratories, 33, Savita Vihar, Somalwada, Wardha Road, Nagpur-440 015.

West Coast Herbochem Pvt. Ltd., 23A/101, Nalanda, C.S.T. Road, Kurla West, Mumbai-400 070.

Wockhardt Ltd., Ready Money Terrace, 2nd Floor, 167, Dr. Annie Besant Road, Worli, Mumbai-400 018.

Zandu Pharmaceutical Works Ltd., 70, Gokhale Road (South), Mumbai-440 025.

11.7.10. New Delhi

ICAR Institutes

Division of Agrochemicals, Indian Agricultural Research Institute (IARI), New Delhi-110 012.

Division of Entomology, Indian Agricultural Research Institute (IARI), New Delhi-110 012, E-mail: iari@dBt.ernet.in

Division of Plant Pathology, Indian Agricultural Research Institute (IARI), New Delhi-110 012.

Central Government Organizations

Defence Institute of Physiology & Applied Sciences, Delhi Cantonment, New Delhi-110 022.

Private Organizations

Biotech International Ltd., VIPPS Centre, 2 Local Shopping Centre, Block-EFGH, Masjid Moth, Greater Kailash-II, New Delhi-110 048, Phone: 011-29220546/47/7359, Fax: 011-29229166/3083, E-mail: info@biotech-int.com.

Hindustan Insecticides Ltd., Scope Complex, Core-6, II Floor, Lodi Road, New Delhi-110 003.

Nimba Foods & Agrochemicals Division, 14-A/10, Western Extension, Pusa Road, New Delhi-110 055.

Pest Control (India) Pvt. Ltd., 7, Jantar Mantar Road, Post Box No. 581, New Delhi-110 001, Phone: 011-2336 8772, Fax: 011-2336 8771, E-mail: pcilmkt@bolnet.in / psd_delhi@pcil.co.in (Contact person: Mr. Naresh Patti, Mobile: 011 3257 9897).

Pesto-Chem India Ltd., CM-3, Ansal Dilkush, Industrial Complex, GT Karnal Road, Azadpur, New Delhi-110 033.

11.7.11. Orissa

Private Organizations

Pest Control (India) Pvt. Ltd., 440, Saheed Nagar, Bhubaneswar-751 007, Phone: 0674-, 2540235, 2544274, Fax: 0674-2544274, E-mail: anjan_samantray@yahoo.com (Contact Person: Mr. Anjan Samantray, Mobile: 98611 49280).

11.7.12. Punjab

Private Organizations

Pest Control (India) Pvt. Ltd., SCO No. 9, Cabin No. 5 (Basement), Sector-26, Madhya Marg, Chandigarh-160 019, Phone: 0172-2790230, Mobile: 0172 3105264, E-mail: psd_chandigarh@pcil.co.in.

11.7.13. Rajasthan

State Agricultural Universities

Biocontrol Laboratory, Agricultural Research Station (ARS), Rajasthan Agricultural University (RAU), Sri Ganganagar-335 001.

State Government Organizations

Deputy Director of Agriculture (Agronomy), ATC, Regional IPM Lab, Tabiji Farm, Ajmeer, Phone: 0145-2440652.

Deputy Director of Agriculture (Agronomy), ATC, Regional IPM Lab, Malikpur, Bharatpur, Phone: 05644-223337.

Deputy Director of Agriculture (Agronomy), ATC, Regional IPM Lab, Rampura, Jodhpur, Phone: 02962-222006.

Deputy Director of Agriculture (Agronomy), ATC, Regional IPM Lab, Cjjatrapura, Bundi, Phone: 0747-2442368.

Deputy Director of Agriculture (Agronomy), ATC, Regional IPM Lab, Farm Chittorgarh, Phone: 01472-241319.

Deputy Director of Agriculture (Agronomy), ATC, Regional IPM Lab, Krishi Farm, Hanumangarh Town, Phone: 01552-231291.

Deputy Director of Agriculture (Plant Pathology), IPM Laboratory, Dahod Road, Banaswara, Phone: 02962-243054.

Joint Director of Agriculture (Plant Pathology), State Biocontrol Lab, Durgapura, Jaipur, Phone: 0141-2550830.

Private Organizations

Pest Control (India) Pvt. Ltd., Shankar Sadan, Ist Floor, Adinath Marg, Opp. SMS Hospital, C-Scheme, Jaipur-302 004, Phone: 0141-236 8004, 237 8346, Fax: 0141-237 8346, E-mail: pcijpr@datainfosys.net (Contact person: Mr. Rajan Prakash, Mobile: 0141 312 2470).

Pesticides India Ltd., PO Box No. 20, Udaisagar Road, Udaipur-313 001.

11.7.14. Tamil Nadu

State Agricultural Universities

Centre for Plant Protection Studies (CPPS), Tamil Nadu Agricultural University (TNAU), Coimbatore-641003, Phone: 0422-431222, Fax: 0422-431672.

State Government Organizations

Bio-control Lab., State Seed Farm, Papparapatty Lab., Pennagaram Taluk, Dharmapuri District.

Bio-control Lab., JDA Office Complex, Tadagam Road, Coimbatore-641 013, Phone: 0422-2430766.

Bio-control Lab., ADA Office Complex, Seelanai-Chanpatty, Salem, Phone: 0427-2280214.

Bio-control Lab., Collectorate Office Complex, Villupuram-605 602, Phone: 04146-222291.

Bio-control Lab., ADA Office Complex, Puttur, Trichy-620 017, Phone: 0431-2793241.

Bio-control Lab., JDA Office Complex, Kancheepuram, At Panjupettai-651 501, Kancheepuram District, Phone: 04112-222977.

Bio-control Lab., State Seed Farm Complex, Kuruppanaichenpalyam Post, Bhavani, Erode.

Bio-control Lab., ADA Office Complex, Vinayagapuram, Therkkutheru Post, Melur Taluk, Madurai District.

Bio-control Lab., JDA Office, Tirunelveli Complex, Nirubar Colony, Tirunelveli-626 002.

Bio-control Lab., JDA Office, Thaniavur Complex, Kattuthottam Marianaman Koil Post, Tanjavur.

Bio-control Lab., Panchayat Union Complex, Namakkal-637 001.

Private Organizations

Agro Links, Trichy.

Agronule Industries, Trichy-632 217.

Agro Research Laboratories, 10/224, Avinashi Road, Coimbatore-641 018.

A.V. Thomas Pvt. Ltd., Anna Salai, Chennai-600 032.

Bacto Agro Culture Pvt. Ltd., Bharatha Nagar, Kalapally-643 253, The Nilgiris.

Basarass Biocontrol Research, 3/204, Main Road, Eraiyur Pennadanam RS &PO, SA, Vallavar-606 111, Cuddalore Dist., Phone: 4143-222403/037, Fax: 44-24867867, E–mail: basaras@vsnl.net.

Canster Chemicals Ltd., Adayar, Chennai-600 020.

Ecobiocides and Botanicals Pvt. Ltd., 2 East Market Street, Periyakulam Road, Theni-625 531, Phone: 4546-252020, Fax: 4546-251010, Mobile: 094433 42020, E-mail: ecobiocides@hotmail.com, Web site: www.azgro.com.

E.I.D. Parry (India) Ltd., Bio-Products Division, 234, NSC Bose Road, Chennai-600 001.

E.I.D. Parry (India) Ltd., Pesticides Division, "Dare House", P.B.No. 12, Chennai-600 001.

Green Tech Agro Products Pvt. Ltd., 47-D, Parsen Residency, Rajaji Road, Ramnagar, Coimbatore-641 009, Phone: 422-230107, Fax: 422-235237.

Jeypee Bio Techs, 25, Chinnaiah School Street, Virudhunagar-626 001.

Madurai Chemicals & Agro Industries Ltd., Rajaji Road, Madurai-625 104.

Micro Chemicals Ltd., Rajaji Road, Madurai-625 104.

Pest Control (India) Pvt. Ltd., 28, Errabalu Street, Chennai-600 001. Phone: 044-2522 6745/4266, Fax: 044-2522 2815, E-mail: psd_chennai@pcil.co.in (Contact person: Mr. V.M. Sagar, Mobile: 044 3106 1225).

Phyto Products Ltd., Thiruthuripoondi, S.A. District.

Prabhakar Oil Mills, Trichy-632 217.

Rajendra Foundation for Agricultural Research & Rural Development, Chittar-638 311,

Kesarimangalam Post, Bhavani Taluk, Erode Dist., Phone: 04256-39258, 39858.

Sakthi Biocontrol & Bio-fertilizer Centre, 9/11 Sathy Main Road, Kallipatti Gobi-638 505, Erode Dist., Phone: 04285-263464, 04554-231832.

Shri Durga Agro Services, 10/1, Kongu Nagar, Kalveerapalyam, Bharathiyar University Post, Coimbatore-641 046.

Shrishti Bio Products Pvt. Ltd., 331, Chinnasamy Naidu Road, New Siddapudur, Coimbatore-641 044.

Sinagro Industries India, 178, 9th Street, Mangalanagar, Porur, Chennai-600 116.

Southern Petrochemicals Industries Corporation Ltd. (SPIC), Bio-Tech Division, Agro Industrial Complex, 97, Mount Road, Chennai-600 032.

Sri Krishna Company, Race Course Road, Coimbatore-641 018.

Sun Agro Industries India, 178, 9th Cross Street, Mangala Nagar, Porur, Chennai-600 016.

Tamil Nadu Co-operative Sugar Federation Ltd., Main Biocontrol Research Laboratories, 2E/1, Rajeshwari Vedachalam Street, Chengalpattu-603 001.

TAP Industries, 7, Guna Complex, Tirumogeer Road, Y. Othathadai, Chennai-600 107.

T. Stanes & Co., 8/23-24, Race Course Road, Post Box No. 3709, Coimbatore-641 018.

Tuticorn Alkali Chemicals & Fertilizers Ltd., 553, Anna Salai, Teynampet, Chennai-600 018.

11.7.15. Uttarakhand

State Agricultural Universities

Department of Plant Pathology, GB Pant University of Agriculture & Technology (GBPUA&T), Pantnagar-263 145.

Private Organizations

Pest Control (India) Pvt. Ltd., 37/1, Ballapur Road, Near Hotel Surabhi, Dehradun-248 001, Phone: 0135-275 8568 (Contact person: Mr. Randeep Singh).

Sri Ram Solvent Extraction Plant (Neem), Jaspur.

11.7.16. Uttar Pradesh

Private Organizations

Amitul Agro Chem Pvt. Ltd., New Snehapuri Colony P.O., Gorakhpur-273 001.

Bioved Research Society, 103/42, MLN Road, Allahabad-211 002.

Crop Health Bio-Product Research Centre, R-12/50, Rajanagar, Ghaziabad-201 003.

Pest Control (India) Pvt. Ltd., Ratanadeep Complex, Ist Floor, Chauki Chauraha, Post Box 89, Bareilly-243 001, Phone: 0581-247 5032 / 245 3059, E-mail: sandeepkumar@pcil.co.in (Contact person: Mr. Sandeep Kumar, Mobile: 0581 310 3763).

Pest Control (India) Pvt. Ltd., 306, Lekhraj Khazana, Faizabad Road, Indira Nagar, Lucknow-226 016, Phone: 0522-238 7360 / 234 2606, E-mail: girish.lohumi@ pcil.co.in (Contact person: Mr. G.C. Lahumi, Mobile: 0522 324 4071).

Pest Control (India) Pvt. Ltd., Ramchandani Palace, IInd Floor, 183, The Mall, Post Box No. 271, Kanpur-208 001, E-mail: sanjay.mishra@pcil.co.in (Contact person: Mr. S.K. Mishra, Mobile: 0512 311 2751).

11.7.17. West Bengal

Private Organizations

Pest Control (India) Pvt. Ltd., 23, Mirza Ghalib Street, Post Box No. 9001, Kolkata-700 016, Phone: 033-2252 1184/2824, Fax: 033-2252 3652, E-mail: kolkata_branch@pcil.co.in (Contact person: Mr. Sanjay Choubey, Mobile: 033 3107 5887).

Vinayak Fats & Proteins Pvt. Ltd., Agri. In-put Division, 206, Bentinck Chambers, 37A, Bentinck Street, Kolkata-700 069.

REFERENCES

Abuzar, S. and Haseeb, A. 2006. Efficacy of carbofuran, bavistin, neem, *Trichoderma harzianum* and *Aspergillus niger* against *Meloidogyne incognita* and *Fusarium oxysporum* f. sp. *vasinfectum* disease complex on okra. *Indian Journal of Nematology* **36**: 299-300.

Acharya, A. and Padhi, N.N. 1987. Population threshold of *Rotylenchulus reniformis* in relation to its disease intensity on betel vine. *International Nematology Network Newsletter* **4**: 21-23.

Adams, S. 1997. Seeing red colored mulch starves nematodes. *Agriculture Research*, October, p. 18.

Ahmed, J.A. and Choudhury, B.N. 2004. Management of root-knot nematode, *Meloidogyne incognita* using organic amendments in French bean. *Indian Journal of Nematology* **24**: 137-139.

Ahuja, S. 1983. Studies on the variability in infection of root-knot nematode *Meloidogyne incognita* on potato. *Third Nematology Symposium,* Himachal Pradesh Agricultural University, Solan, p. 4.

Ahuja, S. and Mukhopadhyaya, M.C. 1984. Efficacy of some agricultural residues as compared to nematicides/insecticides in the control of root-knot nematode (*Meloidogyne incognita*) infestation in okra (*Abelmoschus esculentus*). *Proceedings of the Symposium on Utilization of Agricultural Wastes*, pp. 131-135.

Alam, M.M., Ali, Q.G., Masood, A. and Khan, A.M. 1976. *Indian Journal of Experimental Biology* **14**: 517-518.

Alam, M.M., Khan, A.M. and Saxena, S.K. 1973. Some new host records of root-knot nematode, *Meloidogyne incognita* (Kofoid & White, 1919) Chitwood, 1949. *Indian Journal of Nematology* **3**: 159-160.

Alam, M.M., Khan, A.M. and Saxena, S.K. 1977. Influence of different cropping sequences on soil population of plant parasitic nematodes. *Nematologia Mediterranea* **5**: 71-78.

Alam, M.M., Saxena, S.K. and Khan, A.M. 1977. Influence of interculture of marigold and margosa with some vegetable crops on plant growth and nematode population. *Acta Botanica Indica* **5**: 33-39.

Ali, S.S. 1982. Occurrence of root-knot nematodes in cardamom plantations of Tamil Nadu. *Proc. of Fifth Symp. on Plantation Crops*, Kasaragod, Kerala, India, pp. 615-620.

Ali, S.S. 1984. Preliminary observations on the effect of some systemic nematicides and neem oil cakes in a cardamom field infected with root-knot nematodes (Abstr.). *Proceedings of the Sixth Symposium on Plantation Crops*, Kottayam, Kerala, India, pp. 215-223.

Ali, S.S. 1986. Root-knot nematode problem in cardamom and its management. *Proceedings of the Second Group Discussion on the Nematological Problems of Plantation Crops,* Central Coffee Res. Inst., Balehonnur, pp. 10-12.

Ali, S.S. and Koshy, P.K. 1982. Occurrence and distribution of root-knot nematodes in cardamom plantations of Kerala. *Nematologia Mediterranea* **10**: 107-110.

Ali, S.S. and Venugopal, M.N. 1992. *Nematologia Mediterranea* **20**: 65-66.

Ali, S.S. and Venugopal, M.N. 1993. *Current Nematology* **4**: 19-24.

Amalraj, S.F.A. and Rao, G.R. 1992. Potato cyst nematode pathogens in the southern hills of India. *Indian Journal of Nematology* **22**: 82-85.

Anandraj, M., Ramana, K.V. and Sarma, Y.R. 1991. In *Mycorrhizal Symbiosis and Plant Growth* (D.J. Bagyaraj and A. Manjunath, eds.), pp. 110-112. Univ. of Agri. Sci., Bangalore.

Anita, B. and Subramanian, S. 1998. Management of the reniform nematode *Rotylenchulus reniformis* in tomato. In *Nematology - Challenges & Opportunities in 21st Century* (Usha K. Mehta, ed.), pp. 249-250. Sugarcane Breeding Inst., Coimbatore.

Anita, B. and Vadivelu, S. 1997. Management of root-knot nematode, *Meloidogyne hapla* on scented geranium, *Pelargonium graveolans*. *Indian Journal of Nematology* **27**: 123-125.

Anon, 1985. Kufri Swarna, a High Yielding Late Blight and Cyst Nematode Resistant Potato Cultivar Suitable for Cultivation in Tamil Nadu Hills. No. 21(E), Central Potato Research Institute, Shimla, 4 pp.

Anon, 1987. Improvement of betel vine cultivation in Mahoba. *Progress Report 1986-87.* National Botanical Res. Inst., Lucknow, India.

Anon, 1989a. Biennial Report (1987-89) of AICRP on Plant Parasitic Nematodes with Integrated Approach for Their Control, Indian Agri. Res. Inst., New Delhi, 80pp.

Anon, 1989b. Consolidated Biennial Report (1987-89) of AICRP on Plant Parasitic Nematodes with Integrated Approach for Their Control. Dept. of Nematology, Haryana Agri. Univ., Hisar, 35 pp.

Anon, 1989c. QRT Report of All India Co-ordinated Research Project on Plant Parasitic Nematodes with Integrated Approach for their Control (1984-88). Indian Council of Agricultural Research, New Delhi, 103 pp.

Anon. 1990a. Final Report of Research Scheme on Nematode Problems in Some Selected Temperate Fruit Crops and Their Management. Y. S. Parmar University of Horticulture & Forestry, Solan, 54 pp.

Anon. 1990b. Annual Report of the Central Tuber Crops Research Institute, Trivandrum.

Anon. 1993a. *QRT Report of AICRP on Nematodes (1989-93)*, Dept. of Nematol., Haryana Agri. Univ., Hisar, India, 149 pp.

Anon. 1993b. Biennial Report (1991-93) of AICRP on Plant Parasitic Nematodes with Integrated Approach for Their Control, Dept. of Nematology, Haryana Agri. Univ., Hisar, 80 pp.

Anusuya, K. and Vadivelu, S. 2002. Mutualistic symbiosis of VAM fungi on plant growth promoting rhizobacteria (PGPR) on micro propagated carnation. *J. Microbiology World* **4**: 105-107.

Atu, U.G., Odurukwe, S.O. and Ogbuji, R.O. 1983. Root-knot nematode damage to *Dioscorea rotundata. Plant Disease* **67**: 814-815.

Atwal, A.S. and Mangar, A. 1969. Repellent action of root exudates of *Sesamum orientale* against the root-knot nematode, *Meloidogyne incognita* (Heteroderidae: Nematoda). *Indian Journal of Entomology* **31**: 286.

Ayala, A. and Acosta, N. 1971. Observations on yam (*Dioscorea alata*) nematodes. *Nematropica* **1**: 39-40.

Ayala, A., Acosta, N. and Adsuar, J.A. 1971. A preliminary report on the response of *Carica papaya* to foliar application of two systemic nematicides. *Nematropica* **1**: 10.

Ayala, A., Roman, J. and Gonzalez, J. 1967. Pangola grass as a rotation crop for pineapple nematode control. *Journal of Agriculture Science, University of Puerto Rico* **51**: 94-96.

Ayyar, P.N.K. 1926. A preliminary note on the root nematode, *Heterodera radicicola* Mitter and its economic importance in South India. *Science and Culture* **22**: 391-393.

Baghel, P.P.S. 1992. Nematode pests of grapes, peach and papaya. In *"Nematode Pests of Crops"* (D.S. Bhatti and R.K. Walia, eds.), pp. 126-138. CBS Publishers & Distributors, New Delhi.

Baghel, P.P.S. 1995. Nematode problems in citrus. In *Nematode Pest Management-An Appraisal of Eco-friendly Approach* (G. Swarup, D.R. Dasgupta and J.S. Gill, eds.), pp. 203-210.

Baghel, P.P.S. and Bhatti, D.S. 1982. Vertical and horizontal distribution of phytonematodes associated with citrus. *Indian Journal of Nematology* **12**: 339-344.

Baghel, P.P.S. and Bhatti, D.S. 1983a. Evaluation of pesticides for the control of phytonematodes on citrus. *Third Nematology Symposium*, Himachal Pradesh Agricultural University, Solan, pp. 38-39.

Baghel, P.P.S. and Bhatti, D.S. 1983b. Relative efficacy of nematicides for control of phytonematodes on grapevine varieties. *Third Nematology Symposium*, Himachal Pradesh Agricultural University, Solan, p. 39.

Baghel, P.P.S., Bhatti, D.S. and Chauhan, K.S. 1980. *Haryana Journal of Horticultural Science* **9**: 136-137.

Baghel, P.P.S., Bhatti, D.S. and Jalali, B.L. 1990. *Proceedings of the National Conference on Mycorrhiza,* Haryana Agri. Univ., Hisar, p. 118.

Baghel, P.P.S. and Gupta, D.C. 1986. Effect of intercropping on root-knot nematode (*Meloidogyne javanica*) infecting grapevine (var. Pearlette). *Indian Journal of Nematology* **16**: 283-285.

Bagyaraj, D.J., Manjunath, A. and Reddy, D.D.R. 1979. Interaction of vesicular arbuscular mycorrhiza with root-knot nematodes in tomato. *Plant and Soil* **51**: 397-403.

Bahl, N. and Prasad. 1985. *Indian Botanical Reporter* **4**: 174.

Bajaj, H.K. 1989. In *Phyto and Soil Inhabiting Nemic Fauna of Haryana,* Department of Nematology, Haryana Agricultural University, Hisar, India.

Bajaj, H.K. and Jain, R.K. 2001. CCS Haryana Agriculture University, Hisar, Haryana. In *Indian Nematology–Problems and Perspectives* (S.C. Dhawan, H.S. Gaur, Pankaj, K.K. Kaushal, S. Ganguly, G. Chawla, R.V. Singh and Anil Sirohi, eds.), pp. 78-95. Divn of Nematol., Indian Agri. Res. Inst., New Delhi.

Bajaj, H.K., Kanwar, R.S., Baghel, P.P.S. and Bhatti, D.S. 1988. The Motia (*Jasminum sambac*) - a new host record of citrus nematode, *Tylenchulus semipenetrans* Cobb, 1913. *Journal of Research, Haryana Agricultural University* **18**: 146-147.

Bala, S.K. and Sukul, N.C. 1987. Systemic nematicidal effect of eugenol. *Nematropica* **17**: 219-222.

Balasubramanian, P. and Ramakrishnan, C. 1983. *Nematologia Mediterranea* **11**: 203-204.

Balasubramanian, M. and Rangaswami, G. 1964. Studies on host range and histopathology of root-knot infections caused by *Meloidogyne javanica. Indian Phytopathology* **17**: 126-132.

Bansal, R.K. and Verma, V.K. 2002. Antagonistic efficacy of *Azotobacter chroococcum* against *Meloidogyne javanica* infecting brinjal. *Indian Journal of Nematology* **32**: 132-134.

Barber, C.A. 1901. *Department Land Records and Agriculture.* Madras Agri. Branch 2, Bull. No. 45, pp.227-234.

Barbercheck, M.E. and Broembsen, S.L. von. 1986. *Plant Disease* **70**: 945-950.

Basu, S.D. and Gope, B. 1985. Control of root-knot nematode *Meloidogyne incognita* (Kofoid and White) Chitwood in tea seedlings with some systemic nematicides. *Two Leaves and a Bud* **32**: 31-35.

Bergeson, G.B. 1963. Influence of *Pratylenchus penetrans* alone and in combination with *Verticillium albo-utrum* on growth of peppermint. *Phytopathology* **53**: 1164-1166.

Bhagawati, B and Goswami, B. K. 2000. Interaction of *M. incognita* and *Fusarium oxysporum* f. sp. *lycopersici* on tomato. *Indian Journal of Nematology* **30**: 93-94.

Bhagawati, B., Goswami, B. K. and Singh, C.S. 2000. Management of disease complex of tomato caused by *M. incognita* and *Fusarium oxysporum* f. sp. *lycopersici* through bioagents. *Indian Journal of Nematology* **30**: 16-22.

Bhagawati, B. and Phukan, P.N. 1990. Chemical control of *Meloidogyne incognita* on pea. *Indian Journal of Nematology* **20**: 73-83.

Bhattacharya, R.K. and Rao, V.N.M. 1984. Effect of soil covers and soil moisture regimes on nematode population in soil and in root of banana. *Journal of Research Assam Agricultural University* **5**:206-209.

Bhatti, D.S. and Jain, R.K. 1977. Estimation of loss in okra, tomato and brinjal yield due to *Meloidogyne incognita. Indian Journal of Nematology* **7**: 37-41.

Bhosle, B.B., Sehgal, M., Puri, S.N., Sardana, H.R. and Singh, D.K. 2006. Efficacy of organic amendment in management of root-knot nematode, *Meloidogyne incognita* on okra. *Indian Journal of Nematology* **36**: 37-40.

Bindra, O.S., Chhabra, H.K., Chadha, K.L. and Mehrotra, N.K. 1967. A study on the correlation of citrus nematode population with decline of citrus. *Journal of Research, Punjab Agricultural University* **4**: 543-546.

Bird, G.W. 1981. Integrated nematode management for plant protection. In *Plant Parasitic Nematodes, Vol. 3* (B.M. Zuckerman & R.A. Rohde, eds.), pp. 355-375. Academic Press, New York.

Blair, G.P. 1966. The use of immature nuts of *Cocos nucifera* for studies on *Rhadinaphelenchus cocophilus. Nematologica* **11**: 590-592.

Borah, A. and Phukan, P.N. 2000. Effect of VAM fungus, Glomus fasciculatum and root-knot nematode, Meloidogyne incognita on brinjal. Journal of the Agricultural Sciences Society of North-Eastern India **13**: 212-214.

Borah, A. and Phukan, P.N. 2004. Comparative efficacy of *Glomus fasciculatum* with neem cake and carbofuran for the management of *Meloidogyne incognita* on brinjal. *Indian Journal of Nematology* **24**: 129-132.

Borkakaty, D. 1993. Management of Root-knot Nematode, Meloidogyne incognita (Kofoid and White, 1919) Chitwood, 1949 on Brinjal by Integrated Approach. M. Sc. (Agri.) thesis, Assam Agri. Univ., Jorhat.

Butler, E.J. 1906. The wilt disease of pigeon-pea and pepper. *Agriculture Journal of India* **1**: 25-36.

Cannayane, I. and Sivakumar, C.V. 2001. Bare root dip treatment with *Paecilomyces lilacinus* against *Meloidogyne incognita* in tomato. *Annals of Plant Protection Sciences* **9**: 351-354.

Castillo, M.B. 1985. Some studies on the use of organic soil amendments for nematode control. *Philippine Agriculturist* **68**: 76-93.

Caveness, F.E. 1981. Root-knot nematodes on cassava. In Proceedings of the Third Research Planning Conference on Root-knot Nematodes, Meloidogyne spp., Region IV and V, Ibadan, Nigeria.

Cayrol, J.C. 1967. E'tude de cycle e'volutif d' *Aphelenchoides composticola.* Nematologica **13**: 23-32.

Cayrol, J.C. 1974. *Proc. Symp. XII Intnl. De Nematologia*, Society of European Nematologists, Grenada, Spain, pp 21-22.

Chahal, P.P.K. and Chahal, V.P.S. 1988. Biological control of root-knot nematode with *Azotobacter chroococcum* on brinjal. In *Advances in Plant Nematology*, (M.A.

Maqbool, A. Morgan Golden, A. Gaffar and L.R. Krusberg, eds.), pp. 257-263. Natl. Nematol. Centre, Univ. of Karachi, Karachi, Pakistan.

Chahal, P.P.K. and Chahal, V.P.S. 2003. *Azotobacter* for the control of plant parasitic nematodes and for other benefits. In *Nematode Management in Plants* (P.C. Trivedi, ed.), pp. 241-250. Scientific Publishers (India), Jodhpur.

Chahal, V.P.S. and Chahal, P.P.K. 2003. *Bacillus thuringiensis* for the control of *Meloidogyne incognita*. In *Nematode Management in Plants* (P.C. Trivedi, ed.), pp. 251-257. Scientific Publishers (India), Jodhpur.

Chandawani, G.H. and Reddy, T.S.N. 1967. The host range of root-knot nematode, *Meloidogyne javanica* in tobacco nurseries at Rajahmundry, Andhra Pradesh. *Indian Phytopathology* **20**: 383-384.

Channabasappa, B.S., Krishnappa, K. and Reddy, B.M.R. 1995. Utilization of ecofriendly biological agents and biocomponents in the integrated management of *Radopholus similis* on banana. *Natl. Symp. on Nematode problems of India – An Appraisal of the Nematode Management with Ecofriendly Approaches and Biocomponents,* Indian Agri. Res. Inst., New Delhi.

Charles, J.S.K. 1989. *Bioecology of the Cyst Nematode Infecting Banana.* Ph. D. thesis, Kerala Agricultural University, Trichur, 122 pp.

Charles, J.S.K. and Kurian, K.J. 1979. Studies on nematode incidence in ginger. *Proceedings of the Second Symposium on Plantation Crops,* pp. 50-57.

Charles, J.S.K. and Venkitesan, T.S. 1984. New hosts of *Heterodera oryzicola* Rao & Jayaprakash, 1978 in Kerala, India. *Indian Journal of Nematology* **14**: 181-182.

Charles, J.S.K. and Venkitesan, T.S. 1993. *Status Report on the Nematological Investigations on Banana in Kerala Agricultural University.* Kerala Agricultural University, Vellanikkara, 43 pp.

Charles, J.S.K., Venkitesan, T.S. and Thomas, Y. 1983. Field screening of banana against the burrowing nematode, *Radopholus similis* (Cobb) Thorne. *Third Nematology Symposium,* Himachal Pradesh Agricultural University, Solan, p. 28.

Charles, J.S.K., Venkitesan, T.S., Thomas, Y. and Varkey, P.A. 1985. Correlation of plant growth components to bunch weight in banana infested with burrowing nematode, *Radopholus similis* (Cobb). *Indian Journal of Nematology* **15** : 186-190.

Chawla, M.L., Samathanam, G.J. and Sharma, S.B. 1980. Occurrence of citrus nematode (*Tylenchulus semipenetrans* Cobb, 1913) on roots of mangosteen (*Garcinia mangostena*). *Indian Journal of Nematology* **10**: 240-242.

Chawla, M.L. and Sharma, S.B. 1984. Horizontal and vertical distribution of citrus nematode, *Tylenchulus semipenetrans. Indian Journal of Nematology* **14**: 193-195.

Chhabra, H.K. 1972. Citrus nematode and its control. *Pesticides* **6**(2): 88-89.

Chhabra, H.K. 1978. Distribution pattern of *Tylenchulus semipenetrans* and determination of suitable distance and depth for soil sampling. *Indian Journal of Nematology* **7**: 66-68.

Chhabra, H.K. and Bindra, O.S. 1971. Screening of nematicides against the citrus nematode, *Tylenchulus semipenetrans* Cobb. *Indian Journal of Entomology* **33**: 472-473.

Chhabra, H.K. and Bindra, O.S. 1974. Screening of citrus rootstocks against the citrus nematode, *Tylenchulus semipenetrans* Cobb, 1913 in the Punjab. *Indian Journal of Horticulture* **31**: 194-195.

Chhabra, H.K., Bindra, O.S. and Singh, I. 1977. Further studies on the control of the citrus nematode, *Tylenchulus semipenetrans* Cobb, 1913 infecting *Citrus maxima* Merrill (Abstr.). *Proceedings of the International Symposium on Citriculture*, Bangalore, P. 38.

Chitwood, B.G. and Toung, M.C. 1960. Host parasite interactions of the Asiatic pyroid citrus nematode. *Plant Disease Reporter* **48**: 848-854.

Chona, B.L., Sethi, C.L., Swarup, G. and Gill, J.S. 1965. Citrus nematode (*Tylenchulus semipenetrans* Cobb, 1913) associated with die-back disease of citrus. *Indian Journal of Nematology* **22**: 370-373.

Cohn, E. 1965. The development of the citrus nematode on some of its hosts. *Nematologica* **11**: 593-600.

Connick, W.J. Jr. 1988. Formulation of living biological control agents with alginate. *ACS Symposium Series* No. **371**: 241-250.

Dahiya, R.S., Bansal, R.K. and Paruthi, I.J. 1998. Efficacy of mustard cake and poultry manure for the management of *Meloidogyne javanica* in bottle gourd and okra. *Proc. of National Symposium on Rational Approaches in Nematode Management for Sustainable Agriculture*, Nematol. Soc. of India, New Delhi, pp. 100-102.

Dalal, M.R. and Vats, R. 1998. Estimation of loss in pea (*Pisum sativum* L.) due to *Rotylenchulus reniformis*. In *Nematology – Challenges and Opportunities in 21ˢᵗ Century*, (Usha K. Mehta, ed.), pp. 8-9. Sugarcane Breeding Institute, Coimbatore.

Darekar, K.S. and Mahse, N.L. 1988. Assessment of yield losses due to root-knot nematode, *Meloidogyne incognita* Race 3 in tomato, brinjal and bitter gourd. *Internatioanl Nematology Network Newsletter* **5**(4): 7-9.

Darekar, K.S. and Patil, N.G. 1985. Journal of Maharashtra Agricultural Universities **10**: 104.

Darling, H.M. 1959. *Plant Disease Reporter* **23**: 239-242.

Das, A.K. 1994. *Assessment of Yield Loss Due to* Meloidogyne incognita *on French Bean and its Management*. M. Sc. (Agri.) thesis, Assam Agri. Univ., Jorhat.

Das, N. and Sinha, A.K. 2005. Integrated management of root-knot nematode (*Meloidogyne incognita*) on okra (*Abelmoschus esculentus* (L.) Moench). *Indian Journal of Nematology* **35**: 175-182.

Davide, R.G. and Zorilla, R.A. 1986. Evaluation of a fungus, *Paecilomyces lilacinus* for the biological control of root-knot nematodes *Meloidogyne incognita* on okra. *International Nematology Network Newsletter* **3**(3): 32-33.

Davide, R.G. and Zorilla, R.A. 1995. *Biocontrol* **1**: 63-75.

Decker, H., Casamayor, G.R. and Bosch, D. 1967. Observaciones sobre la presencia del nematode *Scutellonema bradys* en el tuberculo die name en la provincea de Oriente (Cuba). *Centro, Boletin de Ciencias y Tecnologia, Universidad Central de Las Villas* **2**: 67-75.

Deshmukh, M.R., Karmakar, S.P. and Patil, S.G. 2004. Screening of grape varieties for resistance to root-knot nematode (*Meloidogyne incognita*). *Indian Journal of Nematology* **34**: 224-226.

Devarajan, K. and Rajendran, G. 2001. Effect of fungus *Paecilomyces lilacinus* (Thom.) Samson on the burrowing nematode *Radopholus similis* (Cobb) Thorne in banana. *Pest Management in Horticultural Ecosystems* **7**: 171-173.

Devarajan, K. and Rajendran, G. 2002. Effect of fungal egg parasite, *Paecilomyces lilacinus* (Thom.) Samson on *Meloidogyne incognita* in banana. *Indian Journal of Nematology* **32**: 98-101.

Devi, Geetanjali. 1993. Pathogenicity, Crop Loss Assessment and Management of Meloidogyne incognita (Kofoid and White, 1919) Chitwood, 1949 on Carrot (Dacus carota L.). M. Sc. (Agri.) thesis, Assam Agri. Univ., Jorhat.

Dhawan, S.C. and Sethi, C.L. 1977. Inter-relationship between root-knot nematode, *Meloidogyne incognita* and little leaf of brinjal. *Indian Phytopathology* **30**: 55-63.

D'Souza, G.I., Kumar, A.C., Viswanathan, P.R.K. and Shamanna, H.V. 1970. Relative distribution and prevalence of plant parasitic nematodes in the coffee tracts of South-Western India. *Indian Coffee* **34**: 329-330 & 342.

DuCharme, E.P. 1959. Morphogenesis and histopathology of lesions induced on citrus roots by *Radopholus similis*. *Phytopathology* **49**: 388-395.

DuCharme, E.P. and Suit, R.F. 1956. Immunity of the 'Lychee' from the burrowing nematode. *Proceedings of Florida State Horticultural Society* **68**: 270-272.

Duncan, L.W. and Cohn, E. 1986. Nematode parasites of citrus. In *Plant Parasitic Nematodes in Subtropical and Tropical Agriculture* (M. Luc, R.A. Sikora and J. Bridge, eds.), pp. 321-346. CAB International, Wallingford, UK.

Eapen, S.J. 1987. Plant parasitic nematode problem in small cardamom. *Proceedings of the Third Group Discussion on the Nematological Problems of Plantation Crops*, Sugarcane Breeding Institute, Coimbatore, pp. 43-44.

Eapen, S.J. 1995. *Investigations on Plant Parasitic Nematodes Associated with Cardamom*. Final Report of the Project Nema I (813), Indian Inst. of Spices Research, Calicut, 39 pp.

Eapen, S.J., Geetha, S.M. and Leemoll, M. 1987. A note on the association of *Radopholus similis* with the yellow disease of betel vine. *Indian Journal of Nematology* **17**: 137-138.

Eapen, S.J., Ramana, K.V. and Sarma, Y.R. 1997. Evaluation of *Pseudomonas fluorescens* isolates for control of *Meloidogyne incognita* in black pepper (*Piper nigrum* L.). In *Biotechnology of Spices, Medicinal & Aromatic Plants* (S. Edison, K.V. Ramana, B. Sasikumar, K.N. Babu and S.J. Eapen, eds.), pp. 129-133. Indian Inst. of Spices Res., Calicut

Eapen, S.J. and Venugopal, M.N. 1995. Field evaluation of *Paecilomyces lilacinus* and *Trichoderma* spp. in cardamom nurseries for the control of root-knot nematodes and rhizome rot disease. *National Symp. on Nematode Problems of India – An Appraisal of the Nematode Mangmt. with Eco-friendly Approaches and Biocomponents,* Indian Agri. Res. Inst., New Delhi.

Edward, J.C., Misra, S.L., Rai, B.B. and Peter, E. 1969. *Allahabad Farmer* **43**: 175-179.

Edward, J.C., Misra, S.L. and Singh, G.R. 1965. Qualitative and quantitative studies of plant parasitic nematodes associated with the rhizosphere of some citrus plants. *Punjab Horticultural Journal* **5**: 29-36.

Edward, J.C. and Rai, B.B. 1970. Plant parasitic nematodes associated with hill oranges (*Citrus reticulata* Blanco) in Sikkim. *Allahabad Farmer* **44**: 251-254.

Egunjobi, O.A. 1985. Effect of cocoa pod husks soil amendment on cowpea infestation by *Meloidogyne* spp. *Pakistan Journal of Nematology* **3**: 99-103.

Ekanayake, H.M.R.K. and Jayasundara, N.J. 1994. Effect of *Paecilomyces lilacinus* and *Beauvaria bassiana* in controlling *Meloidogyne incognita* on tomato in Sri Lanka. *Nematologia Mediterranea* **22**: 87-88.

Elgoorani, M.A., Abo-Eldahab, M.K. and Mchair, F.F. 1976. *Phytopathology* **64**: 271-272.

Esser, R.P. 1969. *Rhadinaphelenchus cocophilus* a potential threat to Florida palms. *Nematology Circular No. 9,* Division of Plant Industry, Florida Department of Agriculture, Gainesville, Florida, USA.

Fassuliotis, G. and Sparrow, A.H. 1955. Preliminary report of x-ray studies on the golden nematode. *Plant Disease Reporter* **39**: 572.

Faulkner, L.R. and Skotland, C.B. 1965. Interaction of *Verticillium dahliae* and *Pratylenchus miniyus* in *Verticillium* wilt of peppermint. *Phytopathology* **55**: 583-586.

Ferraz, S. and Lear, B. 1976. Interaction of four plant parasitic nematodes and *Fusarium oxysporum* f. sp. *dianthi* on carnation. *Experientiae* **22**: 272-277.

Ford, H.W. 1967. Burrowing nematode resistant rootstocks as biological barriers in citrus groves. *Proceedings of Florida State Horticultural Society* **80**: 56-59.

Ford, H.W. and Feder, W.A. 1964. *Three Citrus Rootstocks Recommended for the Trial in Spreading Decline Areas.* Univ. of Florida Agri Expt. Stn. Circular S-151, 8pp.

Freitas, O.M.B.L. De and Moura, R.M. 1986. Response of cultivars of cassava (*Manihot esculenta* Crantz.) to *Meloidogyne incognita* and *M. javanica* (Nematoda: Heteroderidae) and comparison with hydrogen cyanide content. *Nematologia Mediterranea* **10**: 109-131.

Garabedian, S. and Van Gundy, S.D. 1983. Use of avermectins for the control of *Meloidogyne incognita* on tomatoes. *Journal of Nematology* **15**: 503-510.

Gaspin, R.M. and Valdez, R.B. 1979. Pathogenicity of *Meloidogyne* spp. and *Rotylenchulus reniformis* on sweet potato. *Annals of Tropical Research* **1**: 20-25.

Gaur, H.S. and Sehgal, M. 1988. A comparative study of the effect of moisture stress and period of storage on the survival of *Meloidogyne incognita, Rotylenchulus reniformis* and *Tylenchulus semipenetrans* in soil. *International Nematology Network Newsletter* **5**(3): 5-8.

Gautam, A., Siddiqui, Z.A. and Mahmood, I. 1995. Integrated management of *Meloidogyne incognita* on tomato. *Nematologia Mediterranea* **23**: 245-247.

Giamalava, M.J., Martin, W.J. and Hernandez, T.P. 1960. *Phytopathology* **50**: 575.

Giannakou, I.O., Pembroke, B., Gowen, S.R. and Doloumpala, S. 1999. Effects of fluctuating temperatures and different host plants on development of *Pasteuria penetrans* in *Meloidogyne javanica*. *Journal of Nematology* **31**: 313-318.

Gill, J.S. 1981. Fighting the foliar nematode of chrysanthemum. *Indian Horticulture* **25**: 21-22.

Gill, J.S. and Sharma, N.K. 1976. Additional hosts of the foliar nematode, *Aphelenchoides ritzemabosi* from India. *Indian Journal of Nematology* **6**: 169-171.

Gill, J.S. and Walia, R.K. 1980. Control of chrysanthemum foliar nematode, *Aphelenchoides ritzemabosi* on *Zinnia elegans*. *Indian Journal of Nematology* **10**: 86-88.

Gnanapragasam, N.C. 1982. Effect of potassium fertilization and of soil temperature on the incidence and pathogenicity of the root-lesion nematode, *Pratylenchus loosi* Loof on tea. *Tea Quarterly* **51**: 169-174.

Gnanapragasam, N.C. 1987. Integrated strategies to manage nematodes in tea. *Proceedings of the International Conference on Tea Quality and Human Health*, Hangzhou, China, pp. 117-183.

Gnanapragasam, N.C. 1991. Influence of soil amendments in reducing pathogenicity to tea by the root lesion nematode, *Pratylenchus loosi*. *Proceedings of the International Symposium on Tea Science*, Shizuoka, Japan, pp. 684-688.

Gnanapragasam, N.C., Dharmasena, W.A.M. and Liyanage, D.D. 1989. Development of age resistance in young tea to different species of *Meloidogyne*. *International Nematology Network Newsletter* **6**: 4.

Gnanapragasam, N.C. and Herath, U.B. 1990. Pathogenicity of the burrowing nematode, *Radopholus similis* to young tea (susceptible clone, TRI 2025) at different density. *Sri Lanka Journal* **58**: 83-86.

Gnanapragasam, N.C. and Manuelpillai, M.E.K. 1990. Influence of inoculum levels and temperature on the population build-up and pathogenicity of the root-lesion nematode, *Pratylenchus loosi* Loof on tea. *Tea Quarterly* **53**: 19-22.

Good, J.M. 1964. Effect of soil application and sealing methods on the efficacy of row application of several soil nematicides for controlling root-knot nematodes, weeds and *Fusarium* wilt. *Plant Disease Reporter* **48**: 199-203.

Gopinatha, K.V., Nanje Gowda, D.N. and Nagesh, M. 2002. Management of root-knot nematode *Meloidogyne incognita* on tomato using bioagent *Verticillium chlamydosporium*, neem cake, marigold and carbofuran. *Indian Journal of Nematology* **32**: 179-181.

Goswami, B.K. and Chenulu, V.V. 1974. Interaction of root-knot nematode, *Meloidogyne incognita* and tobacco mosaic virus in tomato. *Indian Journal of Nematology* **4**: 69-80.

Goswami, B.K. and Singh, H. 2001. Biomanagement of disease complex caused by *Meloidogyne incognita* and *Fusarium oxysporum* on tomato. *National Conference on Centenary of Nematol. in India – Appraisal & Future Plans*, Indian Agri. Res. Inst., New Delhi, p.169.

Goswami, B.K., Uma Rao and Singh, S. 1998. Potentiality of some fungal bioagents against root-knot nematode, *Meloidogyne incognita* infecting tomato. *Proceedings of First National Symposium on Pest Management in Horticultural Crops*, Bangalore, pp. 304-307.

Govindaiah, Dandin, S.B., Philip, T. and Datia, R.K. 1990. Effect of marigold (*Tagetes patula*) intercropping against *Meloidogyne incognita* infecting mulberry. *Indian Journal of Nematology* **20**: 96-99.

Govindaiah, Dandin, S.B. and Sharma, D.D. 1991. Pathogenicity and avoidable leaf yield loss due to *Meloidogyne incognita* in mulberry (*Morus alba* L.). *Indian Journal of Nematology* **21**: 52-57.

Gowda, D.A., Swamy, B.C.N. and Setty, K.G.H. 1982. Nematophagous fungi from cultivated soils in Karnataka. *Journal of Soil Biology and Ecology* **2**: 95-96.

Gowdar, S.B. and Kulkarni, S. 1999. Compatibility effect of antagonists and seed dressers against *Fusarium udum* – the causal agent of pigeon pea wilt. *Karnataka Journal of Agriculture Science* **12**: 197-199.

Grewal, P.S. 1989. Effect of leaf-matter incorporation on *Aphelenchoides composticola* (Nematode), mycofloral composition, mushroom compost quality and yield of *Agaricus bisporus*. *Annals of Applied Biology* **115**: 299-312.

Grewal, P.S. and Grewal, S.K. 1988. Solar heat pasteurization of casing soil and chicken manure for mushroom cultivation. *Mushroom Journal of Tropics* **8**: 35-45.

Grewal, P.S. and Sohi, H.S. 1988. A new and cheaper technique for the rapid multiplication of *Arthrobotrys conoides* and its potential as a bio-nematicide in mushroom culture. *Current Science* **57**: 44-46.

Griffith, R. 1977. A species of bacterium pathogenic to the palm weevil, *Rhynchophorus palmarum*. 1976 Ann. Rept., Red Ring Res. Divn., Minst. of Agri., Land and Fisheries, Trinidad and Tobago, pp. 31-35.

Griffith, R. 1987. Red ring disease of coconut palm. *Plant Disease* **71**: 193-196.

Griffith, R. and Koshy, P.K. 1989. Nematode parasites of coconut and other palms. In *Plant Parasitic Nematodes in Subtropical and Tropical Agriculture* (M. Luc, R.A. Sikora & J. Bridge, eds.), pp. 363-386. CAB International, Wallingford, Oxon, U.K.

Gunasekaran, C.R., Vadivelu, S. and Jayaraj, S. 1987. Experiments on nematodes of turmeric–A review. *Proceedings of the Third Group Discussions on the Nematological Problems of Plantation Crops*, Sugarcane Breeding Institute, Coimbatore, pp. 45-46.

Gupta, D.C. and Mukhopadhyaya, M.C. 1971. Effect of N, P and K on the root-knot nematode, *Meloidogyne javanica* (Treub.) Chitwood. *Science and Culture* **37**: 246-247.

Gupta, J.C. and Atwal, A.S. 1972. Biology and ecology of *Hoplolaimus indicus* (Hoplolaiminae: Nematoda). 3. Influence of population density on tomato and citrus plants. *Proceedings of XIth International Symposium on Nematology*, Reading, U. K., p.26.

Gupta, N.K. and Gupta, J.C. 1966. On a nematode species of the genus *Hoplolaimus* Daday, 1905 from a citrus plant in the Punjab. *Research Bull., Punjab University of Science* 17: 211-213.

Hackney, R.W. and Dickerson, O.J. 1975. Marigold, castor bean and chrysanthemum as controls of *Meloidogyne incognita* and *Pratylenchus alleni*. *Journal of Nematology* 7: 84-90.

Hagley, E.A.C. 1962. The palm weevil *Rhynchophorus palmarum* L. a probable vector of red ring disease of coconuts. *Nature* 193: 499.

Hague, N.G.M. 1972. Nematode diseases of flower bulbs, glasshouse crops and ornamentals. In *Economic Nematology* (J.M. Webster, ed.), pp. 409-434. Academic Press, London.

Handa, D.K. and Mathur, B.N. 1981. Control of root-knot of chillies with nematicides. *Second Nematology Symposium*, Tamil Nadu Agricultural University, Coimbatore, p. 48.

Haq, S., Singh, R.P. and Saxena, S.K. 1985. Control of the root-knot nematode *Meloidogyne incognita* on tomato with soft coal fly ash. *International Nematology Network Newsletter* 2(3): 4-6.

Haque, M.M. and Gaur, H.S. 1985. Effect of multiple cropping sequences on the dynamics of nematode population and crop performance. *Indian Journal of Nematology* 15: 262-263.

Hare, W.W. 1951. *Phytopathology* 41: 16.

Hasan, N. and Jain, R.K. 1992. Field application of *Paecilomyces lilacinus* in combination with certain organic matters for controlling *Meloidogyne incognita* infecting cowpea followed by Lucerne. *First Afro-Asian Nematol. Symp.*, Aligarh Muslim Univ., Aligarh, pp. 10-11.

Hasan, N. and Jain, R.K. 1998. Nematological research in Uttar Pradesh: An overview. In *Phytonematology in India* (P.C. Trivedi, ed.), pp. 150-169. C.B.S. Publishers and Distributors, New Delhi.

Haseeb, A. 1992. Nematode pests of medicinal and aromatic plants. In *"Nematode Pests of Crops"* (D. S. Bhatti and R. K. Walia, eds.), pp. 239-249. CBS Publishers & Distributors, Delhi, 381 pp.

Haseeb, A. and Butool, F. 1991. Evaluation of different pesticides and oilcake on the control of root-knot nematode (*Meloidogyne incognita*) infesting davana (*Artemisia pallens*). *Current Nematology* 2: 201-204.

Haseeb, A., Butool, F. and Pandey, R. 1988. Influence of different initial population densities of *Meloidogyne incognita* on the growth and oil yield of *Ocimum basilicum* cv. French. *Indian Journal of Mycology and Plant Pathology* 6: 176-179.

Haseeb, A., Kumar, V., Shukla, P.K. and Ahmad, A. 2006. Effect of different bioinoculants, organic amendments and pesticides on the management of

Meloidogyne incognita – Fusarium solani disease complex on tomato cv. K-25. *Indian Journal of Nematology* **36**: 65-69.

Haseeb, A. and Pandey, R. 1987. Incidence of root-knot nematodes in medicinal and aromatic plants - new host records. *Nematropica* **17**: 209-212.

Haseeb, A. and Pandey, R. 1989a. Root-knot disease of Henbane, *Hyscyamus niger* - a new disease record. *Tropical Pest Management* **35**: 212-213.

Haseeb, A. and Pandey, R. 1989b. Observations on *Meloidogyne* spp. affecting Japanese mint - new disease records. *Nematropica* **19**: 93-97.

Haseeb, A. and Pandey, R. 1990. Root-knot nematodes - a constraint to cultivation of Davana, *Artemisia pallens*. *Tropical Pest Management* **36**: 317-319.

Haseeb, A., Shukla, P.K., Kumar, V. and Khan, R.U. 2004. Comparative efficacy of bio-control agents, neem seed powder and pesticides against *Meloidogyne incognita* and *Fusarium oxysporum* on brinjal. *National Symposium on Paradigms in Nematological Research for Biodynamic Farming*, Univ. of Agri. Sci., Bangalore, p. 80.

Hashmi, S. and Hashmi, G. 1990. Crop rotation and *Meloidogyne javanica* in vegetable production. *International Nematology Network Newsletter* **7** (1): 19-21.

Hassan, A. 1988. Interaction between *Pratylenchus coffeae* and *Pythium aphanidermatum* and/or *Rhizoctonia solani* on chrysanthemum. *Zeist Phytopathology* **123** : 227-232.

Hassan, M.G. and Sobita Devi. 2004. Efficacy of bioagents against root-knot nematode, (*Meloidogyne incognita*) of tomato. *National Symposium on Paradigms in Nematological Research for Biodynamic Farming*, Univ. of Agri. Sci., Bangalore, p. 88.

Hazarika, K. 1990. *Pathogenicity and Management of* Meloidogyne incognita *on Brinjal* (Solanum melongena). M. Sc. (Ag) thesis, Assam Agri. Univ., Jorhat.

Hazarika, K., Dutta, P.K. and Saikia, L. 1998. Field efficacy of *Paecilomyces lilacinus* against root-knot nematode in betel vine. *Journal of Phytological Research* **11**: 171-173.

Hemavathi, M.R. 2002. *Evaluation of Bioagents,* Pasteuria penetrans *and* Glomus fasciculatum *in the Management of Root-Knot Nematode,* Meloidogyne incognita *Infesting Tomato.* M. Sc. (Agri.) thesis, Univ. of Agri. Sci., Bangalore, 102 pp.

Hide, G.A. and Corbett, D.C.M. 1974. Field experiments in the control of *Verticillium dahliae* and *Heterodera rostochiensis* on potatoes. *Annals of Applied Biology* **78**: 295-307.

Hoyle, J.C. 1968. A simple method for poisoning coconut trees. *Journal of Agricultural Science Trinidad* **68**: 459-465.

Huang, C.S. 1966. Host-parasite relationships of the root-knot nematode in edible ginger. *Phytopathology* **56**: 755-759.

Ibrahim, I.K.A. and Rezk, M.A. 1988. The root-knot nematodes – a major problem in crop production in Egypt. In *Advances in Plant Nematology* (M.A. Maqbool, A.M. Golden, A. Gaffar and L.R. Krusberg, eds.), pp. 81-98. Natl. Nematol. Res. Centre, Univ. of Karachi, Karachi, Pakistan.

Ichniohe, M. 1975. Japanese Journal of Nematology **5**: 36-40.

Ichniohe, M. 1980. Studies on the root-knot nematode of black pepper plantation in Amazon. *Annual Report of the Society of Plant Protection, North Japan* **31**: 1-8.

Ingham, R.E., Newcomb, G.B. and Morris, M.A. 1988. Spring nematicide application for control of root lesion and pin nematodes of peppermint, 1987. *Fungicide and Nematicide Test* **43**: 170.

Inomoto, M.M. and Monteiro, A.R. 1989. Thermal treatment of suckers of 'Giant Cavendish' banana for the eradication of plant-parasitic nematodes. *Nematologia Brasileira* **13**: 139-150.

Inserra, R.N., Vovlas, N. and O'Bannon, J. 1980. A classification of *Tylenchulus semipenetrans* biotypes. *Journal of Nematology* **12**: 283-287.

Jaffee, B.A., Gaspard, J.T., Ferris, H. and Muldoon, A. 1988. *Soil Biology and Biochemistry* **20**: 631-636.

Jagdale, G.B., Pawar, A.B. and Darekar, K.S. 1985a. Pathogenicity of *Meloidogyne incognita* on betel vine (*Piper betle* L.). *Indian Journal of Nematology* **15**: 244.

Jagdale, G.B., Pawar, A.B. and Darekar, K.S. 1985b. Effect of organic amendments on root-knot nematodes infecting betel vine. *International Nematology Network Newsletter* **2**(3): 7-10.

Jain, R.K. and Bhatti, D.S. 1978. Bare root treatment with systemics for controlling root-knot nematode in tomato transplants. *Indian Journal of Nematology* **8**: 19-24.

Jain, R.K. and Bhatti, D.S. 1981. Systemics as seed treatment for the control of *Meloidogyne javanica* in okra. *Second Nematology Symposium*, Tamil Nadu Agricultural University, Coimbatore, p. 44.

Jain, R.K. and Bhatti, D.S. 1985. Effect of summer ploughing alone and in combination with some other effective practices on the incidence of root-knot nematode (*Meloidogyne javanica*) on tomato (Cultivar HS 101). *Indian Journal of Nematology* **15**: 262.

Jain, R.K. and Bhatti, D.S. 1987. Population development of root-knot nematode (*Meloidogyne javanica*) and tomato yield as influenced by summer ploughing. *Tropical Pest Management* **33**: 122-125.

Jain, R.K. and Gupta, D.C. 1991. Integrated management of root-knot nematode, *Meloidogyne javanica* infecting *Solanum melongena*. *Afro-Asian Journal of Nematology* **1**: 156-160.

Jain, R.K. and Hasan, N. 1984. Increased seed production of cowpea with low dose of phenamiphos. *Pesticides* **18**: 49.

Jain, R.K. and Hasan, N. 1986. Efficacy of "Neem" cake on fodder production, photosynthetic pigments and nematodes associated with oats and its residual effect on cowpea. *Indian Journal of Nematology* **16**: 98-100.

Jain, R.K., Paruthi, I.J. and Gupta, D.C. 1988. Control of root-knot nematode (*Meloidogyne javanica*) in tomato through nursery bed treatment alone and in combination with application of carbofuran at transplanting. *Indian Journal of Nematology* **18**: 340-341.

Jain, R.K., Paruthi, I.J. and Gupta, D.C. 1990. Influence of intercropping on the incidence of root-knot nematode (*Meloidogyne javanica*) in brinjal. *Indian Journal of Nematology* **20**: 236-238.

Jain, R.K., Paruthi, I.J., Gupta, D.C. and Dhankar, B.S. 1986. Appraisal of losses due to root-knot nematode, *Meloidogyne javanica* on okra under field conditions. *Tropical Pest Management* **32**: 341-342.

Jakobsen, I. 1999. Effects of *Pseudomonas fluorescens* DF 57 on growth and P uptake of two arbuscular mycorrhizal fungi in symbiosis with cucumber. *Mycorrhiza* **8**: 319-334.

Jatala, P. 1986. Biological control of nematodes. *Annual Review of Phytopathology* **24**: 453-489.

Jatala, P. 1988. Nematodes in tuber and root crops and their management – Improvement of sweet potato (*Ipomea batatas*) in East Africa with some references of other tuber and root crops. *Report of the Workshop on Sweet Potato Improvement in Africa*. ILRAD, Nairobi, Kenya, UNDP Project CIAT – CIP – IITA, pp. 91-100.

Jatala, P., Kaltenbach, R., Devaux, A.J. and Campos, R. 1980. Field application of *Paecilomyces lilacinus* for controlling *Meloidogyne incognita* on potatoes. *Journal of Nematology* **12**: 226-227.

Jatala, P., Martin, C., Mendoza, H.A., Hidalgo, O.A. and Rincon, R.H. (eds.). 1990. Role of nematodes in the expression of *Pseudomonas solanacearum* and strategies for screening and breeding for combined resistance. *Advances en el Mejoramiento Genetico de la Papa en los Paises del cono sur.*, pp.187-189.

Jayakumar, J., Rajendran, G. and Ramakrishnan, S. 2004. Management of reniform nematode, Rotylenchulus reniformis in okra through Streptomyces avermitilis. National Symposium on Paradigms in Nematological Research for Biodynamic Farming, Univ. of Agri. Sci., Bangalore, p. 98.

Jayakumar, J., Rajendran, G. and Ramakrishnan, S. 2005. Bio-efficacy of avermectin against root-knot nematode, *Meloidogyne incognita* in tomato. *Indian Journal of Nematology* **35**: 151-156.

Jayaraman, V., Rajendran, G. and Muthukrishnan, T.S. 1975. Occurrence of root-knot nematodes in *Polianthes tuberosa* L. in Tamil Nadu. *Indian Journal of Nematology* **5**: 101-102.

Jenkins, W.R. and Coursen, B.W. 1957. The effect of root-knot nematodes, *Meloidogyne incognita acrita* and *M. hapla*, on *Fusarium* wilt of tomato. *Plant Disease Reporter* **41**: 182-186.

Johnson, N. 2000. Studies on Nematode Pests of Cut Flowers with Special Reference to Carnation, Cystera, Gladiolus and Asiatic Lily. Ph. D. thesis, Tamil Nadu Agri. Univ., Coimbatore, 160 pp.

Jonathan, E.I. 1994. *Studies on the Root-knot Nematodes* (Meloidogyne incognita) *(Kofoid and White, 1919) Chitwood, 1949, in Banana* (Musa *sp.) cv. Poovan (AAB)*. Ph. D. Thesis, Tamil Nadu Agri. Univ., Coimbatore, Tamil Nadu, India.

Jonathan, E.I., Arulmozhiyan, R., Muthusamy, S. and Manuel, W.W. 2000. Field application of *Paecilomyces lilacinus* for the control of *Meloidogyne incognita* on betel vine, *Piper betle. Nematologia Mediterranea* **28**: 131-133.

Jonathan, E.I. and Cannayane, I. 2002. Field application of biocontrol agents for management of spiral nematode, *Helicotylenchus multicinctus* in banana (Abstr.). *Global Conference on Banana and Plantain,* Bangalore, p. 193.

Jonathan, E.I., Cannayane, I., Bommiraju, P., Amutha, G. and Samiyappan, R. 2004. Biological control of root-knot nematode, *Meloidogyne incognita* and *Phytophthora* wilt complex in betel vine with plant growth promoting rhizobacteria. *National Symposium on Paradigms in Nematological Research for Biodynamic Farming*, Univ. of Agri. Sci., Bangalore, p. 73-74.

Jonathan, E.I., Nagalakshmi, S. and Padmanabhan, D. 1990. Estimation of yield losses in betel vine due to *Meloidogyne incognita*. *International Nematology Network Newsletter* **7** (4): 26.

Jonathan, E.I., Padmanabhan, D. and Ayyamperumal, A. 1995. Biological control of root-knot nematode on betel vine, *Piper betel* by *Paecilomyces lilacinus*. *Nematologia Mediterranea* **23**: 191-193.

Jonathan, E.I. and Rajendran, G. 2000. Assessment of avoidable yield loss in banana due to root-knot nematode, *Meloidogyne incognita. Indian Journal of Nematology* **30**: 162-164.

Jones, F.C.E. 1961. The potato root eelworm *Heterodera rostochiensis* Woll. in India. *Current Science* **30**: 187.

Jones, F.C.E. 1969. Integrated control of the potato cyst nematode. *Proceedings of Fifth British Insecticide and Fungicide Conference* **3**: 646-656.

Jothi, G., Ramakrishnan, S. and Sundara Babu, R. 2004. Integrated management of *Pratylenchus delattrei* in crossandra. *National Symposium on Paradigms in Nematological Research for Biodynamic Farming,* Univ. of Agri. Sci., Bangalore, p. 97.

Jothi, G. and Sundarababu, R. 2000. Interaction effect of VAM and root-knot nematode on the growth and nutrient content of brinjal. *International Journal of Tropical Plant Diseases* **18**: 91-100.

Kalita, S. and Bora, B.C. 2006. Effect of nematicides and organic amendments in the management of *Meloidogyne incognita* in nursery tea. *Indian Journal of Nematology* **36**: 148-149.

Kamra, A. and Gaur, H.S. 1998. Control of nematodes, fungi and weeds in nursery beds by soil solarization. *Indian Journal of Nematology* **28**: 46-52.

Kandasamy, K. and Sivakumar, C.V. 1981. Effect of *Meloidogyne arenaria* on the germination of chilli (*Capsicum annum* Linn.) seeds. *Second Nematology Symposium,* Tamil Nadu Agricultural University, Coimbatore, p.16.

Kantharaju, V., Prakash Babu, N., Krishnappa, K. and Reddy, B.M.R. 2002. Biodiversity of indigenous isolates of VAM, *Glomus fasciculatum* against *Meloidogyne incognita* race-1 on tomato (Abstr.). *National Symposium on Biodiversity and Management of Nematodes in Cropping Systems for Sustainable Agriculture,* Jaipur, p.3.

Kanwar, R.S. 1990. Studies on Cropping Systems for Management of Root-knot Nematode, Meloidogyne javanica (Treub, 1985) Chitwood, 1949. Ph. D. thesis, Haryana Agri. Univ., Hisar, 118 pp.

Kanwar, R.S. and Bhatti, D.S. 1990. Yield performance of resistant and susceptible tomato cultivars in monoculture and related reproduction of *Meloidogyne javanica. International Nematology Network Newsletter* **7**(3): 16-18.

Kanwar, R.S. and Bhatti, D.S. 1992. Controlling root-knot nematode, *Meloidogyne javanica* through crop rotation. *Current Nematology* **3**: 65-70.

Karuna, K., Krishnappa, K. and Reddy, B.M.R. 2001. Utilization of *Pasteuria penetrans* in the management of root-knot nematode on brinjal. *National Conference on Centenary of Nematology in India- Appraisal and Future Plans*, Indian Agri. Res. Inst., New Delhi, p. 142.

Karuna, K., Ravichandra, N.G., Krishnappa, K. and Sreenivasa, K.R. 2004. Management of root-knot nematode, *Meloidogyne incognita* on banana by biocontrol agents. *National Symposium on Paradigms in Nematological Research for Biodynamic Farming*, Univ. of Agri. Sci., Bangalore, p. 70.

Kaur, D.J. 1987. *Studies on Nematodes Associated with Ginger* (Zingiber officinale Rosc.) *in Himachal Pradesh*. Ph. D. thesis, Himachal Pradesh Univ., Shimla, India, 198 pp.

Kerr, A. and Vythilingam, M.K. 1966. Replanting eelworm infested areas. *Tea Quarterly* **37**: 67-72.

Khan, A.M., Saxena, S.K. and Khan, M.W. 1971. Interaction of *Rhizoctonia solani* Kuhn and *Tylenchorhynchus brassicae* Siddiqi, 1961 in pre-emergence damping-off of cauliflower seedlings. *Indian Journal of Nematology* **1**: 85-86.

Khan, A.M., Saxena, S.K. and Siddiqi, Z.A. 1971. Efficacy of *Tagetes erecta* in reducing root infesting nematodes of tomato and okra. *Indian Phytopathology* **24**: 166-169.

Khan, A.M., Saxena, S.K., Siddiqi, Z.A. and Upadhyay, R.S. 1975. Control of nematodes by crop rotation. *Indian Journal of Nematology* **5**: 214-221.

Khan, A. M., Siddiqi, M. R., Khan, E., Hussain, S. I. and Saxena, S. K. 1964. *List of Stylet Bearing Nematodes Reported from India-I*. Nematology Publication 1, Aligarh Muslim University, Aligarh, U. P., India, 19 pp.

Khan, E. and Nanjappa, C.K. 1972. Four new species of Criconematoidea (Nematoda) from India. *Indian Journal of Nematology* **2**: 59-68.

Khan, M.R. and Ghadipur, M.H. 2004. Management of root-knot disease of some solanaceous vegetables by soil application of fly ash. *National Symposium on Paradigms in Nematological Research for Biodynamic Farming*, Univ. of Agri. Sci., Bangalore, p. 92.

Khan, M.R. and Tarannum, Z. 1999. Effects of field application of various micro-organisms on *Meloidogyne incognita* on tomato. *Nematologia Mediterranea* **27**: 233-238.

Khan, R.M. and Parvatha Reddy, P. 1989. *Jasminum pubescens* hitherto unrecorded host of *Radopholus similis* (Cobb, 1893) Thorne, 1949 from India. *International Nematology Network News Letter* **6**: 19-20.

Khan, R.M. and Parvatha Reddy, P. 1992. Nematode pests of ornamental crops. In *Nematode Pests of Crops* (D. S. Bhatti and R. K. Walia, eds.), pp. 250-257. CBS Publishers & Distributors, Delhi, 381 pp.

Khan, R.M. and Parvatha Reddy, P. 1994. Nematode problems of ornamental crops and their management. In *Floriculture-Technology, Trades and Trends* (J. Prakash and K.R. Bhandary, eds.), pp.468-472. Oxford & IBH Publishing Co. Pvt. Ltd., New Delhi, 667 pp.

Khan, R.M., Saxena, S.K., Khan, M.W. and Khan, A.M. 1980. *Indian Journal of Nematology* 10: 242-244.

Khan, T., Khan, S.T., Fazal, M. and Siddiqi, Z.A. 1997. Biological control of *Meloidogyne incognita* and *Fusarium solani* disease complex in papaya using *Paecilomyces lilacinus* and *Trichoderma harzianum*. *International Journal of Nematology* 7: 127-132.

Khan, T.A. 1991. *Studies on a Disease Complex of Papaya caused by* Meloidogyne incognita *and* Fusarium solani. Final Tech. Rept., DST Project, Aligarh Muslim Univ., Aligarh, 47 pp.

Khan, T.A. and Khan, S.T. 1998. Nematode diseases of papaya. In *Nematode Diseases in Plants* (P.C. Trivedi, ed.), pp. 344-360. CBS Publishers and Distributors, New Delhi.

Khan, T.A., Khan, S.T. and Saxena S.K. 1992. Effect of interaction of *Meloidogyne incognita* and *Fusarium solani* on the relative response of papaya cvs. *First Afro-Asian Nematology Symposium*, Aligarh Muslim Univ., Aligarh, p. 14.

Khan, T.A. and Saxena, S.K. 1993. Effect of Fusarium solani on the life cycle of Meloidogyne incognita on papaya. National Seminar on Diagnosis and Management of Plant Diseases, Lucknow, p. 47.

Khanna, A.S. and Sharma, N.K. 1988. *Indian Phytopathology* **41**: 472-473.

Khanna, A.S. and Sharma, N.K. 1990. Efficacy of nematophagous fungi against myceliophagus nematode *in vitro*. *Proc. National Seminar on Bioagents in Nematode Management*, Indian Agri. Res. Inst., New Delhi, pp. 83-84.

Khanna, A.S. and Sharma, N.K. 1994. Management of *Aphelenchoides composticola* infesting white button mushroom by neem cake. *Indian Journal of Hill Farming* 7: 114-115.

Khuntia, N. and Das, S. N. 1969. Plant parasitic nematodes associated with fruit trees in Orissa (Abstract). *All India Nematology Symposium,* Indian Agri. Res. Inst., New Delhi, pp. 22-23.

Kiran Kumar, K.C., Krishnappa, K., Karuna, K., Ravichandra, N.G. and Sreenivas, K.R. 2004. Evaluation of proposed management package of growth, development and yield of tomato infested by root-knot nematode. *National Symposium on Paradigms in Nematological Research for Biodynamic Farming,* Univ. of Agri. Sci., Bangalore, p. 94.

Kolodge, C., Radewald, J.D. and Shibaya, F. 1987. Revised host range and studies on life cycle of *Longidorus africanus*. *Journal of Nematology* **19**: 77-81.

Koshy, P.K. 1986. The burrowing nematode, *Radopholus similis* (Cobb, 1893) Thorne, 1949. In *Plant Parasitic Nematodes of India-Problems and Progress*. (G. Swarup and D.R. Dasgupta, eds.), pp. 223-248. Indian Agri. Res. Inst., New Delhi, India.

Koshy, P.K. 2002. Nematode pests of spices and condiments and their control. In *Nematode Control in Crops* (M.M. Alam & N. Sharma, eds.), pp. 187-200. International Book Distributing Company, Lucknow, India.

Koshy, P.K. and Geetha, S.M. 1992. Nematode pests of palms and cocoa. In *Nematode Pests of Crops* (D.S. Bhatti and R.K. Walia, eds.), pp. 214-227. CBS Publishers & Distributors, Delhi.

Koshy, P.K. and Jasy, T. 1990. Effect of the burrowing nematode, Radopholus similis on sweet potato (Ipomea batatas L.) (Abstr.). National Symposium on Recent Advances in the Production and Utilization of Tropical Tuber Crops, ISRC, Central Tuber Crops Res. Inst., Thiruvanathapuram, India, p. 47.

Koshy, P.K. and Jasy, T. 1991. Host preference of the burrowing nematode, *Radopholus similis* populations from India. *Indian Journal of Nematology* **21**: 39-51.

Koshy, P.K., Nair, R.C.P., Sosamma, V.K. and Sundararaju, P. 1976. On the incidence of root-knot nematode in cardamom nurseries. *Indian Journal of Nematology* **6**: 174-175.

Koshy, P.K. and Sosamma, V.K. 1975. Host-range of the burrowing nematode *Radopholus similis* (Cobb, 1893) Thorne, 1949. *Indian Journal of Nematology* **5**: 255-257.

Koshy, P.K. and Sosamma, V.K. 1977. Identification of the physiological races of *Radopholus similis* populations infesting coconut, areca nut and banana in Kerala, South India. *Indian Phytopathology* **30**: 415-416.

Koshy, P.K. and Sosamma, V.K. 1978. Studies on the population fluctuations of *Radopholus similis* in coconut and areca nut roots. *Indian Phytopathology* **31**: 180-185.

Koshy, P.K. and Sosamma, V.K. 1979. Control of the burrowing nematode, *Radopholus similis* (Cobb, 1893) Thorne, 1949 on coconut seedlings with DBCP. *Indian Journal of Nematology* **9**: 32-33.

Koshy, P.K. and Sosamma, V.K. 1987. Pathogenicity of *Radopholus similis* on coconut (*Cocos nucifera* L.) seedlings under greenhouse and field conditions. *Indian Journal of Nematology* **17**: 108-118.

Koshy, P.K., Sosamma, V.K. and Faleiro, J.R. 1987. Occurrence of *Heterodera oryzicola* on banana in Goa. *Indian Journal of Nematology* **17**: 334.

Koshy, P.K., Sosamma, V.K. and Samuel, R. 1999. Interactive effect of vesicular arbascular mycorrhizae (VAM) with *Radopholus similis* on coconut seedlings. In *Nematology – Challenges and Opportunities in 21st Century*, (Usha K. Mehta, ed.), pp. 106-110. Sugarcane Breeding Institute, Coimbatore.

Koshy, P.K., Sosamma, V.K. and Sundararaju, P. 1977. Screening of plants used as pepper standards against root-knot nematode. *Indian Phytopathology* **30**: 128-129.

Koshy, P.K., Sosamma, V.K. and Sundararaju, P. 1979. Reactions of thirty one areca germplasm collections to the burrowing nematode, *Radopholus* similis. *Plant Disease Reporter* **63**: 433-435.

Koshy, P.K., Sundararaju, P. and Sosamma, V.K. 1978. Occurrence and distribution of *Radopholus similis* (Cobb, 1893) Thorne, 1949 in South India. *Indian Journal of Nematology* **8**: 49-58.

Krishnamurthy, G.V.G. and Elias, N.A. 1967. Host range of *Meloidogyne incognita* causing root-knot on tobacco in Hunsur, Mysore State. *Indian Phytopathology* **20**: 374-377.

Krishnamurthy Rao, B.H. and Thammiraju, N.B. 1975. Nematodes associated with citrus. In *"Citrus Decline in Andhra Pradesh"* (G.S. Reddy and V. D. Murthy, eds.), pp. 55-58. Andhra Pradesh Agri. Univ., Hyderabad.

Krishna Rao, A.B.V., Padhi, N.N. and Acharya, A. 1987. Effect of different nematicides, oilcakes and urea in the control of *Rotylenchulus* in okra. *Indian Journal of Nematology* **17**: 171-173.

Kumar, A.C. 1982. Evaluation of Nemacur 5G against coffee root-lesion nematode, *Pratylenchus coffeae. Journal of Coffee Research* **12**: 1-7.

Kumar, A.C. 1984. The symptoms and diagnosis of the disorder 'spreading decline' (Cannoncadoo dieback) with a note on spread and control of the causal agent. *Journal of Coffee Research* **14**: 156-159.

Kumar, A.C. 1985. Investigations on the three species of *Hemicriconemoides* (Nematoda) associated with 'Crinkle leaf' disorder in coffee. *Journal of Coffee Research* **15**: 90-98.

Kumar, A.C. 1988. Nematode problem of coffee and its management. *Indian Coffee* **52**: 12-19.

Kumar, A.C., Viswanathan, P.R.K. and D'Souza, G.I. 1971. A study of plant parasitic nematodes of certain commercial crops in coffee tracts of South India. *Indian Coffee* **36**: 1-3.

Kumar, S. and Sivakumar, C.V. 1981. Disease complex involving *Rotylenchulus reniformis* and *Rhizoctonia solani* in okra. *Second Nematology Symposium*, Tamil Nadu Agricultural University, Coimbatore, p.12.

Kumari, R., Verma, K.K., Dhindsa, K.S. and Bhatti, D.S. 1986. Datura, Ipomea, Tagetes and Lawsonia as control of *Tylenchulus semipenetrans* and *Anguina tritici. Indian Journal of Nematology* **16**: 226-240.

Kumutha, K. 2001. Symbiotic Influence of AM Fungi and Rhizobacteria on Biochemical and Nutritional Changes in Mulberry (Morus alba). Ph. D. thesis, Tamil Nadu Agri. Univ., Coimbatore.

Lakshmana Murthy, K.V. 1983. *Studies on Burrowing Nematode,* Radopholus similis *(Cobb, 1983) Thorne, 1949 on Banana in Karnataka.* M. Sc. (Agri.) thesis, Univ. of Agri. Sci., Bangalore.

Lakshmanan, P.L. and Sivakumar, C.V. 1981. Control of *Rotylenchulus reniformis* in Bhendi by integrated method. *Second Nematology Symposium*, Tamil Nadu Agricultural University, Coimbatore, p.54.

Lal, B., Rana, B.P. and Dahia, R.S. 2002. Efficacy of antagonistic fungi on root-knot nematode, Meloidogyne incognita in okra. Proceedings of the National Symposium on Biodiversity and Management of Nematodes in Cropping Systems for Sustainable Agriculture, Rajasthan Agri. Univ., Bikaner.

Lal, B.S. and Ansari, M.N.A. 1960. Field studies on the root-knot nematodes (*Meloidogyne* spp.) (Nematoda: Heteroderidae). *Science and Culture* **26**: 279-281.

Lamberti, F. 1981. Combating nematode vectors of viruses. *Plant Disease* **65**: 113-117.

Laqman Khan, M. 2001. Dr. Y.S. Paramar University of Horticulture and Forestry, Solan, Himachal Pradesh. In *Indian Nematology-Progress and Prospectives.* (S.C. Dhawan *et al.*, eds.), pp. 153-158. Division of Nematology, Indian Agri. Res. Inst., New Delhi.

Loganathan, M., Swarnakumari, N., Sivakumar, M., Prakasam, V., Mohan, L., Ramraj, B. and Samiyappan, R. 2001. Biological suppression of fungal nematode complex diseases of major cruciferous vegetables. *South Indian Horticulture* **49**: Special issue.

Loos, C.A. 1961. Eradication of the burrowing nematode, *Radopholus similis*, from bananas. *Plant Disease Reporter* **45**: 457-461.

Lordello, L.G.E. 1968. *Nematodes das Plantas Cultivadas.* Livaria Nobel SA, Sao Paulo, Brazil.

Loubser, J.T. 1985. *Deciduous Fruit Grower* **35**: 285-290.

Loubser, J.T. and Klerk, C.A.D.E. 1985. *South African Journal of Enology and Viticulture* **6**: 31-33.

Lownsbery, B.F. 1959. Studies on the nematode, *Criconemoides xenoplax* on peach. *Plant Disease Reporter* **43**: 913-917.

Lownsbery, B.F., English, H., Noel, G.R. and Schick. 1977. *Journal of Nematology* **9**: 221-224.

Maas, P.W.T. 1970. Tentative list of plant parasitic nematodes in Surinam, with description of two new species of Hemicycliophorinae. *Bulletin of Landbouwproefstation Surinam* No. 87, 9 pp.

Maheswari, T.U. and Mani, A. 1988. Combined efficacy of Pasteuria penetrans and Paecilomyces lilacinus on the biocontrol of Meloidogyne javanica on tomato. International Nematology Network Newsletter 5(3): 10-11.

Maheswari, T.U., Mani, A. and Rao, P.K. 1987. Combined efficacy of the bacterial spore parasite *Pasteuria penetrans* (Thorne, 1940) and nematicides in the control of *Meloidogyne javanica* on tomato. *Journal of Biological Control* **1**: 53-57.

Mai, W.F., Brodie, B.B., Harrison, M.B. and Jatala, P. 1981. Nematodes In *Compendium of Potato Diseases* (W.J. Hooker, ed.), pp. 93-101. American Phytopathological Society.

Malaguti, G. 1953. Pudricon del cogello de la palmera de accite Africana (*Elaeis guineensis* Jacq.). *Agr. Trop. Venezuela* **3**: 13-21.

Mammen, K.V. 1974. On a wilt disease of betel vine in Kerala caused by root-knot nematode. *Agricultural Research Journal of Kerala* **12**: 76.

Mangat, B.P.S. and Sharma, N.K. 1981. Influence of host nutrition on multiplication and development of citrus nematode. *Indian Phytopathology* **34**: 90-91.

Mani, A. 1986. Occurrence of *Meloidogyne javanica* on citrus in Andhra Pradesh (India). *International Nematology Network Newsletter* **3**(2): 9-10.

Mani, A. 1988a. Studies on the bacterial parasite *Pasteuria penetrans:* I. Spore viability after storage. II. Culture on citrus nematode *Tylenchulus semipenetrans. International Nematology Network Newsletter* **5** (2): 24-25.

Mani, A. 1988b. Effect of interculture of marigold and mustard with acid lime on citrus nematode, *Tylenchulus semipenetrans. International Nematology Network Newsletter* **5** (4): 14-15.

Mani, A. 1990. Control of citrus nematode, *Tylenchulus semipenetrans*, by bare-root dip treatment of acid lime seedlings in pesticides. *International Nematology Network Newsletter* **6** (3): 20-21.

Mani, A. 1994. Occurrence and distribution of *Tylenchulus semipenetrans* in Andhra Pradesh. *Indian Journal of Nematology* **24**: 106-111.

Mani, A., Murty, I.R., Rao, P.K. and Prasad, V.D. 1989. Growth of *Paecilomyces lilacinus* on natural substrates and its efficacy against citrus nematode, *Tylenchulus semipenetrans. Journal of Biological Control* **3**: 59-61.

Mani, A. and Murty, V.D. 1986. Citrus nematode and root-knot nematodes: Distribution, crop loss, symptoms and management. In *"Plant Parasitic Nematodes - Problems and Progress"* (G. Swarup and D.R. Dasgupta, eds.), pp. 249-260. Indian Agricultural Research Institute, New Delhi.

Mani, A. and Murty, V.D. 1989. *National Symposium on Insect Pests and Disease Management on Citrus.* Jawaharlal Nehru Krishi Viswa Vidyalaya, Chindwara.

Mani, A., Naidu, P.H. and Madhavachari, S. 1987. Occurrence and control of *Meloidogyne incognita* on turmeric in Andhra Pradesh, India. *International Nematology Network Newsletter* **4**: 13-18.

Mani, A. and Reddy, G.S. 1986. Studies on the varietal reaction of some species of *Citrus* and *Poncirus* to *Tylenchulus semipenetrans* Cobb. *Indian Journal of Nematology* **16**: 267-268.

Mani, M.P. 1996. *Effect of* Pasteuria penetrans *(Thorne) Sayre & Starr and* Pseudomonas fluorescens *(Migula) Against* Meloidogyne incognita *(Kofoid & White) Chitwood in Grapevine (Vitis vinefera* Linn.). M. Sc. (Agri.) thesis, Tamil Nadu Agri. Univ., Coimbatore, 46 pp.

Mani, M.P., Rajeswari, S. and Sivakumar, C.V. 1998. Management of the potato cyst nematodes, *Globodera* spp. through plant rhizosphere bacterium *Pseudomonas fluorescens* Migula. *Journal of Biological Control* **12**: 131-134.

Mani, M.P., Sivakumar, C.V. and Ramakrishnan, S. 1999. Status of *Pasteuria penetrans* in root-knot nematode infested vineyards of Tamil Nadu. *Indian Journal of Nematology* **29**: 104.

Mankau, R. 1975. *Bacillus penetrans* n. comb. causing a virulent disease of plant parasitic nematodes. *Journal of Invertebrate Pathology* **26**: 333-339.

Manzoor, S., Sinha, A.K. and Bora, B.C. 2002. Management of citrus nematode, *Tylenchulus semipenetrans* on Khasi mandarin by *Paecilomyces lilacinus. Indian Journal of Nematology* **32**: 153-155.

Martin, W.J. 1970. Elimination of root-knot and reniform nematodes and scurf infection from rootlets of sweet potato plants by hot water treatment. *Plant Disease Reporter* **54**: 1056-1058.

McGuire, J.M. and Good, M.J. 1970. The effect of benomyl on *Xiphinema americanum* and tobacco ring spot virus infection. *Phytopathology* **60**: 1150-1151.

McElroy, F.D. and Van Gundy, S.D. 1968. *Phytopathology* **58**: 1558-1565.

Meagher, J.W. and Jenkins, P.T. 1970. Interaction of *Meloidogyne hapla* and *Verticillium dahliae* and chemical control of wilt in strawberry. *Australian Journal of Experimental Agriculture and Animal Husbandry* **10**: 493-496.

Medhane, N.S., Jagdale, G.B., Pawar, A.B. and Darekar, K.S. 1985. Effect of *Tagetes erecta* on root-knot nematodes infecting betel vine. *International Nematology Network Newsletter* **2**(3): 11-12.

Menon, K.K. 1949. The survey of pollu and root diseases of pepper. *Indian Journal of Agricultural Science* **19**: 89-136.

Midha, R.L. and Trivedi, P.C. 1991. Estimation of losses caused by *Meloidogyne incognita* on coriander, cumin and fennel. *Current Nematology* **2**: 159-162.

Midha, R.L., Trivedi, P.C. and Yadav, B.D. 2001. *Paecilomyces* spp. as biocontrol agents against *Meloidogyne incognita* on coriander. *Natl. Cong. on Centenary of Nematol. in India – Appraisal & Future Plans*, Indian Agri. Res. Inst., New Delhi, pp.146-147.

Misra, S.L. and Edward, J.C. 1977. Observations on histopathology of some of the more important rootstocks for commercial citrus affected by citrus nematodes. *Allahabad Farmer* **53**: 7-18.

Mishra and Shukla, B.N. 1997. Interaction between Glomus fasciculatum and Meloidogyne incognita on tomato. Journal of Mycology and Plant Pathology **27**: 199-202.

Mobin, M. and Khan, A.M. 1969. Effect of organic amendments on the population of rhizosphere fungi and nematodes around the roots of some fruit trees (Abstr.). *All India Nematol. Symp.*, Indian Agri. Res. Inst., New Delhi, p. 68.

Mohandas, C. 2001. Nematode diseases of tuber crops and their management. *Natl. Cong. on Centenary of Nematol. in India – Appraisal & Future Plans*, Indian Agri. Res. Inst., New Delhi, pp. 35-36.

Mohandas, C. and Palaniswami, M.S. 1990a. Resistance in sweet potato (*Ipomea batatas* L.) to *Meloidogyne incognita* (Kofoid and White) Chitwood in India. *Journal of Root Crops* **16**: 148-149.

Mohandas, C. and Palaniswami, M.S. 1990b. Effect of Meloidogyne incognita on three popular cultivars of Colocasia esculenta (Abstr.). National Symposium on Recent Advances in the Production and Utilization of Tropical Tuber Crops, ISRC, Central Tuber Crops Res. Inst., Thiruvanathapuram, India.

Mohandas, C., Palaniswami, M.S. and Potty, V.P. 1990. Survey, identification and pathogenicity of nematodes in tuber crops. *Annual Report of Central Tuber Crops Research Institute (1989-90)*, Thiruvanthapuram, pp. 72-73.

Mohandas, C. and Ramana, K.V. 1987. Slow wilt disease of black pepper and its control. *Indian Cocoa, Areca nut and Spices Journal* **11**: 10-11.

Mohandas, C. and Ramana, K.V. 1987. *Third Group Discussion on the Nematological Problems of Plantation Crops*, Sugarcane Breeding Institute, Coimbatore, pp. 43-44.

Mohandas, C. and Ramana, K.V. 1991. Pathogenicity of *Meloidogyne incognita* and *Radopholus similis* on black pepper (*Piper nigrum* L.). *Journal of Plantation Crops* **19**: 41-43.

Mohanty, K.C., Sahoo, N.K. and Ray, S. 1972. Occurrence of *Radopholus similis* (Cobb, 1893) Thorne, 1949 on banana and pepper in wide areas of Orissa, India. *Afro-Asian Nematology Network Newsletter* **1**(1): 25-26.

Morgan-Jones, G. and Rodriguez-Kabana, R. 1985. Phytonematode Pathology: Fungal modes of action – A perspective. *Nematropica* **15**: 107-114.

Mortensten, J.A. 1985. Proceedings of Florida State Horticulture Society **98**: 166-169.

Mukhopadhyaya, M.C. 1970a. Nematodes associated with fruit trees and vegetable crops in Himachal Pradesh. *Journal of Research, Punjab Agricultural University* **7**: 656-658.

Mukhopadhyaya, M.C. 1970b. Efficacy of D-D, EDB, DBCP and VC-13 in controlling plant parasitic nematodes. *Journal of Research, Punjab Agricultural University* **7**: 659-662.

Mukhopadhyaya, M.C. and Dalal, M.R. 1971. Effect of two nematicides on *Tylenchulus semipenetrans* and on sweet lime yield. *Indian Journal of Nematology* **1**: 95-97.

Mukhopadhyaya, M.C. and Suryanarayana, D. 1969. Citrus decline in Haryana : Role of *Tylenchulus semipenetrans* and its control. *Indian Phytopathology* **22**: 495-497.

Muraleedharan, N. 1991. Non-arthropod pests – Nematodes and rodents. In *Pest Management in Tea* (Director, ed.), pp. 77-81. United Planter's Association of Southern India, Tea Res. Inst.

Mustafa, U. and Khan, M.R. 2004. Management of disease complex caused by *Fusarium oxysporum* f. sp. *gladioli* and *Meloidogyne incognita* on gladiolus. *National Symposium on Paradigms in Nematological Research for Biodynamic Farming*, Univ. of Agri. Sci., Bangalore, p. 91-92.

Muthukrishnan, T.S., Chandrasekaran, J., Lakshmanan, P.L. and Chinnarajan, A.M. 1975. Occurrence of the nematode *Hemicycliophora labiata* Colbran 1960 on roses in Tamil Nadu. *South Indian Horticulture* **23**: 75-76.

Muthukrishnan, T.S. and Sivakumar, C.V. 1973. Some observations on nematodes of citrus, banana and papaya in Tamil Nadu. *Proc. Workshop on Fruit Research*, Coimbatore.

Muthukrishnan, T.S., Vadivelu, S. and Rajendran, G. 1977. *Longidorus africanus* and its pathogenic effect on *Crossandra undulaefolia*. *Indian Journal of Nematology* **7**: 69-72.

Nadakal, A.M. and Thomas, A.N. 1967. Observations on nematodes associated with dry rot of *Dioscorea alata* L. *Science and Culture* **33**: 142-143.

Nagesh, M., Hussaini, S.S., Ramanujam, B. and Chidanandaswamy, B.S. 2006. Management of *Meloidogyne incognita* and *Fusarium oxysporum* f. sp. *lycopersici* wilt complex using antagonistic fungi in tomato. *Nematologia Mediterranea* **34**: 63-68.

Nagesh, M., Hussaini, S.S., Singh, S.P. and Biswas, S.R. 2003. Management of root-knot nematode, *Meloidogyne incognita* (Kofoid & White) Chitwood in chrysanthemum using *Paecilomyces lilacinus* (Thom.) Samson in combination with neem cake. *Journal of Biological Control* **17**: 125-131.

Nagesh, M. and Janakiram, T. 2004. Root-knot nematode problem in polyhouse roses and its management using dazomet, neem cake and *Pochonia chlamydosporia* (*Verticillium chlamydosporium*). *Journal of Ornamental Horticulture* **7**: 147-152.

Nagesh, M., Janakiram, T., Rao, T.M. and Parvatha Reddy, P. 1995. Evaluation of China aster varieties/pure lines for resistance to *Meloidogyne incognita* (Kofoid & White, 1919) Chitwood, 1949. *Pest Management in Horticultural Ecosystems* **1**: 101-104.

Nagesh, M., Meenakshi, S., Murthy, N. and Parvatha Reddy, P. 1995. Screening of some tuberose cultivars/hybrids for resistance against *Meloidogyne incognita* race-1. *Indian Perfumer* **39**: 138-140.

Nagesh, M. and Parvatha Reddy, P. 1995. Comparative efficacy of *Paecilomyces lilacinus* (Thom.) Samson and *Verticillium lecanii* (A. Zimmerman) Viegas in combination with botanicals against *Meloidogyne incognita* (Kofoid and White, 1919) Chitwood, 1949 infecting *Crossandra undulaefolia* L. *Journal of Biological Control* **9**: 109-112.

Nagesh, M. and Parvatha Reddy, P. 1996a. Management of *Meloidogyne incognita* on carnation and gerbera in commercial polyhouses. *Second International Crop Science Congress*, New Delhi, p.349.

Nagesh, M. and Parvatha Reddy, P. 1996b. Eco-friendly management of *Meloidogyne incognita* on *Crossandra undulaefolia* using vesicular arbuscular mycorrhiza, *Glomus mosseae* and oil cakes. *Mycorrhiza News* **9**(1): 12-14.

Nagesh, M. and Parvatha Reddy, P. 2000. Crop loss estimation in carnation and gerbera due to the root-knot nematode, *Meloidogyne incognita* (Kofoid & White) Chitwood. *Pest Management in Horticultural Ecosystems* **6**: 158-159.

Nagesh, M. and Parvatha Reddy, P. 2005. Management of carnation and gerbera to control the root-knot nematode, *Meloidogyne incognita*, in commercial polyhouses. *Nematologia Mediterranea* **33**: 157-162.

Nagesh, M., Parvatha Reddy, P., Janakiram, T. and Rao, T.M. 1999. Sequential biochemical changes in roots of *Callistiphus chinensis* lines resistant and susceptible to *Meloidogyne incognita* race 1. *Nematologia Mediterranea* **27**: 39-42.

Nagesh, M., Parvatha Reddy, P. and Rama, N. 2001. Pathogenicity of selected antagonistic soil fungi on *Meloidogyne incognita* (Kofoid & white) eggs and egg masses under *in vitro* and *in vivo* conditions. *Journal of Biological Control* **15**: 63-68.

Nagesh, M., Parvatha Reddy, P. and Ramachandran, N. 1998. Integrated management of *Meloidogyne incognita* and *Fusarium oxysporum* f sp. *gladioli* in gladiolus using antagonistic fungi and neem cake. In *Nematology - Challenges & Opportunities in 21ˢᵗ Century* (Usha K. Mehta, ed.), pp. 263-266. Sugarcane Breeding Inst., Coimbatore.

Nagesh, M., Parvatha Reddy, P. and Rao, M.S. 1995. Management of Meloidogyne incognita on Crossandra undulaefolia using Paecilomyces lilacinus and Pasteuria penetrans formulations. National Symposium on Perspectives in Eco-friendly Approaches to Plant Protection. Indian Agri. Res. Inst., New Delhi, pp. 47-48.

Nagesh, M., Parvatha Reddy, P. and Rao, M.S. 1997. Integrated management of *Meloidogyne incognita* on tuberose using *Paecilomyces lilacinus* in combination with plant extracts. *Nematologia Mediterranea* **25**: 3-7.

Nagesh, M., Parvatha Reddy, P. and Rao, M.S. 1998. Comparative efficacy of VAM fungi in combination with neem cake against *Meloidogyne incognita* on *Crossandra undulaefolia*. *Mycorrhiza News* **11**: 11-13.

Nagesh, M., Parvatha Reddy, P., Vijaya Kumar, M.V. and Nagaraju, B.M. 1998. Management of *Meloidogyne incognita* on *Polianthes tuberosa* using parasitic fungus, *Paecilomyces lilacinus* in combination with oil cakes. In *"Advances in IPM for Horticultural Crops"* (P. Parvatha Reddy, N. K. Krishna Kumar and A. Verghese, eds.), pp. 339-344. Association for Advancement of Pest Management in Hortil. Ecosystems, Bangalore.

Nagesh, M., Parvatha Reddy, P., Vijaya Kumar, M.V. and Nagaraju, B.M. 1999. Studies on correlation between *Glomus fasciculatum* spore density, root colonization and *Meloidogyne incognita* infection on *Lycopersicon esculentum*. *Journal of Plant Disease and Protection* **106**: 82-87.

Naik, D. 2004. Biotechnological Approaches for the Management of Wilt Disease Complex in Capsicum (Capsicum annum L.) and Egg plant (Solanum melongena) with Special Emphasis on Biological Control. Ph. D. thesis, Kuvempu Univ., Shimoga.

Nair, K.K.R. 1979. Studies on the chemical control of banana nematodes. *Agricultural Research Journal of Kerala* **17**: 232-235.

Nair, M.R.G.K. 1965. On the occurrence of the citrus nematode, *Tylenchulus semipenetrans* Cobb, in Kerala. *Agriculture Research Journal Kerala* **3**: 105.

Nair, M.R.G.K., Das, N..M. and Menon, M.R. 1966. On the occurrence of the burrowing nematode *Radopholus similis* (Cobb, 1893) Thorne, 1949 on banana in Kerala. *Indian Journal of Entomology* **28**: 553-554.

Nakat, R.V., Acharya, A., Jonathan, E.I., Hazarika, K., Jha, S., Singh, U.S., Wasnikar, A.R., Subba Rao, D.V. and Sitaramaiah, K. 1995. Evaluation of *Paecilomyces lilacinus* for the control of root-knot nematodes *Meloidogyne incognita* on betel vine. *National Symp. on Nematode Problems of India – An*

Appraisal of the Nematode Mangmt. with Eco-friendly Approaches and Biocomponents, Indian Agri. Res. Inst., New Delhi, p. 44.

Nand, S., Meher, H.C. and Singh, G. 1992. New host of Meloidogyne incognita and Tylenchulus semipenetrans. Indian Journal of Nematology **22**: 149.

Nanjappa, C.K., Panda, M. and Seshadri, A.R. 1978. Effects of *Meloidogyne incognita* and *Rotylenchulus reniformis* on the emergence and survival of seedlings. *Third International Congress on Plant Pathology*, Germany.

Nath, A., Sharma, N.K., Bhardwaj, S. and Thapa, C.N. 1982. Nematicidal property of garlic. *Nematologica* **28**: 253-255.

Nath, R., Khan, M.N., Wanshi, R.S.K. and Dwivedi, R.P. 1984. Influence of root-knot nematode, *Meloidogyne javanica* on pre and post-emergence damping off of tomato. *Indian Journal of Nematology* **14**: 135-140.

Nath, R.P., Sinha, B.K. and Haider, M.G. 1976. Studies on the nematodes of vegetables in Bihar-II. Combined effect of *Meloidogyne incognita* and *Ozonium texanum* var. *parasiticum* on germination of eggplant. *Indian Journal of Nematology* **6**: 177-179.

Nayak, G.M. 1986. *Studies on Papaya in Relation to Certain Plant Diseases*. M. Sc. Thesis, University of Agricultural Sciences, Bangalore, 106 pp.

Nayar, K., Mathew, J., Kuriyan, J., Gnanamanickam, S.S. and Mahadevan, A. (eds.). 1988. Role and association of root-knot nematode (*Meloidogyne incognita*) in the bacterial wilt of brinjal incited by *Pseudomonas solanacearum*. In *Advances in Research on Pant Pathogenic Bacteria*. Based on the proceedings of the *National Symposium on Phytobacteriology*, University of Madras, Madras, India, pp. 45-48.

Netscher, C. 1983. Control of *Meloidogyne incognita* in vegetable production by crop rotation in Ivory Coast. *Acta Horticulture* **152**: 219-225.

Ngundo, B.W. 1977. Screening of bean cultivars for resistance to *Meloidogyne* spp. in Kenia. *Plant Disease Reporter* **61**: 991-993.

Niknam, G.R. and Dhawan, S.C. 2001a. Induction of systemic resistance by *Bacillus subtilis* isolate Bs1 against *Rotylenchulus reniformis* in tomato. *Natl. Cong. on Centenary of Nematol. in India – Appraisal & Future Plans*, Indian Agri. Res. Inst., New Delhi, pp.143-144.

Niknam, G.R. and Dhawan, S.C. 2001b. Effect of seed bacterization, soil drench and bare root-dip application methods of *Pseudomonas fluorescens* isolate Pf1 on the suppression of *Rotylenchulus reniformis* infecting tomato. *Natl. Cong. on Centenary of Nematol. in India – Appraisal & Future Plans*, Indian Agri. Res. Inst., New Delhi, p. 144.

Nirmal Johnson, S.B. and Devarajan, K. 2004. Effect of *Pseudomonas fluorescens* on banana nematodes. *National Symposium on Paradigms in Nematological Research for Biodynamic Farming*, Univ. of Agri. Sci., Bangalore, p. 81-82.

Nirula, K.K. 1959. Root-knot nematode on *Colocasia*. *Current Science* **28**: 125-126.

Nirula, K.K. 1963. Collateral host plants of root-knot nematodes. *Current Science* **32**: 25-221-222.

Nirula, K.K. and Bassi, K.K. 1965. Thermotherapy for root-knot nematode, *Meloidogyne incognita* in potato tubers. *Indian Potato Journal* **7**: 9-11.

Nisha, M.S. and Sheela, M.S. 2004. Rhizome treatment – A low cost technology for the management of nematodes in Kacholam, an important medicinal plant. *National Symposium on Paradigms in Nematological Research for Biodynamic Farming,* Univ. of Agri. Sci., Bangalore, p. 62-63.

Nisha, M.S and Sheela, M.S. 2006. Bio-management of *Meloidogyne incognita* on Coleus, *Solenostemon rotundifolius* by integrating solarization, *Paecilomyces lilacinus, Bacillus macerans* and neem cake. *Indian Journal of Nematology* **36**: 136-138.

Nisha, M.S., Sheela, M.S. and Beena, R. 2001. Effect of mulching for the management of plant parasitic nematodes associated with kacholam, *Kaempferia galangal* L. *Proceedings of the Second National Symposium on Integrated Pest Management in Horticultural Crops: New Molecules, Biopesticides and Environment*, Bangalore, pp. 145-147.

O'Bannon, J.H. and Tomerlin, A.T. 1973. Citrus tree decline caused by *Pratylenchus coffeae. Journal of Nematology* **5**: 311-316.

Ohkawa, K. and Saigusa, T. 1981. Resistance of rose rootstocks to *Meloidogyne hapla, Pratylenchus penetrans* and *P. vulnus. Horticultural Science* **16**: 559-560.

Oostenbrink, M. 1972. Evaluation and integration of nematode control methods. In *Economic Nematology* (J.M. Webster, ed.), pp. 497-514. Academic Press, New York.

Oostendorp, M. and Sikora, R.A. 1989. Seed treatment with antagonistic rhizobacteria for the suppression of *Heterodera schachtii* early root infection of sugar beet. *Review of Nematology* **12**: 77-83.

Palanisamy, S. and Sivakumar, C.V. 1981. Assessment of avoidable yield loss in cowpea, black gram, maize and finger millet. *Second Nematology Symposium,* Tamil Nadu Agricultural University, Coimbatore, p. 58.

Pandey, R. 1990. Pathogenicity and reproduction of *Meloidogyne incognita* on *Hyoscyamnus albus* and rhizosphere association of other phytoparasitic nematodes. *International Nematology Network Newsletter* **7**: 34-37.

Pandey, R. 1994a. Bionomics of phytonematodes in relation to medicinal and aromatic plants. *Indian Journal of Plant Pathology* **12**: 16-23.

Pandey, R. 1994b. Comparative performance of oil seed cakes and pesticides in management of root-knot nematode disease of davana. *Nematologia Mediterranea* **22**: 17-19.

Pandey, R. 2001. Influence of Meloidogyne incognita, Pratylenchus thornei and Tylenchorhynchus vulgaris on growth and oil yield of menthol mint. Indian Journal of Nematology **31**: 111-114.

Pandey, R. 2002. Application of botanicals for management of root-knot nematode disease of *Ammi majus* L. *Indian Journal of Nematology* **32**: 198-200.

Pandey, R. 2003. Mint nematology – Current status and future needs. In *Advances in Nematology* (P.C. Trivedi, ed.), pp. 155-166. Scientific Publishers (India), Jodhpur.

Pandey, R. 2005a. Field application of bio-organics in the management of *Meloidogyne incognita* in *Mentha arvensis*. *Nematologia Mediterranea* **33**: 51-54.

Pandey, R. 2005b. Management of *Meloidogyne incognita* in *Artemisia pallens* with bioagents. *Phytoparasitica* **33**: 304-308.

Pandey, R., Gupta, M.L., Singh, H.N. and Kumar, S. 1999. The influence of vesicular arbuscular mycorrhizal fungi alone and IPM combination with *Meloidogyne incognita* on *Hyoscyamnus niger*. *Bioresource Technology* **69**: 275-278.

Pandey, R., Haseeb, A. and Husain, A. 1992. Distribution, pathogenicity and management of *Meloidogyne incognita* on *Mentha arvensis* cv. MAS-1. *Afro-Asian Journal of Nematology* **2**: 27-34.

Pandey, R. and Kalra, A. 2005a. Chemical activators: A novel and sustainable management approach for *Meloidogyne incognita* (Kofoid and White) Chitwood in *Chamomile recutita* L. *Archives of Phytopathology and Plant Protection* **38**: 107-111.

Pandey, R. and Kalra, A. 2005b. Root-knot disease of ashwagandha *Withania somnifera* and its eco-friendly cost effective management. *Journal of Mycology and Plant Pathology* **33**: 240-245.

Pandey, R., Kalra, A.,Gupta, M.L and Sharma, P. 2003. Phytonematodes: Major pest of MAPs. In *Proceedings of First National Interactive Meet on Medicinal and Aromatic Plants* (A.K. Mathur, S. Dwivedi, D.D. Patra, G.D. Bagchi, N.S. Sangwan, A. Sharma and S.P.S. Khanuja, eds.), pp. 188-197. CIMAP, Lucknow, India.

Pandey, R., Kalra, A., Gupta, M.L and Sharma, S.C. 2002. *Bacopa monnieri* damaged by root-knot nematode and remedial measures. *Journal of Tropical Medicinal Plants*

Pandey, R., Kalra, A., Singh, H.N., Gupta, M.L., Singh, H.B. and Kumar, S. 2000. Suppressive response of bioagent, organic material and pesticide on population development of *Meloidogyne incognita* on *Matricartia chamomilla*. *Journal of Spices and Aromatic Plant Sciences* **22**: 649-651.

Pandey, R. and Patra, N.K. 2001. Screening mint (*Mentha* spp.) accessions against root-knot nematode infection. *Journal of Spices and Aromatic Crops* **10**: 55-56.

Pandey, R., Singh, H.B. and Gupta, M.L. 1997. Antagonistic impact of vesicular arbuscular mycorrhizae (VAM) on *Meloidogyne incognita* population development in Japanese mint. *International Journal of Tropical Plant Disease* **15**: 237-245.

Pandey, R., Singh, H.B., Gupta, M.L., Singh, H.N., Sharma, S. and Kumar, S. 1998. Efficacy of VAM fungi, green manure, organic materials and pesticides in the management of root-knot nematode in menthol mint. *Proc. of National Symp. on Rational Approaches in Nematode Mangmt. for Sustainable Agri* (S.C. Dhawan and K.K. Kaushal, eds.), pp. 47-50. Gujarat Agri. Univ., Anand.

Pandey, R., Singh, H.B. and Kalra, A. 2000. Microbial management of plant parasitic nematodes. *International Journal of Tropical Plant Disease* **18**: 1-23.

Pandey, R., Singh, H.B. and Kumar, S. 1999. Pathogenicity of *Meloidogyne incognita* (Kofoid & White) Chitwood on *Matricaria chamomilla* L. *Journal of Spices and Aromatic Crops* **8**: 201-203.

Pandey, R.C. and Dwivedi, B.K. 2001. Study on the effect of biocontrol agents against root-knot disease of brinjal. *Current Nematology* **12**: 73-74.

Pandey, R.C. and Trivedi, P.C. 1992. Biological control of Meloidogyne incognita by Paecilomyces lilacinus in Capsicum annum. Indian Phytopathology **45**: 134-135.

Pani, A.K. and Das, S.N. 1972. Studies on etiological complexes in plant disease. I. Association of root-knot nematodes in bacterial wilt of tomato. *Journal of Research, Orissa University of Agriculture & Technology* **2**: 54-59.

Parvatha Reddy, P. 1984. Efficacy of seed treatment with three nematicides for the control of *Meloidogyne incognita* infecting cowpea, French bean and peas. *Indian Journal of Nematology* **14**: 39-40.

Parvatha Reddy, P. 1985a. Chemical control of root-knot nematodes infecting peas. *Indian Journal of Nematology* **15**: 120-121.

Parvatha Reddy, P. 1985b. Estimation of crop losses in peas due to *Meloidogyne incognita. Indian Journal of Nematology* **15**: 226.

Parvatha Reddy, P. 1986. Chemical control of plant parasitic nematodes infecting citrus and grapevine. *Indian Journal of Horticulture* **43**: 303-305.

Parvatha Reddy, P. 1988. *Nematode Management in Horticultural Crops.* Tech. Bull. No. 4, Indian Inst. of Hortil. Res., Bangalore, 52 pp.

Parvatha Reddy, P. and Agarwal, P.K. 1987. Resistance in citrus rootstocks to the citrus nematode, *Tylenchulus semipenetrans. Indian Journal of Horticulture* **44**: 111-114.

Parvatha Reddy, P., Iyer, C.P.A. and Subramanyam, M.D. 1988. Evaluation of papaya cultivars and hybrids against *Meloidogyne incognita. Indian Journal of Nematology* **18**: 381-382.

Parvatha Reddy, P. and Khan, R.M. 1987. First report on the occurrence of *Radopholus similis* (Cobb, 1893) Thorne, 1949 in Andhra Pradesh, India. *International Nematology Network Newsletter* **4**(1): 11.

Parvatha Reddy, P. and Khan, R.M. 1988. Evaluation of *Paecilomyces lilacinus* for the biological control of *Rotylenchulus reniformis* infecting tomato as compared with carbofuran. *Nematologia Mediterranea* **16**: 113-116.

Parvatha Reddy, P. and Khan, R.M. 1989. Evaluation of biocontrol agent *Paecilomyces lilacinus* and carbofuran for the management of *Rotylenchulus renfiromis* infecting brinjal. *Pakistan Journal of Nematology* **7**: 55-59.

Parvatha Reddy, P. and Khan, R.M. 1991. Integrated management of root-knot nematodes infecting okra. *Current Nematology* **2**:115-116.

Parvatha Reddy, P., Khan, R.M. and Agarwal, P.K. 1987. Selection of citrus rootstocks and hybrids resistant to the citrus nematode, *Tylenchulus semipenetrans. Pakistan Journal of Nematology* **5**(2): 69-72.

Parvatha Reddy, P., Khan, R.M. and Agarwal, P.K. 1989. Screening of banana germplasm against the burrowing nematode, *Radopholus similis. Indian Journal of Horticulture* **46**: 276- 278.

Parvatha Reddy, P., Khan, R.M. and Rao, M.S. 1991. Integrated management of the citrus nematode, *Tylenchulus semipenetrans* using oil cakes and *Paecilomyces lilacinus. Afro-Asian Journal of Nematology* **1**: 221-222.

Parvatha Reddy, P., Khan, R.M. and Rao, M.S. 1993. Integrated management of the citrus nematode, *Tylenchulus semipenetrans* using neem cake and the parasitic fungus, *Paecilomyces lilacinus. Proceedings of World Neem Conference,* Bangalore, pp. 647-650.

Parvatha Reddy, P. and Nagesh, M. 1995. Nematode problems of ornamental plants in India. In *Nematode Pest Management – An Appraisal of Eco-friendly Approaches* (G. Swarup, D.R. Dasgupta and J.S. Gill, eds.), pp. 244-251. Nematol. Soc. of India, New Delhi.

Parvatha Reddy, P. and Nagesh, M. 2000. Integrated management of the citrus nematode using bacterial (*Pasteuria penetrans)* and fungal (*Paecilomyces lilacinus)* biocontrol agents. *Proceedings of International Symposium on Citriculture*, Nagpur, pp. 825-829.

Parvatha Reddy, P. and Nagesh, M. 2002. *Avermectins – Isolation, Fermentation, Preliminary Charecterization and Screening for Nematicidal Activity.* Tech. Bull.17, Indian Inst. of Hortil. Research, Bangalore, 28 pp.

Parvatha Reddy, P., Nagesh, M. and Devappa, V. 1997. Effect of integration of *Pasteuria penetrans, Paecilomyces lilacinus* and neem cake for the management of root-knot nematodes infecting tomato. *Pest Management in Horticultural Ecosystems* **3**: 100-104.

Parvatha Reddy, P., Nagesh, M. and Devappa, V. 1998. Management of *Meloidogyne incognita* on tomato by integration of *Trichoderma harzianum, Glomus fasciculatum* and neem cake. In *"Advances in IPM for Horticultural Crops"* (P. Parvatha Reddy, N. K. Krishna Kumar & A. Verghese, eds.), pp. 349-352. Association for Advancement of Pest Management in Hortil. Ecosystems, Bangalore.

Parvatha Reddy, P., Nagesh, M., Devappa, V. and Vijaya Kumar, M.V. 1998. Management of *Meloidogyne incognita* on tomato by integrating endomycorrhiza, *Glomus mosseae* with oil cakes under nursery and field conditions. *Journal of Plant Disease and Protection* **105**: 53-57.

Parvatha Reddy, P., Nagesh, M., Rao, M.S. and Devappa, V. 1997. Integrated management of the burrowing nematode *Radopholus similis* using endomycorrhiza, *Glomus mosseae* and oil cakes. *Pest Management in Horticultural Ecosystems* **3**: 25-30.

Parvatha Reddy, P., Nagesh, M., Rao, M.S. and Rama, N. 2000. Management of *Tylenchulus semipenetrans* by integration of *Pseudomonas fluorescens* with oil cakes. *Proc. International Symposium on Citriculture*, Nagpur, pp. 830-833.

Parvatha Reddy, P., Rao, M.S., Mohandas, S. and Nagesh, M. 1995. Integrated management of the citrus nematode, *Tylenchulus semipenetrans* Cobb using VA mycorrhiza, *Glomus fasciculatum* (Thaxt.) Gerd & Trappe and oil cakes. *Pest Management in Horticultural Ecosystems* **1**: 37-41.

Parvatha Reddy, P., Rao, M.S. and Nagesh, M. 1996a. Management of the citrus nematode, *Tylenchulus semipenetrans* by integration of *Trichoderma harzianum* with oil cakes. *Nematologia Mediterranea* **24**: 265-267.

Parvatha Reddy, P., Rao, M.S. and Nagesh, M. 1996b. Management of the citrus nematode on acid lime by integration of parasitic fungi and oil cakes. *Pest Management in Horticultural Ecosystems* **2**: 15-18.

Parvatha Reddy, P., Rao, M.S. and Nagesh, M. 1996c. Integrated management of the citrus nematode, *Tylenchulus semipenetrans* using pesticides and parasitic fungus, *Paecilomyces lilacinus. Pest Management in Horticultural Ecosystems* **2**: 61-63.

Parvatha Reddy, P., Rao, M.S. and Nagesh, M. 1996d. Crop loss estimation in banana due to the burrowing nematode, *Radopholus similis. Pest Management in Horticultural Ecosystems* **2:** 85-86.

Parvatha Reddy, P., Rao, M.S. and Nagesh, M. 1998. Effect of bare-root dip treatment of tomato seedlings in plant leaf extracts mixed with *Paecilomyces lilacinus* spores for the management of *Meloidogyne incognita*. In *"Advances in IPM for Horticultural Crops"* (P. Parvatha Reddy, N. K. Krishna Kumar and A. Verghese, eds.), pp. 334-338. Association for Advancement of Pest Management in Hortil. Ecosystems, Bangalore.

Parvatha Reddy, P., Rao, M.S. and Nagesh, M. 1999. Eco-friendly management of *Meloidogyne incognita* on tomato by integration of *Verticillium chlamydosporium* with neem and calotropis leaves. *Journal of Plant Disease and Protection* **106**: 530-533.

Parvatha Reddy, P., Rao, M.S. and Nagesh, M. 2002. Integrated management of burrowing nematode (*Radopholus similis)* using endomycorrhiza (*Glomus mosseae)* and oil cakes. In *Banana – Improvement, Production and Utilization* (H.P. Singh and K.L. Chadha, eds.), pp. 344-348. AIPUB, Trichy.

Parvatha Reddy, P. and Singh, D.B. 1978a. The association of the citrus nematode with grape roots in a commercial orchard. *Current Science* **47**: 640-641.

Parvatha Reddy, P. and Singh, D.B. 1978b. Evaluating the reaction of some species and varieties of *Citrus* and *Poncirus* to the citrus nematode. *Indian Journal of Nematology* **8**: 82-84.

Parvatha Reddy, P. and Singh, D.B. 1979. Control of root-knot nematode, *Meloidogyne incognita* on tomato by chemical bare-root dips. *Indian Journal of Nematology* **9**: 40-43.

Parvatha Reddy, P. and Singh, D.B. 1980. Plant parasitic nematodes associated with banana in South India. *Proceedings of National Seminar on Banana Production Technology*, Tamil Nadu Agricultural University, Coimbatore, pp. 161-163.

Parvatha Reddy, P. and Singh, D.B. 1981. Assessment of avoidable yield loss in okra, brinjal, French bean and cowpea due to root-knot nematodes. *Third International Symposium on Plant Pathology*, New Delhi, pp.93-94.

Parvatha Reddy, P., Singh, D.B. and Nagalingam, B. 1981. Fighting citrus nematodes in Andhra Pradesh. *Indian Horticulture* **26**(3): 24-25.

Parvatha Reddy, P., Singh, D.B. and Ram Kishun. 1979. Effect of root-knot nematodes on the susceptibility of Pusa Purple Cluster brinjal to bacterial wilt. *Current Science* **48**: 915-916.

Parvatha Reddy, P., Singh, D.B. and Rao, V.R. 1979. The root-knot nematode on Gladiolus from India. *Current Science* **48**: 82.

Parvatha Reddy, P., Singh, D.B. and Sharma, S.R. 1979. Interaction of *Meloidogyne incognita* and *Rhizoctonia solani* in a root rot disease complex of French bean. *Indian Phytopathology* **32**: 651-652.

Paruthi, L.J., Jain, R.K. and Gupta, D.C. 1987. Effect of different periods of degradation of subabool leaves alone and in combination with nematicides in okra. *Indian Journal of Nematology* **17**: 30-32.

Patel, H.V., Patel, D.J. and Patel, B.A. 1989. Reaction of papaya varieties to *Rotylenchulus reniformis*. *International Nematology Network Newsletter* **6** (3): 24.

Patel, B.A., Patel, R.G., Patel, H.R., Vyas, R.V. and Patel, B.N. 2001. Management of root-knot nematodes in turmeric. *Natl. Cong. on Centenary of Nematol. in India – Appraisal & Future Plans*, Indian Agri. Res. Inst., New Delhi, p.165.

Patel, D.J., Makadia, B.M. and Shah, H.M. 1982. Occurrence of root-knot on turmeric (*Curcuma longa* L.) and its chemical control. *Indian Journal of Nematology* **12**: 168-171.

Pathak, K.N. and Keshari, N. 2004. Interaction of *Meloidogyne incognita* with *Fusarium oxysporum* f. sp. *conglutinans* on cauliflower. *Indian Journal of Nematology* **34**: 85-87.

Patnaik, P.R., Mohanty, K.C., Mahapatra, S.N. and Sahoo, S. 2004. Evaluation of chilli varieties against root-knot nematode, *Meloidogyne incognita*. *National Symposium on Paradigms in Nematological Research for Biodynamic Farming*, Univ. of Agri. Sci., Bangalore, p. 100.

Perveen, S., Ehteshamul-Haque, S. and Ghaffar, A. 1998. Efficacy of *Pseudomonas aeruginosa* and *Paecilomyces lilacinus* in the control of root rot-root knot disease complex on some vegetables. *Nematologia Mediterranea* **26**: 209-212.

Perwez, M.S., Rehman, M.F. and Haider, S.R. 1988. Effect of *Tagetes erecta* on *Meloidogyne javanica* infecting lettuce. *International Nematology Network Newsletter* **5**(3): 18-19.

Pillai, K.S. 1976. Nematicidal control of root-knot nematode on *Coleus parviflorus*. *Journal of Root Crops* **2**: 60-63.

Pinochet, J., Calvet, C., Camprubi, A. and Fernandez, C. 1995. Growth and nutritional response of Nemared peach rootstock infected with *Pratylenchus vulnus* and the mycorrhizal fungus, *Glomus mosseae*.

Pinochet, J. and Rowe, P.R. 1979. Progress in breeding for resistance to *Radopholus similis* in banana. *Nematropica* **9**: 76-78.

Ponte, J.J.D.A. 1980. *Meloidogyne* - importanica and control no nordeste. *Sociedade Brasileira Nematologia* **4**: 1-14.

Poornima, K., Kumar, N. and Balamohan, T.N. 2004. Biological control of lesion nematode in banana. *National Symposium on Paradigms in Nematological Research for Biodynamic Farming*, Univ. of Agri. Sci., Bangalore, p. 99.

Poornima, K. and Vadivelu, S. 1998. Pathogenicity of *Meloidogyne incognita* to turmeric (*Curcuma longa* L.). In *Nematology-Challenges and Opportunities in 21st Century* (Usha K. Mehta, ed.), pp. 29-33. Sugarcane Breeding Institute, Coimbatore.

Poucher, C., Ford, H.W., Suit, R.F. and DuCharme, E.P. 1967. *Burrowing Nematode in Citrus*. Florida Dept. of Agri., Divn. of Plant Industry, Bull. No. 7, 63 pp.

Prakash Babu, N. 2000. *Role of VAM in the Management of* Radopholus similis *on Banana*. M. Sc. (Agri.) thesis, University of Agricultural Sciences, Bangalore, 89 pp.

Prakash Babu, N., Kantharaju, V., Krishnappa, K. and Reddy, B.M.R. 2002. Biodiversity in indigenous isolates of vesicular arbuscular mycorrhiza, *Glomus fasciculatum* against *Radopholus similis* on banana (Abstr.). *Symposium on Biodiversity and Management of Nematodes in Cropping Systems for Sustainable Agriculture,* Jaipur, p. 108.

Prasad, K.S.K. 1988. Nematology in 2000 AD as I see it in relation to potato cyst nematode. *National Nematology Symposium*, New Delhi, pp. 26-29.

Prasad, K.S.K. 1989. Nematological problems and progress of research on potato in India. *Fourth Group Meeting on Nematological Problems of Plantation Crops,* University of Agricultural Sciences, Bangalore.

Prasad, K.S.K. 1990. *Helicotylenchus dihystera*, a potential pest of potato. *Journal of Indian Potato Association* **17**: 204-205.

Prasad, K.S.K. and Krishnappa, K. 1981. Effect of bare-root dip treatment with pesticides on *Rotylenchulus reniformis* affecting brinjal. *Indian Journal of Nematology* **11**:154-158.

Prasad, K.S.K., Raj, D. and Sharma, R.K. 1983. Effect of planting time on *Meloidogyne incognita* infesting potato at Shimla Hills. *Third Nematology Symposium*, Himachal Pradesh Agricultural University, Solan, pp. 5-6.

Prasad, K.S.K. and Sharma, R.K. 1985. Parasitic nematodes associated with potato and the pathogenicity of *Quinisulcius capitatus* on variety Kufri Jyoti. *Journal of Indian Potato Association* **12**: 56-62.

Prasad, M.B.N.V., Parvatha Reddy, P., Rao, M.S. and Rekha, A. 1998. Evaluation of some intergeneric citrus rootstock hybrids for resistance to the citrus nematode *Tylenchulus semipenetrans*. In "*Advances in IPM for Horticultural Crops*" (P. Parvatha Reddy, N. K. Krishna Kumar and A. Verghese, eds.), pp. 328-331. Association for Advancement of Pest Management in Hortil. Ecosystems, Bangalore.

Prasad, M.B.N.V., Sawant, S.D., Parvatha Reddy, P., Palaniappan, R., Srinivasa Rao, N. K. and Rekha, A. 1999. Citrus rootstock hybrids resistant to *Phytophthora* and citrus nematodes and also tolerant to salinity and draught. *Proceedings of National Symposium on Citriculture*, Nagpur, pp. 59-62.

Prasad, P.R.K 1978. *Studies on the Root-knot Nematodes Infecting Patchouli* (Pogostemon cablin *Bench*). M. Sc. (Agri.) thesis, Univ. of Agri. Sci., Bangalore.

Prasad, P.R.K. and Reddy, D.D.R. 1984. Pathogenicity and analysis of crop losses on patchouli due to *Meloidogyne incognita*. *Indian Journal of Nematology* **14**: 36-38.

Prasad, S.K. 1960. The problem of plant parasitic nematodes in the Indian Union. *Indian Journal of Entomology* **22**: 301-304.

Prasad, S.K. and Chawla, M.L. 1965. Observations on the population fluctuations of citrus nematode, *Tylenchulus semipenetrans* Cobb, 1913. *Indian Journal of Entomology* **27**: 450-455.

Prasad, S.K. and Dasgupta, D.R. 1964. Plant parasitic nematodes associated with roses in Delhi. *Indian Journal of Entomology* **26**: 12-122.

Prasad, S.K., Dasgupta, D.R. and Mukhopadhyaya, M.C. 1964. Nematodes associated with commercial crops in Northern India. I. Uttar Pradesh. *Indian Journal of Nematology* **3**: 8-23.

Raj, B.T. and Nirula, K.K. 1969. Trials on the eradication of root-knot nematodes from potato fields with DD (Abstr.). *All India Nematology Symposium*, Indian Agri. Res. Inst., New Delhi, India.

Rajagopalan, P. and Chinnarajan, A.M. 1976. Plant parasitic nematodes associated with banana in Tamil Nadu. *South Indian Horticulture* **24**: 69-72.

Rajagopalan, P. and Naganathan, T.G. 1977a. Studies on nematode parasites of grapevine. *Tamil Nadu Agricultural University Annual Report* **6**: 129.

Rajagopalan, P. and Naganathan, T.G. 1977b. Studies on nematode parasites of banana. *Tamil Nadu Agricultural University Annual Report* **6**: 131.

Rajani, T.S., Sheela, M.S. and Sivaprasad, P. 1998. Management of nematodes associated with kacholam, *Kaempferia galanga* L. In *Advances in IPM of Horticultural Crops* (P. Parvatha Reddy, N.K. Krishna Kumar and A. Verghese, eds.), pp. 326-327. Association for Advancement of Pest Management in Horticultural Ecosystems, Bangalore.

Rajendran, B. and Rajendran, G. 1979. Report of Meloidogyne incognita in Jasminum flexile. South Indian Horticulture **27**: 70.

Rajendran, G. and Naganathan, T.G. 1981. Evaluation of systemic chemicals for the control of *Rotylenchulus reniformis* on papaya. *Indian Journal of Nematology* **11**: 130.

Rajendran, G., Naganathan, T.G. and Sivagami, V. 1979. Studies on banana nematodes. *Indian Journal of Nematology* **9**: 54.

Rajendran, G., Vadivelu, S. and Muthukrishnan, T.S. 1976. Pathogenicity of *Meloidogyne incognita* on crossandra. *Indian Journal of Nematology* **6**: 115-116.

Ram, B. and Baheti, B.L. 2004. Management of reniform nematode, *Rotylenchulus reniformis* on cowpea through seed and soil treatment with plant products. *Indian Journal of Nematology* **34**: 193-195.

Ram, S. and Gupta, D.C. 2001. Studies on the effect of intercropping of onion on the growth of tomato and population dynamics of root-knot nematode, *Meloidogyne javanica. National Conference on Centenary of Nematology in India – Appraisal & Future Plans*, Indian Agri. Res. Inst., New Delhi, p. 94.

Ramakrishnan, C. and Balasubramanian, P. 1981. Control of *Meloidogyne incognita* in chilli *(Capsicum annum* L.) nursery with nematicides. *Second Nematology Symposium*, Tamil Nadu Agricultural University, Coimbatore, p. 54.

Ramana, K.V. 1986. Slow wilt disease of black pepper and the role of plant parasitic nematodes in its etiology (Abstr.). *Proceeding of the Second Group Discussion on Nematological Problems of Plantation Crops*, Central Coffee Research Station, Balehonnur, pp. 17-21.

Ramana, K.V., Eapen, S.J. and Sarma, Y.R. 1998. Effect of *Meloidogyne incognita, Pythium aphanidermatum* and *Ralstonia solanacearum* alone and in combinations in ginger. In *Nematology-Challenges and Opportunities in 21st Century* (Usha K. Mehta, ed.), pp. 87-93. Sugarcane Breeding Institute, Coimbatore.

Ramana, K.V. and Mohandas, C. 1987. Plant parasitic nematodes associated with black pepper (*Piper nigrum* L.) in Kerala. *Indian Journal of Nematology* **17**: 62-66.

Ramana, K.V., Mohandas, C. and Balakrishnan, R. 1987. Role of plant parasitic nematodes in the slow wilt disease complex of black pepper (*Piper nigrum* L.) in Kerala. *Indian Journal of Nematology* **17**:225-230.

Ramappa, H.K. 1988. *Studies on Nematode Parasites of French Bean and Their Control*. M. Sc. (Agri.) thesis, Univ. of Agri. Sci., Bangalore.

Rameshchand and Gill, J.S. 2002. Evaluation of various application methods of *Pasteuria penetrans* against *Meloidogyne incognita* in tomato. *Indian Journal of Nematology* **32**: 23-26.

Rangaswamy, S.D., Parvatha Reddy, P., Nagesh, M. and Nanje Gowda, D.N. 1999. Integrated management of *Meloidogyne incognita* using *Pasteuria penetrans, Trichoderma viride* and oil cakes in tomato. *Pest Management in Horticultural Ecosystems* **5**: 122-126.

Rangaswamy, S.D., Parvatha Reddy, P. and Nanje Gowda, D.N. 1999. Management of root-knot nematode, *Meloidogyne incognita* in tomato by intercropping with marigold and mustard. *Pest Management in Horticultural Ecosystems* **5**: 118-121.

Rangaswamy, S.D., Parvatha Reddy, P. and Narayana Swamy, B.C. 1995. Management of *Meloidogyne incognita* in tomato nursery by growing trap/antagonistic crops in rotation. *Current Nematology* **6**: 9-12.

Rao, M.S., Gowen, S.R., Pembroke, B. and Parvatha Reddy, P. 1997. Relationship of *Pasteuria penetrans* spore encumbrance on juveniles of *Meloidogyne incognita* and their infection in adults. *Nematologia Mediterranea* **25**: 129-131.

Rao, M.S., Kerry, B.R., Gowen, S.R., Bourne, J.M. and Parvatha Reddy, P. 1997. Management of *Meloidogyne incognita* in tomato nurseries by integration of *Glomus deserticola* with *Verticillium chlamydosporium*. *Journal of Plant Disease and Protection* **104**: 419-422.

Rao, M.S., Mohandas, S., Parvatha Reddy, P. and Khan, R.M. 1993. Synergistic effect of endomycorrhizal fungus (*Glomus fasciculatum*) and biocontrol fungus (*Paecilomyces lilacinus*) in the management of root-knot nematode infecting tomato. In "*Soil Organisms and Sustainability*" (D. Rajagopal, R. D. Kale and K. Bano, eds.), pp. 119-123. Indian Society of Soil Biology and Ecology, Univ. of Agri. Sci., Bangalore.

Rao, M.S. and Naik, D. 2003. Effect of Trichoderma harzianum and Paecilomyces lilacinus on Meloidogyne incognita on papaya (Carica papaya L.) nursery seedlings. Pest Management in Horticultural Ecosystems **9**: 155-160.

Rao, M.S., Naik, D., Shylaja, M. and Parvatha Reddy, P. 2002. Prospects for the management of the root-knot nematode and bacterial wilt disease complex in capsicum by integration of fungal and bacterial biocontrol agents. *Proceedings of International Conference on Vegetables,* Bangalore, pp. 347-351.

Rao, M.S. and Parvatha Reddy, P. 1992c. Innovative approaches of utilization of biocontrol agents-*Paecilomyces lilacinus* and *Verticillium chlamydosporium* against root-knot nematode on tomato (Abstr.). *First Afro-Asian Nematology Symposium,* Aligarh Muslim University, Aligarh, pp. 25-26.

Rao, M.S. and Parvatha Reddy, P. 1993a. Effective use of 'Welgro' and 'RD-9 Repelin' for the management of root-knot nematode on tomato. In *"Neem and Environment"* (R. P. Singh, M. S. Chari, A. K. Raheja and W. Kraus, eds.), **2**: 651-656. Oxford & IBH Publishing Co. Pvt. Ltd., New Delhi.

Rao, M.S. and Parvatha Reddy, P. 1993b. Interactive effect of *Verticillium chlamydosporium* and castor cake on the control of root-knot nematode on eggplant. In *"Soil Organisms and Sustainability"* (D. Rajagopal, R.D. Kale and K. Bano, eds.), pp. 114-118. Indian Society of Soil Biology and Ecology, Univ. of Agri. Sci., Bangalore.

Rao, M.S. and Parvatha Reddy, P. 1994. A method for conveying *Paecilomyces lilacinus* to soil for the management of root-knot nematodes on eggplant. *Nematologia Mediterranea* **22**: 265-267.

Rao, M.S. and Parvatha Reddy, P. 2001. Control of *Meloidogyne incognita* on eggplant using *Glomus mosseae* integrated with *Paecilomyces lilacinus* and neem cake. *Nematologia Mediterranea* **29**: 153-157.

Rao, M.S., Parvatha Reddy, P. and Mohandas, S. 1995. Studies on development of a biorational management strategy against root-knot nematode attacking tomato. *National Symposium on Nematode Problems of India – An Appraisal of the Nematode Management with Eco-friendly Approaches* (G. Swarup, D.R. Dasgupta and J.S. Gill, eds.), p. 14. Nematol. Soc. of India, New Delhi.

Rao, M.S., Parvatha Reddy, P. and Mohandas, S. 1996. Effect of integration of *Calotropis procera* leaf and *Glomus fasciculatum* on the management of *Meloidogyne incognita* infesting tomato. *Nematologia Mediterranea* **24**: 59-61.

Rao, M.S., Parvatha Reddy, P. and Nagesh, M. 1997a. Integrated management of *Meloidogyne incognita* on okra by castor cake suspension and *Paecilomyces lilacinus. Nematologia Mediterranea* **25**: 17-19.

Rao, M.S., Parvatha Reddy, P. and Nagesh, M. 1997b. Integration of *Paecilomyces lilacinus* with neem leaf suspension for the management of root-knot nematodes on eggplant. *Nematologia Mediterranea* **25**: 249-252.

Rao, M.S., Parvatha Reddy, P. and Nagesh, M. 1997c. Management of root-knot nematode, *Meloidogyne incognita* on tomato by integration of *Trichoderma harzianum* with neem cake. *Journal of Plant Disease and Protection* **104**: 423-425.

Rao, M.S., Parvatha Reddy, P. and Nagesh, M. 1998a. Integrated management of Meloidogyne incognita on tomato using Verticillium chlamydosporium and Pasteuria penetrans. Pest Management in Horticultural Ecosystems **4**: 32-35.

Rao, M.S., Parvatha Reddy, P. and Nagesh, M. 1998b. Integrated management of *Meloidogyne incognita* on tomato using oil cakes and a bioagent, *Verticillium chlamydosporium.* In *"Advances in IPM for Horticultural Crops"* (P. Parvatha Reddy, N. K. Krishna Kumar and A. Verghese, eds.), pp. 345-348. Association for Advancement of Pest Management in Hortil. Ecosystems, Bangalore.

Rao, M.S., Parvatha Reddy, P. and Nagesh, M. 1998c. Evaluation of plant based formulations of *Trichoderma harzianum* for the management of *Meloidogyne incognita* on eggplant. *Nematologia Mediterranea* **26**: 59-62.

Rao, M.S., Parvatha Reddy, P. and Nagesh, M. 1998d. Evaluation of *Paecilomyces lilacinus* cultured on neem cake extract for the management of root-knot nematodes on eggplant. *Pest Management in Horticultural Ecosystems* **4**:116-119.

Rao, M.S., Parvatha Reddy, P. and Nagesh, M. 1999. Bare root-dip treatment of tomato seedlings in calotropis or castor leaf extracts mixed with *Paecilomyces lilacinus* spores for the management of *Meloidogyne incognita. Nematologia Mediterranea* **27**: 323-326.

Rao, M.S., Parvatha Reddy, P. and Nagesh, M. 2000. Management of *Meloidogyne incognita* on tomato by integrating *Glomus mosseae* with *Pasteuria penetrans* under field conditions. *Pest Management in Horticultural Ecosystems* **6**: 130-134.

Rao, M.S., Parvatha Reddy, P., Somasekhar, N. and Nagesh, M. 1997. Management of root-knot nematodes, *Meloidogyne incognita* in tomato nursery by integration of endomycorrhiza, *Glomus fasciculatum* with castor cake. *Pest Management in Horticultural Ecosystems* **3**: 31-35.

Rao, M.S., Parvatha Reddy, P. and Sukhada, M. 1998. Biointensive management of *Meloidogyne incognita* on eggplant by integrating *Paeciloymces lilacinus* and *Glomus mosseae. Nematologia Mediterranea* **26**: 213-216.

Rao, M.S., Parvatha Reddy, P., Sukhada, M., Nagesh, M. and Pankaj. 1998. Management of root-knot nematodes on eggplant by integrating endomycorrhiza (*Glomus fasciculatum*) and castor (*Ricinus communis*) cake. *Nematologia Mediterranea* **26**: 217-219.

Rao, M.S., Reddy, P.P. and Mohandas, S. 1995. Effect of integration of endomycorrhiza (*Glomus mosseae*) and neem cake on the control of root-knot nematode on tomato. *Journal of Plant Disease and Protection* **102**: 526-529.

Rao, M.S. and Shylaja, M. 2004. Role of *Pseudomonas fluorescens* (Migula) in induction of systemic resistance (ISR) and managing *Rotylenchulus reniformis* (Linford and Oliveira) on carrot. *Pest Management in Horticultural Ecosystems* **10**: 87-93.

Rao, M.S., Tewari, R.P. and Pandey, M. 1992. Occurrence of mushroom nematodes in South India. *Mushroom Research* **1**: 135-136.

Rao, V.R. and Singh, D.B. 1978. Chemical control of root-knot nematodes in okra with the aid of nematicides. *Indian Journal of Entomology* **35**: 151-154.

Rashid, A. and Khan, A.M. 1975. Studies on the lesion nematode, *Pratylenchus coffeae* associated with chrysanthemum. *Indian Phytopathology* **28**: 22-28.

Rashid, A., Khan, F.A. and Khan, A.M. 1973. Plant parasitic nematodes associated with vegetables, fruits, cereals and other crops in North India. I. Uttar Pradesh. *Indian Journal of Nematology* **3**: 8-23.

Raveendran, V. and Nadakal, A.M. 1975. An additional list of plants infected by the root-knot nematode, *Meloidogyne incognita* (Kofoid & White, 1919) Chitwood, 1949 in Kerala. *Indian Journal of Nematology* **5**: 126-127.

Ravi, K., Nanje Gowda, D.N. and Parvatha Reddy, P. 2001. Integrated management of disease complex in banana involving borrowing nematode and Panama wilt – an ecofriendly approach. In *IPM in Horticultural Crops: Emerging Trends in the New Millenium* (A. Verghese and P. Parvatha Reddy, eds.), p. 154. Association for Advancement of Pest Management in Horticultural Ecosystems, Bangalore.

Ravichandra, N.G. and Krishnappa, K. 1985. Effect of various treatments, both individually and in integration, in controlling the burrowing nematode, *Radopholus similis*, infesting banana. *Indian Journal of Nematology* **15**: 62-65.

Ravi Dutt and Bhatti, D.S. 1986. Determination of effective doses and time of application of nematicides and castor leaves for controlling *Meloidogyne javanica* in tomato. *Indian Journal of Nematology* **16**: 8-11.

Ray, S. and Parija, A.K. 1987. Community analysis of nematodes in the rhizosphere of banana in Puri district, Orissa. *Indian Journal of Nematology* **17**: 159-161.

Ray, S., Sahoo, N.K. and Mohanty, K. 1990. Plant parasitic nematodes associated with tuber crops in Orissa (Abstr.). *National Symposium on Recent Advances In the Production and Utilization of Tropical Tuber Crops,* ISRC, Central Tuber Crops Res. Inst., Thiruvanthapuram.

Reddy, D.B. 1977. *Technical Document,* Plant Protection Committee for the South East Asia and Pacific Region, FAO, Bangkok, pp. 108-114.

Reddy, D.D.R. 1985. Analysis of crop losses in tomato due to *Meloidogyne incognita*. *Indian Journal of Nematology* **15**: 55-59.

Reddy, D.D.R. and Seshadri, A.R. 1972. Studies on some systemic nematicides. II. Further studies on the action of thionazin and aldicarb on *Meloidogyne incognita* and *Rotylenchulus reniformis*. *Indian Journal of Nematology* **2**: 182-190.

Reddy, D.D.R. and Seshadri, A.R.1975. Elimination of root-knot nematode infestation from tomato seedlings by chemical bare-root dips or soil application. *Indian Journal of Nematology* **5**: 170-175.

Reka Arya and Saxena, S.K. 1998. *Trichothecium roseum* together with *Rhizoctonia solani* and *Meloidogyne incognita* on germination of seeds of tomato cv. Pusa Ruby. *Indian Journal of Nematology* **28**: 217-221.

Rhoade, R.A. and Jenkins, N.R. 1957. *Phytopathology* **47**: 29.

Rhoades, H.L. 1984. Control of *Pratylenchus scribneri* on spearmint, *Mentha spicata* with nonfumigant nematicides. *Nematropica* **14**: 85-89.

Rom, R.C., Dozier, W.A, Knowles, J.W., Carlton, C.C., Arrington, E.H., Wehunt, E.J., Yadava, U.L., Doud, S.L., Ritchie, D.F., Clayton, C.N., Zehr, E.L., Gamrbell, C.E., Britton, J.A. and Lockwood, D.W. 1985. *Compact Fruit Tree* **18**: 85-91.

Routaray, B.N. and Sahoo, H. 1985. Integrated control of root-knot nematode, *Meloidogyne incognita* with neem cake and granular nematicides on tomato. *Indian Journal of Nematology* **15**: 261.

Rowe, R.C., Davis, J.R., Powelson, M.L. and Rouse, D.I. 1987. Potato early dying: causal agents and management strategies. *Plant Disease* **71**: 482-489.

Sahoo, S., Mohanty, K.C., Sahoo, N.K., Mohapatra, A.K.B. and Mishra, H.N. 2004a. Effect of different cropping sequences on root-knot nematode (*Meloidogyne incognita*) and wilt incidence in tomato. *National Symposium on Paradigms in Nematological Research for Biodynamic Farming,* Univ. of Agri. Sci., Bangalore, p. 37.

Sahoo, S., Mohanty, K.C., Sahoo, N.K., Mohapatra, A.K.B. and Mishra, H.N. 2004b. Comparative efficacy of different eco-friendly approaches for the management of root-knot nematode in okra. *National Symposium on Paradigms in Nematological Research for Biodynamic Farming,* Univ. of Agri. Sci., Bangalore, p. 79.

Saifullah. 1996. Effect of nematophagous fungi on growth of potato and *Globodera rostochiensis* and their environmental tolerance. *Afro-Asian Journal of Nematology* **6**: 128-132.

Saikia, B. 1992. Pathogenicity, Crop Loss Assessment and Biological Control of Meloidogyne incognita (Kofoid and White, 1919) Chitwood, 1949 on Betel Vine (Piper betle L.). M. Sc. (Agri.) thesis, Assam Agri. Univ., Jorhat.

Samiyappan, R. 2003. Biological control of fungal and nematode complex diseases by plant growth promoting rhizobacteria (PGPR). *Winter School on Biological Control of Plant Parasitic Nematodes,* Tamil Nadu Agri. Univ., Coimbatore, pp. 69-77.

Samuel, M. and Mathew, J. 1983. *Indian Phytopathology* **36**: 398-399.

Sangwan, N.K., Verma, K.K., Verma, B.S., Malik, M.S. and Dhindsa, K.S. 1985. Nematicidal activity of essential oils of *Symbopogon* grasses. *Nematologica* **13**: 93-99.

Santhosh Kumar, T. and Sheela, M.S. 2004. Status of nematode pests of Chethikoduveli, *Plumbago rosea* L. *National Symposium on Paradigms in Nematological Research for Biodynamic Farming,* Univ. of Agri. Sci., Bangalore, p. 28-29.

Sarma, Y.R., Nambiar, K.R.N. and Nair, C.P.R. 1974. Brown rot of turmeric. *Journal of Plantation Crops* **2**: 33-34.

Sarwar, M., Parameswaran, T.N., Shanmugam, C. and Narayan, M.R. 1982. Appraisal of Patchouli in India. Root-knot nematode problem and control. *Indian Perfumer* **26** (2-4): 117-121.

Sasser, J.N. 1989. *The Plant-parasitic Nematodes, The Farmers Hidden Enemy.* Dept. of Plant Path., North Carolina State Univ. & Consortium for Intnl. Crop Prot., 115 pp.

Sasser, J.N. and Freckman, D.W. 1987. A world perspective on Nematology: The role of the Society. In *Vistas on Nematology* (J.A. Veech and D.W. Dickson, eds.), pp. 7-14. Society of Nematologists, Hyattsville, Maryland.

Saxena, S.K., Alam, M.M. and Khan, A.M. 1974. Chemotherapeutic control of nematodes with Vydate oxamyl, VC-13 and Dazomet on chilli plants. *Indian Journal of Nematology* **4**: 235-238.

Schieber, E. and Sosa, O.N. 1960. Nematodes on coffee in Guatemala. *Plant Disease Reporter* **44**: 722-723.

Scholte, K. 1990. Causes of differences in growth pattern, yield and quality of potatoes (*Solanum tuberosum* L.) in short rotations on sandy soil as affected by crop rotation, cultivar and application of granular nematicides. *Potato Research* **33**: 181-190.

Seenivasan, N., Parameswaran, S., Sridar, P., Gopalakrishnan, C. and Gnanamurthy, P. 2001. Application of bioagents and neem cake as soil application for the management of root-knot nematode in turmeric. *Natl. Cong. on Centenary of Nematol. in India – Appraisal & Future Plans*, Indian Agri. Res. Inst., New Delhi, pp. 164.

Sehgal, M., Chawla, M.L., Singh, S. and Balakrishnan, K. 1990. Studies on interactive effect of moisture and temperature on *Tylenchulus semipenetrans*. *Current Nematology* **1**: 35-36.

Sen, K. and Dasgupta, M.K. 1977. Additional hosts of the root-knot nematode, *Meloidogyne* spp. from India. *Indian Journal of Nematology* **7**: 74.

Sen, K. and Dasgupta, M.K. 1981. Effect of rice hull ash on root-knot nematodes in tomato. *Tropical Pest Management* **27**: 518-519.

Senthamarai, M., Poornima, K. and Subramanian, S. 2006a. Nematode-fungal disease complex involving *Meloidogyne incognita* and *Macrophomina phaseolina* on *Coleus forskohlii* Briq. *Indian Journal of Nematology* **36**: 181-184.

Senthamarai, M., Poornima, K. and Subramanian, S. 2006b. Bio-management of root-knot nematode, *Meloidogyne incognita* on *Coleus forskohlii* Briq. *Indian Journal of Nematology* **36**: 206-208.

Senthamarai, M., Poornima, K. and Subramanian, S. 2006c. Assessment of avoidable yield loss on *Coleus forskohlii* due to *Meloidogyne incognita*. *Indian Journal of Nematology* **36**: 296-297.

Senthil Kumar, T. and Rajendran, G. 2004. Biocontrol agents for the management of disease complex involving root-knot nematode, *Meloidogyne incognita* and *Fusarium moniliforme* on grapevine (*Vitis vinefera*). *Indian Journal of Nematology* **34**: 49-51.

Seshadri, A.R. 1964. Investigations on the biology and life cycle of *Criconemoides xenoplax* Raski, 1952 (Nematoda: Criconematidae). *Nematologica* **10**: 540-562.

Seshadri, A.R. and Kumaraswami, T. 1963. A preliminary report on the occurrence of root-knot nematodes (*Meloidogyne* spp.) on some important crop plants in Madras State. *Madras Agricultural Journal* **50**: 98-99.

Seth, A. 1984. Ph. D. thesis, Dept. of Entomol. and Apiculture, Himachal Pradesh Krishi Viswa Vidyalaya, Solan, India.

Seth, A. and Sharma, N.K. 1986. Five new species of genus *Aphelenchoides* (Nematoda: Aphelenchida) infesting mushroom in North India. *Indian Journal of Nematology* **16**: 205-215.

Sethi, C.L., Nand, S. and Srivastava, A.N. 1981. Occurrence of *Radopholus similis* (Cobb, 1893) Thorne, 1949 in Gujarat, India. *Indian Journal of Nematology* **11**: 116.

Sethi, C.L. and Swarup, G. 1971. Plant parasitic nematodes of North-Western India. III. The genus, *Pratylenchus*. *Indian Phytopathology* **24**: 410-412.

Shah, J.J. and Raju, E.C. 1977. Histopathology of ginger (*Zingiber officinale*) infected by soil nematode *Meloidogyne* spp. *Phyton* (Austria) **16**: 79-84.

Shahzad, S. and Gaffer, A. 1989. Use of *Paecilomyces lilacinus* in the control of root rot and root-knot complex in okra and mungbean. *Pakistan Journal of Nematology* **7**: 47-53.

Shanthi, A. 2003. *Management of* Radopholus similis, Helicotylenchus multicinctus *and* Pratylenchus coffeae *in Banana.* Ph. D. thesis, Tamil Nadu Agri. Univ., Coimbatore.

Shanthi, A., Rajendran, G. and Sivakumar, M. 2004. Influence of biofertilizers on lesion nematodes in banana. *National Symposium on Paradigms in Nematological Research for Biodynamic Farming,* Univ. of Agri. Sci., Bangalore, p. 97.

Shanthi, A., Rajeswari, S. and Sivakumar, C.V. 1998. Soil application of *Pseudomonas fluorescens* (Migula) for the control of root-knot nematode (*Meloidogyne incognita*) on grapevine (*Vitis vinefera* Linn.). In *Nematology – Challenges and Opportunities in 21st Century,* (Usha K. Mehta, ed.), pp. 203-206. Sugarcane Breeding Institute, Coimbatore.

Shanthi, A. and Sivakumar, C.V. 1995. Biocontrol potential of *Pseudomonas fluorescens* (Migula) against root-knot nematode, *Meloidogyne incognita* (Kofoid and White, 1919) Chitwood, 1949 on tomato. *Journal of Biological Control* **9**: 113-115.

Shanthi, A., Sundarababu, R. and Sivakumar, C.V. 1999. Field evaluation of rhizobacterium, *Pseudomonas fluorescens* for the management of the citrus nematode, *Tylenchulus semipenetrans.* In *Proc. of the Natl. Symp. on Rational Approaches in Nema Mangmt. for Sustainable Agri.,* Indian Agri. Res. Inst., New Delhi, pp. 38-42.

Sharma, A. and Trivedi, P.C. 1989. Influence of inoculum levels of fungus *Paecilomyces lilacinus* (Thom.) Samson on the biocontrol of root-knot nematode, *Meloidogyne incognita* (Chitwood). *International Nematology Network Newsletter* **6**(2): 27-29.

Sharma, A. and Trivedi, P.C. 1994. Interaction of root-knot nematode with VAM on *Abelmoschus esculentus* and their effect on fruit yield. In *Vistas in Seed Biology* (T. Singh and P.C. Trivedi, eds.), Vol. II, pp. 61-81. Printwell, Jaipur.

Sharma, B.B., Sharma, H.C. and Khan, E. 1969. Nematodes associated with sweet orange and mandarin plants in relation to die-back (Abstr.). *All India Nematology Symposium,* New Delhi, p. 16.

Sharma, G.L. 1989. Estimated losses due to root-knot nematodes, *Meloidogyne incognita* and *M. javanica* in pea crop. *International Nematology Network Newsletter* **6**(1): 28-29.

Sharma, G.L., Trivedi, P.C., Sharma, M.K. and Tiagi, B. 1988. Fungus *Paeciloymces lilacinus* as a biocontrol agent of root-knot nematode, *Meloidogyne incognita. The Indian Zoologist* **12**: 259-262.

Sharma, H.K. and Gill, J.S. 1992. Prevalence of races in root-knot nematode, *Meloidogyne incognita* in India. *Indian Journal of Nematology* **22**: 43-49.

Sharma, H.K., Pankaj and Mishra, S.D. 2005. Polyethylene mulching in the management of plant parasitic nematodes. *Indian Journal of Nematology* **35**: 82-85.

Sharma, N.K. and Sharma, S.K. 1977. Spatial distribution of *Tylenchulus semipenetrans* in citrus rhizospheres (Abstr.). *Proceedings of the International Symposium on Citriculture*, Bangalore, p. 37.

Sharma, N.K. and Sethi, C.L. 1976. Reaction of certain cowpea varieties to *Meloidogyne incognita* and *Heterodera cajani. Indian Journal of Nematology* **6**: 99-102.

Sharma, N.K., Thapa, C.D. and Nath, A. 1981. Pathogenicity and identity of myceliophagus nematode infesting *Agaricus bisporus* (Lange) Sing. In Himachal Pradesh (India). *Indian Journal of Nematology* **11**: 230-231.

Sharma, N.K., Thapa, C.D. and Damanjeet. 1984. Estimation of loss in mushroom yield due to myceliophagus nematode, *Aphelenchoides sacchari. Indian Journal of Nematology* **14**: 121-124.

Sharma, N.K., Thapa, C.D. and Damanjeet Kaur. 1985. Pathogenicity and identity of *Ditylenchus myceliophagus* "Indian population" infesting *Agaricus bisporus* (Lange) Sing. in India. *Indian Journal of Nematology* **15**: 233-234.

Sharma, R.D. and Ferraz, E.C.A. 1977. Eficacia de nematicidas sistemicos e nao in Bahia, Brazil. *Nematologia Brasiliera* **2**: 139-147.

Sharma, S.B., Sharma, H.K. and Pankaj. 2002. Nematode problems in India. In *Crop Pest and Disease Management: Challenges for the Millennium* (D. Prasad and S.N. Puri, eds.), pp.267-275. Jyoti Publishers, New Delhi, India.

Sharma, S.K., Singh, I. and Sakhuja, P.K. 1980. Influence of different cropping sequences on the population of root-knot nematode, *Meloidogyne incognita* and the performance of the subsequent mungbean crop. *Indian Journal of Nematology* **10**: 53-58.

Sharpe, R.H., Hesse, C.O., Lownsbery, B.F., Perry, V.G. and Hansen, C.J. 1969. *Journal of American Society of Horticultural Science* **94**: 209-212.

Sheela, M.S., Hebsybai and Anitha, N. 1996. Nematodes associated with medicinal and aromatic plants in Kerala. *Eigth Kerala Science Congress*, pp. 450-452.

Sheela, M.S., Jiji, T. and Nisha, M.S. 2002. Evaluation of different control strategies for the management of nematodes associated with vegetables (brinjal) (Abstr.). *International conference on Vegetables*, Bangalore, p. 268.

Sheela, M.S. and Nisha, M.S. 2004. Impact of biological control agents for the management of root-knot nematode in brinjal. *National Symposium on Paradigms in Nematological Research for Biodynamic Farming*, Univ. of Agri. Sci., Bangalore, p. 63-64.

Sheela, M.S., Nisha, M.S. and Mohandas, C. 2002. Impact of trap cropping and crop rotation for the management of root-knot nematode in vegetables. *Proceedings of International Conference on Vegetables,* Bangalore, pp. 344 -346.

Sheela, M.S., Nisha, M.S. and Mohandas, C. 2004. Eco-friendly management of nematodes associated with Chinese potato (Coleus), *Solanostemon rotundifolius* (Poir) Morton. *National Symposium on Paradigms in Nematological Research for Biodynamic Farming,* Univ. of Agri. Sci., Bangalore, p. 90.

Sheela, M.S. and Rajani, T.S. 1998. Status of phytonematodes as a part of medicinal plants in Kerala. In *Nematology – Challenges and Opportunities in 21ˢᵗ Century* (Usha K. Mehta, ed.), pp.2-3. Sugarcane Breeding Inst., Coimbatore.

Shetty, K.D. and Reddy, D.D.R. 1984. Pathogenicity of *Meloidogyne incognita* on *Solanum khasianum. Indian Journal of Nematology* **14**: 185-186.

Shetty, D.K and Reddy, D.D.R. 1985. Control of root-knot nematodes on *Solanum khasianum* by chemical bare-root dips. *Indian Journal of Nematology* **15**: 227-229.

Shukla, P.K. and Haseeb, A. 1996. Effectiveness of some nematicides and oil cakes in the management of *Pratylenchus thornei* on *Mentha citrata, M. piperita* and *M. spicata. BITE* **57**: 307-309.

Shylaja, M. 2004. Studies on the development of bio-intensive management strategies of nematode induced wilt disease complex in Tuberose (*Polyanthes tuberosa*) and Carnations (*Dianthus caryophyllus*) using biological control agents. *National Symposium on Paradigms in Nematological Research for Biodynamic Farming,* Univ. of Agri. Sci., Bangalore, p. 136-137.

Shylaja, M., Rao, M.S. and Rahiman, A.B. 2004. Studies on the Development of Bio-intensive Disease Management Strategies in Tuberose (Polyanthes tuberosa L.) and Carnations (Dianthus caryophyllus L.) Using Biological Control Agents. Ph. D. thesis, Kuvempu Univ., Shimoga.

Siddiqi, M.A. and Alam, M.M. 1989a. Seed treatment with azadirachtin for the control of the stunt nematode attacking cabbage and cauliflower. *Annals of Applied Biology* **113**: 4-5.

Siddiqi, M.A. and Alam, M.M. 1989b. Control of stunt nematode by bare-root dip in leaf extracts of margosa and Persian lilac. *Pakistan Journal of Nematology* **7**: 33-38.

Siddiqi, M.A. and Saxena, S.K. 1987a. Effect of interculture of margosa and Persian lilac with tomato and eggplant on root-knot and reniform nematodes. *International Nematology Network Newsletter* 4(2): 5-8.

Siddiqi, M.A. and Saxena, S.K. 1987b. Studies on the control of the stunt nematode, *Tylenchorhynchus brassicae* by interculture of margosa and Persian lilac. *International Nematology Network Newsletter* 4(4): 27-29.

Siddiqi, M.R. 1961. Occurrence of the citrus nematode, *Tylenchulus semipenetrans* Cobb. 1913 and the reniform nematode, *Rotylenchulus reniformis* in India. *Proceedings of 48ᵗʰ Indian Science Congress* **3**: 504.

Siddiqi, M.R. 1962. *Nematologica* **8**: 193-200.

Siddiqi, M.R. 1964. Studies on nematode root-rot of citrus in Uttar Pradesh, India. *Proceedings of Zoological Society, Calcutta* **17**: 67-75.

Siddiqui, Z.A., Khan, A.M. and Saxena, S.K. 1972. Host range and varietal resistance of certain crucifers against *Tylenchorhynchus brassicae. Indian Phytopathology* **25**: 275-281.

Siddiqui, Z.A., Khan, A.M. and Saxena, S.K. 1973. Studies on *Tylenchorhynchus brassicae* II - Effect of temperature and moisture on multiplication of nematode. *Indian Journal of Entomology* **26**: 139-147.

Siddiqui, Z.A., Khan, A.M. and Saxena, S.K. 1976. Nematode population and yield of certain vegetables as influenced by oil-cake amendments. *Indian Journal of Nematology* **6**: 179-182.

Siddiqui, Z.A. and Khan, M.W. 1986. A survey of nematodes associated with pomegranate in Libya and evaluation of some systemic nematicides for their control. *Pakistan Journal of Nematology* **4**: 83-90.

Singh, D. 1991. *Studies on the Effect of* Tagetes *species on* Meloidogyne javanica *Infectivity in Brinjal* (Solanum melongena). Ph. D. thesis, Haryana Agri. Univ., Hisar, 83 pp.

Singh, D.B., Parvatha Reddy, P. and Sharma, S.R. 1981. Effect of the root-knot nematode, *Meloidogyne incognita* on *Fusarium* wilt of French beans. *Indian Journal of Nematology* **11**: 84-85.

Singh, D.B., Rao, V.R. and Parvatha Reddy, P. 1978. Evaluation of nematicides for the control of root-knot nematodes on brinjal. *Indian Journal of Nematology* **8**: 64-66.

Singh, M., Samra, R. and Sharma, G.L. 1998. Effect of various soil types on the development of *Pasteuria penetrans* on *Meloidogyne incognita* in brinjal crop. *Indian Journal of Nematology* **28**: 35-40.

Singh, R.S. and Sitaramaiah, K. 1966. Incidence of root-knot of okra and tomato in oil-cake amended soil. *Plant Disease Reporter* **50**: 668-672.

Singh, R.S. and Sitaramaiah, K. 1971. Control of root-knot through organic and inorganic amendments of soil: Effect of saw dust and inorganic nitrogen. *Indian Journal of Nematology* **1**: 80-84.

Singh, R.S. and Sitaramaiah, K. 1973. *Control of Plant Parasitic Nematodes with Organic Amendments of Soil.* Final technical report on effect of organic amendments, green manuring and inorganic fertilizers on root-knot of vegetable crops, G.B. Pant Univ. of Agri. & Technol., Bull. No. 6, 289 pp.

Singh, R.V. and Gill, J.S. 1998. Important nematode problems on vegetables and their management. In *Pathological Problems of Economic Crop Plants and Their Management* (S.M. Paul Khurana, ed.), pp. 557-570. Scientific Publishers (India), Jodhpur, India.

Singh, R.V., Midha, S.K. and Kumar, V. 2003. Major nematode pests of economically important crops in India and their management. In *Advances in Nematology* (P.C. Trivedi, ed.), pp. 233-256. Scientific Publishers (India), Jodhpur.

Sitaramaiah, K. and Naidu, H. 2003. Plant parasitic nematodes-management opportunities and challenges. In *Nematode Management in Plants* (P.C. Trivedi, ed.), pp. 57-80. Scientific Publishers (India), Jodhpur.

Sitaramaiah, K. and Sikora, R.A. 1982. Effect of the mycorrhizal fungus, *Glomus fasciculatum* on the host-parasite relationship of *Rotylenchulus reniformis* in tomato. *Nematologica* **28**: 412-419.

Sivakumar, C.V. 1998. Bacterial antagonists for suppression of plant parasitic nematodes. In *Biological Suppression of Plant Diseases, Phytoparasitic Nematodes and Weeds* (S.P. Singh and S.S. Hussaini, eds.), pp. 128-137. Project Directorate of Biological Control, Bangalore.

Sivakumar, C.V. and Meerazainuddin, M. 1974. Influence of N, P and K on the reniform nematode, *Rotylenchulus reniformis* and its effect on the yield of okra. *Indian Journal of Nematology* **4**: 243-244.

Sivakumar, C.V., Palanisamy, S. and Naganathan, T.G. 1976. Persistence of nematicidal activity in seed treatment of okra with carbofuran and aldicarb sulfone. *Indian Journal of Nematology* **6**: 106-108.

Sivakumar, C.V. and Seshadri, A.R. 1972. Histopathology of infection by the reniform nematode, *Rotylenchulus reniformis* Linford and Oliveira, 1940 on castor, papaya and tomato. *Indian Journal of Nematology* **2**: 173-181.

Sivakumar, M. and Marimuthu, T. 1986. Efficacy of different organic amendments against phytonematodes associated with betel vine. *Indian Journal of Nematology* **16**: 278.

Sivakumar, M. and Marimuthu, T. 1987. Preliminary studies on the effect of soil solarization on phytonematodes of betel vine. *Indian Journal of Nematology* **17**: 58-59.

Sivakumar, M., Natarajan, S. and Balasubramaniam, M. 1987. Nematode pests of betel vine (*Piper betle* L.) and their management. *Proceedings of the Third Group Discussions on the Nematological Problems of Plantation Crops*, Sugarcane Breeding Institute, Coimbatore, pp. 34-35.

Sivakumar, M. and Vadivelu, S. 1999. Management of *Meloidogyne incognita* on grapevine using biocontrol agents, botanicals and biofertilizers. *Pest Management in Horticultural Ecosystems* **5**: 127-131.

Sivapalan, P. 1972. Nematode pests of tea. In *Economic Nematology* (J.M. Webster, ed.), pp. 285-310. Academic Press, New York and London.

Sivaprasad, P. and Sheela, M.S. 1999. Distribution of phytonematodes and VAM fungi in the rhizosphere of pepper and its impact on nematode management. In *Nematology – Challenges and Opportunities in 21st Century* (Usha K. Mehta, ed.), pp. 102-105. Sugarcane Breeding Institute, Coimbatore.

Sobita Devi, L. and Dutta, U. 2002. Effect of *Pseudomonas fluorescens* on root-knot (*Meloidogyne incognita*) on okra plant. *Indian Journal of Nematology* **32**: 215.

Somasekhar, N. and Gill, J.S. 1991. Efficacy of *Pasteuria penetrans* alone and in combination with carbofuran in controlling *Meloidogyne incognita*. *Indian Journal of Nematology* **21**: 61-65.

Sosamma, V.K. 1984. *Studies on the Burrowing Nematode of Coconut*. Ph. D. thesis, Kerala University, Trivandrum, Kerala, India, 166 pp.

Sosamma, V.K., Geetha, V.K. and Koshy, P.K. 1994. Effect of the fungus *Paecilomyces lilacinus* on the burrowing nematode, *Radopholus similis* infesting betel vine. *Indian Journal of Nematology* **24**:50-54.

Sosamma, V.K. and Koshy, P.K. 1978. A note on the association of *Cylindrocarpon* spp. with *Radopholus similis* in coconut. *Indian Phytopathology* **31**: 381-382.

Sosamma, V.K. and Koshy, P.K. 1995. Effect of Pasteuria penetrans and Paecilomyces lilacinus on population build up of root-knot nematode, Meloidogyne incognita on black pepper. National Symposium on Nematode Problems of India – An Appraisal of the Nematode Mangmt. with Eco-friendly Approaches and Biocomponents, Indian Agri. Res. Inst., New Delhi, p. 47.

Sosamma, V.K., Koshy, P.K. and Bhaskar Rao, E.V.V. 1980. Susceptibility of coconut cultivars and hybrids to *Radopholus similis* in the field. *Indian Journal of Nematology* **10**: 250-252.

Sosamma, V.K., Koshy, P.K. and Samuel, R. 1998. Effect of vesicular arbuscular mycorrhizae, *Glomus mosseae* on nematodes infecting banana under coconut based high density multispecies cropping system. In *Nematology – Challenges and Opportunities in 21st Century* (Usha K. Mehta, ed.), pp. 94-97. Sugarcane Breeding Institute, Coimbatore.

Sosamma, V.K., Koshy, P.K. and Sundararaju, P. 1980a. *Nematodes Associated with Cocoa – A review*. Tech. Doc. No. 123, Plant Prot. Committee for the South East Asia and Pacific Region, 14 pp.

Sosamma, V.K., Koshy, P.K. and Sundararaju, P. 1980b. Plant parasitic nematodes associated with cocoa. *Cocoa Growers Bulletin* **29**: 27-30.

Sosamma, V.K., Sundararaju, P. and Koshy, P.K. 1979. Effect of *Radopholus similis* on turmeric. *Indian Journal of Nematology* **9**: 27-31.

Sreeja, P., Nageswari, S. and Mohandas, C. 1998. Reaction of cassava cultivars to root-knot nematode, *Meloidogyne incognita* (Kofoid and White, 1919) Chitwood, 1949. in *Nematology: Challenges and Opportunities in 21st Century* (Usha K. Mehta, ed.), pp. 137-139. Sugarcane Breeding Institute, Coimbatore.

Srinivasan, A. 1974. *Studies on Root Lesion Nematode*, Pratylenchus delattrei. M. Sc. (Agri.) thesis, Tamil Nadu Agricultural University, Coimbatore.

Srinivasan, A. and Muthurksihnan, T.S. 1975. A note on nematode fungal complex in crossandra (*Crossandra undulaefolia*) in Coimbatore. *Current Science* **44**: 743-744.

Srinivasan, A. and Muthukrishnan, T.S. 1976. Control of root lesion nematode *Pratylenchus delattrei* Luc 1958 on crossandra. *Indian Journal of Nematology* **6**: 111-114.

Srinivasan, C.S. and D'souza, G.I. 1965. A note on the occurrence of Brazilian pyroid coffee nematode, *Meloidogyne exigua* Goeldi in South India. *Indian Coffee* **29**(6): 13.

Srivastava, A.S., Verma, R.S. and Pandey, R.C. 1971. Comparative efficacy of different nematicides tested against root-knot nematode infesting *Colocasia antiquorum*. *Labdev Journal of Science and Technology* **9B**: 142-144.

Srivastava, S.S. and Swarup, G. 1986. Biomanagement of root-knot nematode (*Meloidogyne incognita*) through the predatory fungus, *Arthrobotrys conoides*. *National Conference on Plant Parasitic Nematodes of India-Problems and Progress*, Indian Agri. Res. Inst., New Delhi, p. 85.

Stirling, G.R. and White, A.M. 1982. Distribution of a parasite of root-knot nematodes in South Australian vineyards. *Plant Disease* **66**: 52-53.

Subramanian, S. and Selvaraj, P. 1988. Effect of *Tagetes patula* L. leaf extract on *Radopholus similis* (Cobb, 1893) Thorne, 1949. *Indian Journal of Nematology* **18**: 337-338.

Subramanyam, S., Rajendran, G. and Vadivelu, S. 1990. Estimation of loss in tomato due to *Meloidogyne incognita* and *Rotylenchulus reniformis*. *Indian Journal of Nematology* **20**: 239-240.

Sudha, S. 1998. Screening of coconut cultivars against the burrowing nematode, *Radopholus similis*. In *Advances in IPM of Horticultural Crops* (P. Parvatha Reddy, N.K. Krishna Kumar and A. Verghese, eds.), pp. 332-333. Association for Advancement of Pest Management in Horticultural Ecosystems, Bangalore,

Sudha, S. and Sundararaju, P. 1998. Effect of neem oil cake and nematicide for the control of burrowing nematode, *Radopholus similis* in the areca nut based cropping system. In *Nematology - Challenges & Opportunities in 21st Century* (Usha K. Mehta, ed.), pp. 251-257. Sugarcane Breeding Inst., Coimbatore.

Suit, R.F. and DuCharme, E.P. 1953. The burrowing nematode and other parasitic nematodes in relation to spreading decline of citrus. *Plant Disease Reporter* **37**: 379-383.

Sukumaran, S. and Sundararaju, P. 1986. Pathogenicity of *Meloidogyne incognita* on ginger (*Zingiber officinale* Rosc.). *Indian Journal of Nematology* **16**: 258.

Sukumaran, S. and Sundararaju, P. 1986. Effect of root-knot nematode, *Meloidogyne incognita* on the growth of turmeric (Abstr.). *Seventh Symposium of Plantation Crops*, Coonoor, Nilgiris, Tamil Nadu, India, p. 33.

Suma, M.R., Krishnappa, K., Reddy, B.M.R. and Kantharaju, V. 2002. VAM in the management of the burrowing nematode on banana (Abstr.). *Southern Zone Annual Meeting of the Indian Phytopath. Society,* Bangalore, pp. 19-21.

Sumathi, M., Parthiban, S. and Samuel, S.D. 2006. Effect of graded doses of nitrogen and VAM on herbage, oil yield and nematode population on patchouli (*Pogostemon patchouli* Pellet). *Indian Journal of Nematology* **36**: 209-213.

Sundarababu, R. 1992. Nematodes in jasmine. *Tamil Nadu Agricultural University Newsletter* **21**(11): 2.

Sundarababu, R., Mani, M.P. and Sivakumar, C.V. 1999. A new biocontrol agent for plant parasitic nematodes. *Indian Horticulture* **12**: 25.

Sundarababu, R., Sankaranarayanan, C. and Shanthi, A. 1993. Interaction between vesicular arbuscular mycorrhiza and *Meloidogyne javanica* on tomato as influenced by time of inoculation. *Indian Journal of Nematology* **23**: 125-127.

Sundarababu, R. and Suguna, N. 1998. Effective optimum dose of *Glomus mosseae* for the management of *Meloidogyne incognita* on tomato, chillies and bhendi. *International Journal of Tropical Plant Diseases* **16**: 127-131.

Sundarababu, R. and Vadivelu, S. 1988a. Studies on the nematodes associated with rose. *Indian Journal of Nematology* **18**: 139.

Sundarababu, R. and Vadivelu, S. 1988b. Pathogenicity of *Meloidogyne* species to tuberose (*Polyanthes tuberosa* L.). *Indian Journal of Nematology* **18**: 146-148.

Sundararaju, P. and Cannayane, I. 2002. Integrated nematode management on banana in India. *Global Conference on Banana and Plantain*, Bangalore, pp. 73-76.

Sundararaju, P., Cannayane, I. and Sathiamoorthy, S. 2004. *Nematode Management in Banana and Plantains*. Tech. Bull. No. 10, National Res. Centre for Banana, Thiruchirapalli, Kerala, India, 14 pp.

Sundararaju, P., Cannayane, I. and Thangavelu, R. 2002. Biomanagement of *Pratylenchus coffeae* and *Meloidogyne incognita* on banana by using endophytic fungi. *Global Conference on Banana and Plantain*, Bangalore, p. 191

Sundararaju, P. and Koshy, P.K. 1982. Response of Areca collections against *Radopholus similis*. *Indian Journal of Nematology* **12**: 404-405.

Sundararaju, P. and Koshy, P.K. 1986. Control of *Radopholus similis* on areca nut seedlings with aldicarb, aldicarb sulfone, carbofuran and fensulfothion. *Indian Journal of Nematology* **16**: 4-7.

Sundararaju, P. and Koshy, P.K. 1988a. Screening of areca nut cultivars and hybrids for resistance to *Radopholus similis*. *Indian Journal of Nematology* **18**: 376-377.

Sundararaju, P. and Koshy, P.K. 1988b. Histopathology of areca nut roots infected with *Radopholus similis*. *Indian Journal of Nematology* **18**: 378-380.

Sundararaju, P., Koshy, P.K. and Sosamma, V.K. 1979. Plant parasitic nematodes associated with spices. *Journal of Plantation Crops* **7**: 15-26.

Sundararaju, P. and Kumar, V. 2003. Management of root-lesion nematode, *Pratylenchus coffeae* in six commercial cultivars of banana through organic and inorganic amendments. *Infomusa* **12**: 21-25.

Sundararaju, P., Sosamma, V.K. and Koshy, P.K. 1979. Pathogenicity of *Radopholus similis* on ginger. *Indian Journal of Nematology* **9**: 91-94.

Sundaram, H.D. and Arangarasan, V. 1996. Role of vesicular arbuscular mycorrhizal fungi on the growth, nematode (*Meloidogyne incognita*) disease incidence and fruit yield of tomato CO 3 grown in loamy soils of Arcot District (Tamil Nadu). *Mycorrhizae: Biofertilizer for the Future*, pp. 100-103.

Suresh, C.K. and Bagyaraj, D.J. 1984. Interaction between a vesicular arbuscular mycorrhiza and a root-knot nematode and its effect on growth and chemical composition of tomato. *Nematologia Mediterranea* **12**: 31-39.

Susan, M.L., Daniel, R.P., David, C.J., Lyn, C.K., Robert, L.D. and Weili, M. 2002a. *Application of* Burkholderia cepacea *and* Trichoderma virens *Alone and in Combination Against* Meloidogyne incognita. USDA, ARS Publication.

Susan, M.L., Samie, M.I., David, C.J. and Daniel, R.P. 2002b. *Evaluation of* Trichoderma virens *and* Burkholderia cepacea *for Antagonistic Activity Against Root-knot Nematode,* Meloidogyne incognita. USDA, ARS Publication.

Swamy, B.C.N., Rajagopal, B.K. and Anahosur, K.H. 1973. Citrus nematode, a new record from Mysore State. *Mysore Journal of Agricultural Sciences* **7**: 138-139.

Swamy, K.R.M. and Sadashiva, A.T. 2004. *Vegetable Varieties/Hybrids of IIHR.* Tech. Bull. 2, Indian Inst. of Hort. Res., Bangalore, 30 pp.

Swarup, G. and Seshadri, A. R. 1974. Nematology in India - Problems and Progress. In *Current Trends in Plant Pathology* (S.P. Raychaudhuri and J. P. Verma, eds.), pp. 303-311. Dept. of Botany, Univ. of Lucknow, Lucknow.

Talukdar, Rita. 1993. *Management of* Meloidogyne incognita *(Kofoid and White, 1919) Chitwood 1949 on okra* (Abelmoschus esculentus *(L.) Moench*). M. Sc.(Ag) thesis, Assam Agri. Univ., Jorhat.

Tanda, A.S. and Atwal, A.S. 1988. Effect of sesame intercropping against the root-knot nematode (*Meloidogyne incognita*). *Nematologica* **34**: 484-492.

Tarjan, A.C., Jimenez, M.G. and Soria, J. 1973. Increasing yields of cocoa by application of nematocides. *Turialba* **23**: 138-142.

Taylor, A.L. and Sasser, J.N. 1978. *Biology, Identification and Control of Root-knot Nematodes* (Meloidogyne *spp.*). North Carolina State University Graphics, Raleigh, 111 pp.

Taylor, A.L., Sasser, J.N. and Nelson, L.A. 1982. *Relationship of Climate and Soil Characteristics to Geographical Distribution of Meloidogyne Species in Agricultural Soil.* Co-op. Pubn., Dept. of Plant Path., North Carolina State Univ. & U. S. Agency for Intn. Devp., Raleigh, N.C., 65 pp.

Tewari, S.P. and Dave, G.S. 1985. Burrowing nematode, *Radopholus similis* (Cobb) Thorne, 1949 on banana in Madhya Pradesh. *Current Science* **54**: 1286.

Thakar, N.A. and Patel, C.C. 1984. Screening of few cowpea varieties against reniform nematode, *Rotylenchulus reniformis* Linford and Oliveira, 1940. *Indian Journal of Nematology* **14**: 204.

Thangaraju, D. 1983. Distribution of potato cyst nematodes in Kodaikanal Hills, Madurai district, Tamil Nadu. *Indian Journal of Nematology* **13**: 222-223.

Thapa, C.D., Munjal, R.L. and Seth, P.K. 1987. The mushroom nematodes and their control. *Indian Journal of Mushrooms* **3**: 1-11.

Thapa, C.D., Jandaik, C.L. and Sharma, N.K 1987. Comparative resistance in some cultivated edible fungi against two mushroom nematodes occurring in India. *Proceedings of the International Conference on Science and Cultivation Technology of Edible Fungi,* pp. 169-172.

Thapa, C.D., Sharma, N.K and Nath, A. 1981. Mushroom nematodes and their source of infestation in mushroom beds in India. *Mushroom Science* **11**: 443-448.

Thirumalachar, M.J. 1951. Root-knot nematode on potato tubers in Simla. *Current Science* **20**: 20,104.

Thirumala Rao, V. 1956. Stray notes on some crop outbreaks of South India. *Indian Journal of Entomology* **18**: 123-126.

Thorne, G. 1961. *Principles of Nematology*. McGraw-Hill, New York and London, 553 pp.

Tiyagi, S.A., Anver, S., Bano, M. and Siddiqi, M.A. 1986. Feasibility of growing zinnia as a 'mix-crop' along with tomato for control of root-knot and reniform nematodes. *International Nematology Network Newsletter* 3(3): 6-7.

Trivedi, P.C. 1992. Evaluation of a fungus, *Paecilomyces lilacinus* for the biological control of root-knot nematode, *Meloidogyne incognita* on *Solanum melongena*. In *Proc. of the 3rd Intnl. Conf. on Plant Prot. in the Tropics* (F.A.C. Ooi and G.S. Lim, eds.), Vol. 6, pp. 29-33. Genting Highlands, Malaysia.

Trivedi, P.C. and Tiagi, A. 1984. Effect of some plants on *Meloidogyne incognita* population in chilli soil. *Indian Journal of Botanical Science* **63**: 416-419.

Umesh, K.C., Krishnappa, K. and Bagyaraj, D.J. 1988. Interaction of burrowing nematode, *Radopholus similis* and VA mycorrhiza, *Glomus fasciculatum* in banana. *Indian Journal of Nematology* **18**: 6-11.

Upadhyay, K.D. and Dwivedi, K. 1987. Analysis of crop losses in pea and gram due to *Meloidogyne incognita*. *International Nematology Network Newsletter* 4(4): 6-7.

Vadhera, I., Sheila, B.N. and Bhatt, J. 1995. Interaction between reniform nematode (*R. reniformis*) and *Fusarium solani* causing root rot of French bean. *Indian Journal of Agriculture Sciences* **65**: 774-777.

Vadivelu, R.S. and Muthukrishnan, T.S. 1979. Control of root lesion nematode, *Pratylenchus delattrei* on crossandra. *Indian Journal of Nematology* **9**: 86.

Vadivelu, R.S. and Rajendran, G. 1986. Damage potential of the root-knot nematode, *Meloidogyne arenaria* to onion. *International Nematology Network Newsletter* **4**(2): 5-8.

Vadivelu, R.S., Rajendran, G., Naganathan, T.G. and Jayaraj, S. 1987. Studies on the nematode pests of banana and their management - A review. *Third Group Discussion on the Nematological Problems of Plantation Crops*, Sugarcane Breeding Institute, Coimbatore, pp.32-33.

Vadivelu, S., David, B.V. and Rajmohan, N. 1976. *Pesticides* **10**: 39-40.

Van Berkum, J.A. and Seshadri, A.R. 1970. Some important nematode problems in India. *Proceedings of Xth International Symposium of European Society of Nematologists*, Pescara, pp. 135-137.

Van der Vetch, J. 1950. Plant parasitic nematodes. In *Diseases of Cultivated Plants in Indonesia* Vol. I (L.G.E. Karsh van and Jived. Vetch, eds.), pp. 16-45. W. van Wove, S'Gravenhagen.

Van Gundy, S.D. 1957. *Plant Disease Reporter* **41**: 1016-1018.

Van Gundy, S.D. 1958. The life history of the citrus nematode *Tylenchulus semipenetrans*. *Nematologica* **3**: 283-294.

Van Gundy, S.D. 1959. The life history of *Hemicycliophora arenaria* Raski (Nematoda: Criconematidae). *Proceedings of Helminthological Society of Washington* **26**: 67-72.

Van Gundy, S.D. and McElroy, F.D. 1969. *Proceedings of Ist International Citrus Symposium*, **Vol. II**: 985-989.

Van Hoof, H.A. and Seinhorst, J.W. 1962. *Rhadinaphelenchus cocophilus* associated with little leaf of coconut and oil palm. *Tijdschr. Plziekt.* **68**: 251-256.

Varma, M.K., Sharma, H.C. and Pathak, V.N. 1978. Evaluation of systemic granular nematicides against *Tylenchorhynchus* spp. parasitic on cabbage. *Indian Journal of Nematology* **8**: 59-60.

Vats, R. and Dalal, M.R. 1997. Interaction between *R. reniformis* and *Fusarium oxysporum* f. sp. *pisi* on pea. *Annals of Biology, Ludhiana* **13**: 239-242.

Veere Gowda, R., Singh, T.H., Nagesh, M. and Rao, M.S. 2001. Evaluation of carrot (*Dacus carota L.*) genotypes for resistance to root-knot nematode (*Meloidogyne incognita*). *Proceedings of the Second National Symposium on Integrated Pest Management in Horticultural Crops: New Molecules, Biopesticides and Environment*, Bangalore, p. 155.

Venkata Ram, C.S. 1963. The Root-knot Eelworm of Mature Tea Meloidogyne brevicauda, its Importance in South India and an Evaluation of Certain Chemical Treatments for its Control. Tea Scientific Report for 1962-63, United Planters Association of South India, Appendix pp. 31-36.

Venkitesan, T.S. 1972. On the occurrence of plant parasitic nematodes associated with different crops in Cannanore District, Kerala. *Agriculture Journal Kerala* **10**: 179-180.

Venkitesan, T.S. 1976. Studies on the Burrowing Nematode, Radopholus similis (Cobb, 1893) Thorne, 1949 on Pepper (Piper nigrum L.) and its Role in Slow Wilt Disease. Ph. D. thesis, Univ. of Agri. Sci., Bangalore.

Venkitesan, T.S. and Charles, J.S.K. 1983. Sucker dip treatment for banana nematode control (Abstr.). *Third Nematology Symposium*, Himachal Pradesh Agricultural University, Solan, p. 45.

Venkitesan, T.S. and Jacob, A. 1985. Integrated control of root-knot nematode, infecting pepper vines in Kerala (Abstr.). *Indian Journal of Nematology* **15**: 261-262.

Venkitesan, T.S. and Setty, K.G.H. 1977. Pathogenicity of *Radopholus similis* to black pepper (*Piper nigrum*). *Indian Journal of Nematology* **7**: 17-26.

Verma, A.C. 2001. N.D. University of Agriculture and Technology, Kumarganj, Faizabad, Uttar Pradesh. In *Indian Nematology-Progress and Perspectives* (S.C. Dhawan *et al.*, eds.), pp. 121-125. Division of Nematology, Indian Agri. Res. Inst., New Delhi.

Verma, A.C., Singh, H.K. and Khan, N. 2005. Management of root-knot nematode, *Meloidogyne incognita* through antagonistic approaches in pointed gourd. *Indian Journal of Nematology* **35**: 78-79.

Verma, K.K., Gupta, D.C. and Paruthi, I.J. 1999. Preliminary trial on the efficacy of Pseudomonas fluorescens as seed treatment against Meloidogyne incognita in tomato. Proceedings of the National Symposium on Rational Approaches in Nematode Management for Sustainable Agriculture, Indian Agri. Res. Inst., New Delhi, pp. 79-81.

Verma, K.K., Verma, B.S., Sangwan, N.K. and Dhindsa, K.S. 1989. Toxicity of some indigenous plant extracts to root-knot, seed gall and citrus nematode. *Pesticides* **23**(11): 25-27.

Verma, R.R. 1987. *Indian Journal of Nematology* **17**: 123-124.

Vetrivelkalai, P. and Subramanian, S. 2006. Plant parasitic nematode populations as affected by onion based coping sequences. *Indian Journal of Nematology* **36**: 10-14.

Victoria, K.J., Sanchez, P.A. and Barriga, O.R. 1970. Eradication de palms de cocotero afectadas por el anillo rojo (*Rhadinaphelenchus cocophilus* (Cobb, 1919) Goodey, 1960, Nematoda: Aphelenchoididae) mediante la utilization de sustancias quimicas. *Revista Inst. Colomb. Agropec.* **5**: 185-197.

Vidya, K. and Reddy, B.M.R. 1998. Integrated management of *Radopholus similis* infecting banana using plant product, nematicide and biocontrol agents. *Mysore Journal of Agricultural Sciences* **32**: 186-190.

Viswanathan, P.R.K., Kumar, A.C. and D'Souza, G.I. 1974. Nematode association of cardamom in South India. *Cardamom News* **6**: 5 & 19.

Walia, R.K., Bansal, R.K. and Bhatti, D.S. 1991. Effect of *Paecilomyces lilacinus* application time and method in controlling *Meloidogyne javanica* on okra. *Nematologia Mediterranea* **19**: 247-149.

Walia, R.K., Bansal, R.K. and Bhatti, D.S. 1992. Efficacy of bacterium (*Pasteuria penetrans*) and fungus (*Paecilomyces lilacinus*) alone and in combination against root-knot nematode (*Meloidogyne javanica*) infecting brinjal. *First Afro-Asian Nematology Symposium*, Aligarh Muslim Univ., Aligarh, p. 38.

Walia, R.K. and Dalal, M.R. 1994. Efficacy of bacterial parasite *Pasteuria penetrans* application as nursery soil treatment in controlling root-knot nematode *Meloidogyne javanica* on tomato. *Pest Management and Economic Zoology* **2**: 19-21.

Wallace, H.R. 1960. Observations on the behaviour of *Aphelenchoides ritzemabosi* in chrysanthemum leaves. *Nematologica* **5**: 315-321.

Webster, B.N. and Gonzales, J. 1959. Investigaciones sobre la enfermedad 'amillo-rojo en coco soy palma des aceite. IX Conv. de la assoc. Venezolana. el avance de la ciencia, service shell paa el agr. Establa Arajua.

Weischer, B. 1967. *Report to the Government of India on Plant Parasitic Nematodes.* PL 24/52, Report No. TA 2332, FAO, Rome, 15 pp.

Whitehead, A.G. 1968. Taxonomy of *Meloidogyne* (Nematodea: Heteroderidae) with descriptions of four new species. *Transactions of Zoological Society of London* **31**: 263-401.

Winslow, R.D. and Willis, R.J. 1972. Nematode diseases of potato. In *Economic Nematology* (J.M. Webster, ed.), pp. 17-48. Academic Press, London.

Wood, F.C. and Goodey, J.B. 1957. Effects of gamma-ray irradiation on nematodes infesting cultivated mushroom beds. *Nature* **180**: 760-761.

Wyatt, J.E., Fassuliotis, G., Hoffman, J.C. and Deakin, J.R. 1983. "Nemasnap" snap bean. *Hort Science* **18**: 776.

Xiuuan, Y., Yuxian, H., Furn, C. and Zhengliang 2000. Isolation and selection of egg-mass fungi of *Meloidogyne* species in Fujian Province. *Fujian Journal of Agriculture Science* **15**: 12-15.

Zaki, F.A. 1998. Biological control of *Meloidogyne javanica* in tomato by *Paecilomyces lilacinus* and castor leaves. *Indian Journal of Nematology* **28**: 132-139.

Zaki, F.A. and Bhatti, D.S. 1989. Effect of castor (*Ricinus communis*) leaves in combination with different fertilizer doses on *Meloidogyne javanica* infecting tomato. *Indian Journal of Nematology* **19**: 171-176.

Zaki, F.A. and Bhatti, D.S. 1991a. Effect of castor *Ricinus communis* and the biocontrol fungus, *Paecilomyces lilacinus* on *Meloidogyne javanica*. *Nematologica* **36**: 114-122.

Zaki, M.J. and Maqbool, M.A. 1992. *Pakistan Journal of Nematology* **10**: 75-79.

Zuckerman, B.M., Dicklow, M.B. and Acosta, N. 1993. A strain of *Bacillus thuringiensis* for the control of plant parasitic nematodes. *Biocontrol Science and Technology* **3**: 41-46.

APPENDICES

1. Quotable Quotes

"If all the matter in the universe except the nematodes were swept away, our world would still be dimly recognizable............we would find its mountains, hills, valleys, rivers, lakes and oceans represented by a film of nematodes".

– N.A. Cobb (1914)

"Some say roundworms, others thread or eelworms
Sexes are separate, producing eggs and sperms

*They reproduce sexually, have habits strange
A millimeter to a few meters, is their body range*

*Esophagus with nerve ring, lumen triradiate
Body cavity false, hence pseudocoelomate*

*Circulatory organs lacking, yet triploblastic
Respiration through cuticle, which is elastic*

*Amphids and phasmids are their eyes and ears
Plant feeders possess – either stylets or spears".*

– R.S. Kanwar

"Whatever approaches we take to achieve the goal of controlling diseases associated with plant nematodes, knowledge of the nematodes themselves as well as the plants which they parasitize, will increase the chances of success...... In other words, there must always be room for the specialist who studies different aspects of the biology of plant nematodes".

– H.R. Wallace

"The extraction of food from one organism, the host, by another organism, the parasite, inevitably entails some disturbance of the host's integrity".

– H.R. Wallace

"Plants are rarely, if indeed ever, subject to association with only one potential pathogen".

— N.T. Powell

"Each year these minute organisms (nematodes) exact an ever-increasing toll from almost every cultivated acre in the world: a bag of rice in Burma, a pound of tea in Ceylon, a ton of sugar beets in Germany, a bag of potatoes in England, a bale of cotton in Georgia, a bushel of corn in Iowa, a box of apple in New York, a sack of wheat in Kansas, or a crate of oranges in California........which the farmers can ill afford to lose and which the undernourished millions on earth would welcome if made available to them ".

— G. Thorne (1961)

"The most severe nematode problems occur where good host crops are grown too frequently for too long on the same land".

— B.G. Chitwood

"If at all there was one single factor that triggered nematological research in India, I would say it was this discovery (relating to the report of potato cyst nematode from Ooty, Tamil Nadu by Jones, 1961)".

— Abrar M. Khan (1974)

"A convergence of technical, environmental and social forces is moving agriculture towards more non-pesticide pest management alternatives like biological control, host plant resistance and cultural management".

- Michael Fitzner, National IPM Program Leader, USDA Extension Service

2. Nematode Diseases of Horticultural Crops

Radopholus similis – **'Yellows'** of black pepper in Bangka Island, Indonesia.

– **'Slow Wilt'** of black pepper in Kerala.

– **'Black Head'** or **'Toppling'** or **'Rhizome Rot'** of banana.

Radopholus citrophilus – **'Spreading Decline'** of citrus in Florida State, USA.

Pratylenchus coffeae – **'Slump Disease'** of Citrus.

Tylenchulus semipenetrans – **'Slow Decline'** of citrus.

Rhadinaphelenchus cocophilus – **'Red Ring'** of coconut.

Macroposthonia xenoplax – **'Peach Tree Short Life'** disease of peach.

Hemicriconemoides spp. – **"Crinkle Leaf"** disorder of coffee.

Ditylenchus dipsaci – **'Bloat'** in onion.

Ditylenchus destructor – **'Dry Rot'** of potato.

Aphelenchoides besseyi – **'Summer Crimp'** or **'Summer Dwarf'** of chrysanthemum.

Aphelenchoides fragariae – **'Cauliflower Disease'** or **'Spring Dwarf'** or **'Crimp'** or **'Strawberry Bunch'** or **'Red Plant'** of strawberry.

3. Soil Solarization

Soil solarization is a method of killing nematodes by heat. This technique consists of covering moist soil with a plastic film during periods of intense sunshine and heat, thereby capturing radiant heat energy from the sun, causing physical, chemical and biological changes in the soil. The soil temperature is increased (by 5-15°C) to levels lethal to many soil-borne fungi, bacteria, nematodes and weeds. The short wave incident solar radiation penetrates the polyethylene sheet but the long wave radiation is prevented from soil, thus trapping the heat resulting in thermal inactivation and production of heat shock proteins and irreversible heat injury. Solarization also improves plant nutrition by increasing the availability of N and other essential elements. This practice can be profitably used on a small scale and is especially useful for nursery beds.

Tarping with Linear Low Density Poly Ethylene (LLDPE) 100 gauge film during summer for 15 days (in between April 15th to June 15th) in tomato nursery at Anand gave 66 and 92% reduction of root-knot nematodes and weeds, respectively. This practice produced 615% more seedlings over control giving net profit of Rs. 7,661 from 1000 m^2 area with cost: benefit ratio of 1: 5.65.

Soil solarization for 40-45 days using 400 gauge transparent LDPE was proved ideal pre-sowing treatment for the management of nematodes and other soil-borne diseases in cardamom nurseries. Mean soil temperature during the solarization period increased by 8.7°C in solarized beds and the germination was enhanced by 25.5%, while weed growth was suppressed by 82%. *Pythium vexans* was totally eliminated, while populations of *Rhizoctonia solani, Phyllosticta elettaria* and nematodes were suppressed to varying levels. Solarization was also found to stimulate the growth of cardamom seedlings (Eapen, 1995).

White polyethylene mulching for 15 days on beds prepared for planting betel vine revealed high reduction in nematode populations due to increased soil temperature to 44.1°C (Sivakumar and Marimuthu, 1987).

Soil solarization using thin transparent polyethylene mulches for six weeks caused as much as 72.6 and 88.5% decrease in the population densities of *Tylenchorhynchus vulgaris* and *Hoplolaimus indicus*, respectively in nursery beds. Application of mahua cake before solarization usually caused slightly greater reduction in nematode population (Kamra and Gaur, 1998).

4. Mass Production of *Pasteuria penetrans*

Pasteuria penetrans is an obligate parasite of some plant parasitic nematodes including *Meloidogyne* spp. but can not be produced, at the present time, in large numbers *in vitro*. It is a very common parasite on *Meloidogyne* spp. and is often observed attached to the juveniles. The spores can resist both drought and exposure to non-fumigant nematicides. Stirling and Wachtel (1980) were able to produce large numbers of spores by inoculating tomato with infected *Meloidogyne* juveniles. Autoclaved soil should be filled in pots and seedlings of tomato or brinjal are to be transplanted in pots. The upper 5 cm layer of the soil should be infested with a known number of endospores of *P. penetrans*/g soil. Each plant should be inoculated with 1,000 to 2,000 juveniles of root-knot nematode. The plant roots should be harvested 70 days later, dried, weighed and milled into a powder containing *P. penetrans* spores. The number of endospores/g of root/plant should be estimated. This method might be adapted to produce inoculum of the parasite for local use on small farms.

In case of direct seeded crops, seed or soil treatment with known quantity of endospores or root powder infested with endospores is applied. Treated seeds are sown singly in each pot. One week after germination, plants are inoculated with 1 or 2 juveniles/g of soil. Seventy days after germination, data on plant growth parameters, nematode multiplication, % Juveniles and females infected with endospores are recorded.

In transplanted crops, seedling bare root-dip in endospore suspension or soil application with endospores or root powder infested with endospores can be applied and the efficacy of these methods can be evaluated as per method described for direct seeded crops.

5. Mass Production of *Pseudomonas fluorescens*

A loopful of *P. fluorescens* should be transferred to 100 ml of King's B broth in a 250 ml conical flask under aseptic conditions. The flask can be incubated at room temperature for 48 hours. The population of *P. fluorescens* in the broth can be assessed (a minimum of 9×10^9 cfu/ml is necessary).

The pH of the substrate (talc powder) can be adjusted to 7.0 by addition of 150g calcium carbonate/kg of talc powder and sterilized at 1.1 kg/cm^2 for 30 min. for 2 successive days. One kg of the sterilized substrate (talc powder) and 5g of carboxy methyl cellulose should be placed in a polythene bag or any sterile container under aseptic conditions, 400 ml of *P. fluorescens* suspension to be added and mixed thoroughly. The formulation can be stored in a polythene bags for one month. The product should contain a minimum of 2.5×10^8 cfu/g.

The product should have the following quality parameters:

- Fresh product should contain 2.5×10^8 cfu/g.

- Three months after storage at room temperature, the population should be 8-9 x 10^7 cfu/g.

- Storage period is 3-4 months.

- Minimum population load for use is 1×10^8 cfu/g.

- The product should be packed in white polythene bags.

- Moisture content of the final product should not be more than 20%.

- Population in broth should be $9 \pm 2 \times 10^8$ cfu/ml.

The composition of King's B medium is as follows:

Peptone	20.00 g
Dipotassium hydrogen phosphate	1.50 g
Magnesium sulphate	1.50 g
Glycerol	10.00 ml
Agar	15.00 g
Distilled water	1000 ml
Cycloheximide	100 ppm
Ampicillin	50 ppm
Chloramphenicol	12.50 ppm

(King's B broth is prepared without the agar and antibiotics)

6. Mass Production of *Trichoderma viride, T. harzianum*

Among the various fungal biological control agents which are effective against nematode diseases of horticultural crops, *Trichoderma viride* and *T. harzianum* have a good market. They can be easily mass produced on several organic substrates.

6.1. Small Scale

The mycelial discs grown in potato dextrose agar medium was transferred to 250 ml conical flasks containing sterilized molasses yeast medium (Yeast extract – 5 g, Molasses – 30 g, Distilled water – 1 litre sterilized at 1.42 kg/cm² for 20 min.) at 100 ml /flask and incubated for 10 days. The mycelial mat along with the broth was homogenized and mixed with talc powder (500 mesh) at 1:2 ratio. Carboxy methyl cellulose (CMC) was added at 5 g/kg of the product as sticker. The product was shade dried until it attained 12% moisture level. The clumps were broken, homogenized and packed in polythene bags of 100g capacity. The required population level at the time of packing should be 28 x 10^7 cfu/g of the product. The shelf life of the product was 4 months. At the end of expiry, the cfu should not be less than 20×10^6 cfu/g of the product.

6.2. Large Scale

Add 70 ml of molasses yeast medium in each 250 ml conical flasks and plug with non-absorbent cotton wool. Autoclave at 1.42 kg/cm^2 for 1 hr. Transfer 8 mm mycelial disc of *T. viride/T. harzianum* to each flask and incubate for 10 days. This mother culture is used for inoculating the medium in fermentor.

Add 50 litres of molasses yeast medium in 50 litre fermentor, sterilize at 1.42 kg/cm^2 for 20 min. and cool to room temperature. Add 2.5 litres of inoculum (from mother culture) into the sterilized medium and incubate for 10 days. Provide aeration in the fermentor for 4-8 hr/day by agitation. Presence of greenish fungal growth indicates *T. viride/T. harzianum*. This broth culture contains fungal mat, spores and antibiotics. Draw samples 10 days after incubation and assess the spore load in the broth which should be at least 10^8 spores/ml. Mix one litre of broth culture with 2 kg talc powder, add CMC at 5g/kg of the product, air dry (12% moisture content) and sieve. The population should not be less than 20×10^6 cfu/g of the product. It is packed in opaque polythene bags and stored in a cool dry place before use. The product can be safely stored for 4 months without losing its viability.

7. Enrichment of FYM with Bioagents (*Trichoderma harzianum, Paecilomyces lilacinus*)

Commercial formulation of the bioagent (*T. harzianum/P. lilacinus*) (1 kg/ton) was mixed with moist FYM along with 50 kg of neem/pongamia cake. Mixture was spread 15 cm high heap under shade, covered with polyethylene sheet and incubated for 3 weeks at 25 to 30°C and sprayed with water regularly to maintain moisture.

Colonized FYM can be used at 6 g/kg soil in pot culture experiments (12 g/2 kg soil in 15 cm diameter pots). It can also be used for main field treatment at 5 tonnes/ha before sowing of seeds or transplanting of seedlings. It can be applied at 2 kg/plant of banana, acid lime, papaya, pomegranate, grapes and roses along with 400g of neem/pongamia cake once in every 6 months after planting for the management of nematodes and other soil-borne pathogens.

T. harzianum colonized compost significantly enhanced germination and growth of both tomato and okra as compared to non-colonized FYM. Population of *T. harzianum* in colonized compost was several folds higher (48.6×10^{12}) than non- colonized FYM (3×10^4).

Colonization of FYM by *T . harzianum* helps to improve the quality of compost. It significantly enhanced total and water soluble humic matter. A number of macro and micro-nutrients like P, K, S, Zn, Cu and Fe (both total and water soluble content) were significantly higher in *T. harzianum* colonized compost. Six times increase in water soluble humic content due to colonization of *T. harzianum* seems quite important as it is reported to enhance the plant growth.

8. Mass Production of Arbuscular Mycorrhizal Fungi

81. In Glasshouse

Cement pots or tanks are filled with sterilized sand: soil (1: 1 by volume) or soilrite: peat (1: 1 by volume) mixture. A standard starter culture of AMF obtained from any authentic source is placed as a band 3 cm below the soil surface. The seeds of Rhoades grass/ragi/maize are sown on them and covered with soil. Pots are irrigated normally. When the plants are about 10-12 weeks old, the top shoots are cut off and the root system colonized by the fungus along with the soil containing chlamydospores is used as the inoculum.

8.2. On-farm Production

A suitable land (even waste land) is selected near the field where the fungus inoculum has to be used. The land is brought to the fine tilth by ploughing, harrowing and cultivator operations. A raised bed (20 cm height) of soil: sand (1: 1 by volume) mixture is prepared on an even land surface. The raised bed is sterilized by soil solarization for 2 to 3 weeks using polythene sheet. The starter culture of AMF is applied at a depth of 3-4 cm from the top by removing the top soil. The seeds of Rhoades grass/ragi/maize are sown on them and covered with soil. Irrigation is done normally. When the crop attains a height of 10-15 cm, 12 weeks after sowing, the shoot portion is cut off at ground level. The soil is dug up to a depth of 20 cm along with root system is used as the inoculum.

8.3. Method of Inoculation

8.3.1. Fruit Crops

Fruit crops are inoculated in the nursery. AMF inoculum is added to FYM + soil + sand mixture (1: 1: 1 by volume) at the rate of 50 g inoculum/ 500 g soil mixture in polythene bags. Seeds of papaya or rootstocks of mango, citrus, guava or sapota are planted in each bag. Forty five days after sowing, the root system of seedlings would be colonized by the fungus and ready for transplantation in the field (in pits). Before planting, each pit should be inoculated with 250-500 g of the fungus along with FYM and neem cake. Two months after planting, the roots may be tested for colonization by the fungus. The plants get colonized up to 80-90% within 3-8 months. Inoculation of AMF helps in greater uptake of nutrients like P, Zn and Cu. It also increases stem girth, plant height and hastens flower initiation.

8.3.2. Vegetable Crops

Vegetable crops can be inoculated with AMF in the nursery beds. The AMF inoculum consisting of chlamydospores, soil and infected root bits is either spread uniformly in the nursery beds or placed in rows over which seeds of vegetable crops (tomato, chilli, capsicum, brinjal) are sown. The seedlings get colonized by the fungus in about 4-5 weeks after which they are transplanted in the main field. Inoculated plants show better growth and improved nutrient content.

9. Seed Biopriming / Solid Matrix Priming

Seed biopriming (treatment of seeds with a slurry of biocontrol agents mixed with an organic carrier and incubated for 4 days under warm and moist conditions with relative humidity of 80-100% until just prior to radicel emergence) has potential advantages over simple coating of seeds as it results in rapid and uniform seedling emergence. *Trichoderma* required less moisture for growth. *Trichoderma* conidia germinate on the seed surface and form a layer around bioprimed seeds. Such seeds tolerate adverse soil conditions better. Biopriming could also reduce the amount of biocontrol agents that is applied to the seed. Population of *T. harzianum* and *T. virens* on the surface of treated seeds of tomato and brinjal increases from 10^4 cfu per seed up to 10^6-10^7 cfu per seed after biopriming.

10. Liquid Formulation Technology to Extend the Shelf Life of *Pseudomonas fluorescens*

Polyvinyl Pyrrolidone (PVP) (2%) was tested as an effective preservation agent in protecting the cells of *P. fluorescens* for more than a year without affecting the natural properties of bacteria. The antagonistic efficacy of preserved cells was also confirmed *in vitro* and *in vivo* (on black pepper, vanilla and ginger).

11. Formulation of Alginated Pellets of Biocontrol Agents

Alginate is a water soluble polysaccharide gum that has excellent gel-forming properties. Living organisms that are capable of biologically controlling a variety of nematodes can easily be entrapped in a calcium or sodium alginate matrix in aqueous form at room temperature. A pellet formulation of biocontrol agents was prepared by mixing Sodium or Calcium alginate, clay and streptomycin sulfate. The mixture was homogenized in a blender and then mixed with bioagent propagules. The gel beads, granules or seed coatings that are produced by the process are biodegradable. The encapsulation in Ca or Na alginate maintained the viability of the bioagents for 12 months at 7°C and 5 months at room temperature. Antagonistic fungi (species of *Paecilomyces, Trichoderma, Gliocladium, Talaromyces* and *Pencillium*) and bacteria (species of *Pseudomonas* and *Bacillus*) that can control soil-borne pathogens can be formulated in Ca or Na alginate (Connick, 1988). Alginated pellets can be used for soil application at 400 to 800 pellets/sq.m.

12. Production of Bioagent Colonized Seedlings

The horticultural crops are invariably attacked by soil borne pathogens (nematodes, fungi, bacteria) at the seedling stage itself. Securing healthy seedlings of horticultural crops is essential to ensure freedom from soil-borne pathogens, optimum plant stand, good growth and higher yields. The following methods may be adopted to produce bioagent colonized seedlings of horticultural crops:

12.1. Seed Treatment

Seeds have to be treated with talc based formulations of *Pseudomonas fluorescens, Trichoderma harzianum* and *T. viride* at 10g/kg seed before sowing.

12.2. Nursery Bed Treatment

The nursery beds may be treated with *P. fluorescens/ T. harzianum/ T. viride* at 50g/m² along with neem/pongamia cake at 200g/m² before sowing.

12.3. Treatment of Coco Peat or any other Substrate

The coco peat or any other substrate used for producing the nursery seedlings in portrays is treated with *P. fluorescens/ T. harzianum/ T. viride* at 1kg/tonne before filling with the substrate.

12.4. Treatment of Soil Mixture

The soil mixture used for producing the nursery seedlings in polythene bags in a conventional manner is treated with *P. fluorescens/ T. harzianum/ T. viride* at 2kg/tonne along with neem/pongamia cake at 50kg/tonne before sowing.

SUBJECT INDEX